S0-ARM-097

Fundamentals of Television Servicing

Joel Goldberg, Ph.D.

Fundamentals of Television Servicing

PRENTICE-HALL, INC., Englewood Cliffs, New Jersey 07632

Library of Congress Cataloging in Publication Data

Goldberg, Joel (date)
Fundamentals of television servicing.

Includes index.
1. Television—Repairing. I. Title.
TK6642.G63 621.388'87 81-13914
ISBN 0-13-344598-4 AACR2

Editorial/production supervision: Nancy Milnamow and Karen Wagstaff
Interior design: Karen Wagstaff
Cover design: Mario Piazza
Manufacturing buyer: Gordon Osbourne

Printed in the United States of America

10 9 8 7 6 5 4

ISBN 0-13-344598-4

Prentice-Hall International, Inc., *London*
Prentice-Hall of Australia Pty. Limited, *Sydney*
Prentice-Hall of Canada, Ltd., *Toronto*
Prentice-Hall of India Private Limited, *New Delhi*
Prentice-Hall of Japan, Inc., *Tokyo*
Prentice-Hall of Southeast Asia Pte. Ltd., *Singapore*
Whitehall Books Limited, *Wellington, New Zealand*

Contents

Preface

Repair of today's complex television sets can be a very simple process if the technician has a sound knowledge of *how* a set functions. The basic process has not changed much since the first commercial television sets were produced in the early 1950s.

There have always been two types of repair persons in the service industry. One can be classified as a "parts changer." This person does not know exactly where in the set the problem is. Repair is accomplished by changing parts, a system that is partially successful. This type of servicing was especially prevalent in the era of the vacuum tube. Because the vacuum tube was the hardest-working part of a set, the odds were in favor of it failing first. The key to this type of repair was to attempt to determine which of several tubes in the set needed to be replaced. Some people actually replaced all of the tubes when making this type of repair.

Often, persons using this technique were unable to repair a set if the problem was not related to a tube failure. The reason for this is that they did not know, or were unable to relate to, how the set actually functioned. Their knowledge of electronic circuit analysis was minimal. A set requiring circuit analysis repair was often left on the shelf for later repair, or it was sent out to someone else for repair.

This style of repair work was not done by all service technicians. Many technicians could be classified as being of the second type. These technicians *knew* electronic theory. They also *knew* how to utilize testing equipment to diagnose faults. In many cases the knowledgeable service technician did not remove any component from the set until *after* tests were made. This person was able to draw conclusions from observation and testing. These conclusions led to the location of defective components. This style of work enabled the service technician to be more productive and to repair a malfunctioning set correctly.

If television set design had remained static, we would still have these two groups of repair people. But rapid technological change occurred in the industry. The introduction of the transistor changed the entire situation. People who were used to replacing tubes to fix a set were often not able to cope with the new devices. They had no idea how they worked. Often, transistors were soldered into the set rather than being plugged into sockets as tubes were. This required further knowledge on the part of the technician. Removal and testing techniques for transistors are different from those used for vacuum tubes. Persons with minimal technical knowledge and a lack of willingness to learn about technical change were unable to repair these receivers quickly.

Another factor that separated these two groups was the introduction of printed-circuit boards to replace hand-wired sets. Technicians had to learn how to follow wiring on the circuit board. Many of the semitechnically trained persons in the service industry were unable to handle these changes. They left the industry and found employment in other fields.

Persons remaining in the television service industry quickly learned that they had to maintain their technical knowledge as the industry brought in new systems. Set manufacturers learned that they had to establish and offer field training programs for both factory and independent service people. Technological changes are still occurring. Today, most television sets use integrated circuits, which replace individual transistors. Many "modular" sets are being manufactured, also. These use plug-in units, or modules, to replace entire sections of a set. Often, several sections are on a single module. This type of technical change, as well as those changes that are certain to come in the future, place the electronic service technician in an unusual situation.

The successful service technician has to have several qualifications. These include a basic knowledge of electronics, how to use electronic test equipment, and how television sets function. In addition, the technician must be willing to attend technical seminars or return to a technical school periodically in order to maintain a high level of competency.

Television set design technology will continue to change in the coming years. Repair technicians who are still able to function as changes occur will be successful.

Technological changes occur daily. It is impossible to identify each change and to include it in a book of this type. It is better to develop fundamental skills and to be able to apply these skills as the need occurs.

This book is written on the basis of fundamental knowledge. I have spent nine years in the electronic service/parts field as well as seventeen years in teaching electronics and electronic service. Over the years, I have visited many service shops and attended a large number of training seminars. The material presented in this book is a result of my learning experiences over the past twenty-six years.

The book is divided into four major sections. The first section describes how television information is transmitted. The second section relates to how this information is processed in a receiver. The third section is devoted to the use of test equipment. The final section explains how each block, or section, of a set functions. The material is developed using integrated-circuit technology. The methods of diagnosis and repair are based on this type of system. The purpose of the book is to present basic material related to the repair of television sets. It is assumed that persons using the book have a good foundation in basic electronic theory.

Any author writing on a technical subject has to have assistance with materials. I wish to thank the many manufacturers of sets and test equipment who have contributed. They are acknowledged individually for photographs or drawings they have supplied. Support is also required from family and friends, and I acknowledge it with thanks. Specifically, I wish to dedicate this book to my wife, Alice. I could not have considered undertaking such a project without her help, understanding, and support.

Joel Goldberg

Fundamentals of Television Servicing

Chapter 1

Television Signals

People living in a highly technical society quickly learn to expect "things to happen" when a switch is activated. Rarely is thought given as to why these things do work. Little time is spent on determining what has gone into producing the expected result. When one turns on a light switch, one expects the lamp to glow and give off illuminating light rays. No consideration is given to the generating and distributing network required to accomplish this feat.

This is also true for almost all persons who turn on a radio or television set. If the picture, sound, and color are correct, the viewer is content. The ingredients required to accomplish this are either unknown or overlooked. It is only when trouble occurs that we are concerned with the "how and why" of the television signal. It is important that people employed in the television service field have an understanding of this "how and why" before attempting to repair any set.

This chapter explains how the television signal is created. It covers the specific components of the composite signal. Fundamentals of transmission of electromagnetic energy as well as reception are also presented. An understanding of this material will assist the technician when a repair is required.

COMPOSITION OF THE SIGNAL

The composite television signal has several individual components. These are all (except color) required in order to convey intelligence from the broadcast station to the receiver. The components are:

1. Picture information (video)
2. Synchronizing pulses
3. Blanking pulses
4. Synchronizing pulses (color)
5. Picture information (color)
6. Special control information
7. Sound information

Each of these components must be examined and understood in order to service a television set successfully.

All picture information is first converted into electrical information. This is accomplished by use of a television camera and a special tube called a *vidicon.* A cross-sectional view of this system is shown in Figure 1-1. The vidicon has a light-sensitive plate upon which the light image being viewed is focused. This light-sensitive plate is a part of an electronic circuit. The complete circuit contains a power source, the vidicon electron gun, the light-sensitive plate, and a load resistance. The light-sensitive plate is made of hundreds of individual elements. The image from the light source will be focused on the plate. Each element will then change its resistance. The resistance change is based upon the amount of light on each of the individual elements on the plate. The light energy is now used to control the flow of electrical energy. It must be further processed in order to make it useful.

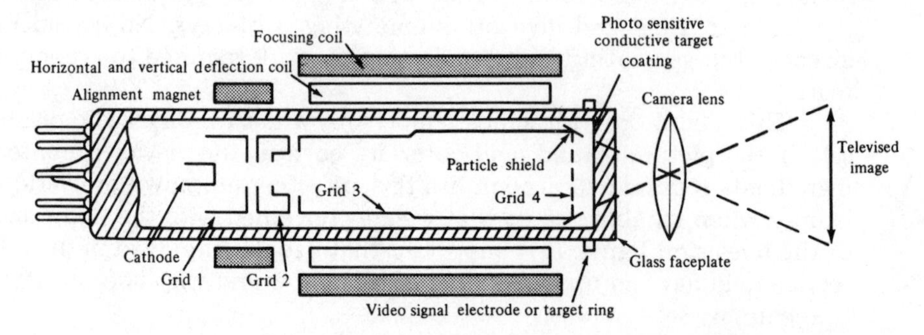

Figure 1-1 Cross-sectional view of the vidicon camera tube. (Clyde N. Herrick, *Television Theory and Servicing: Black/White and Color, 2nd ed.,* 1976. Reprinted with permission of Reston Publishing Co., a Prentice-Hall Co., 11480 Sunset Hills Road, Reston VA 22090)

The method of removing this information is shown in Figure 1-2(a). The system illustrated in Figure 1-1 is shown as an electrical circuit in this illustration. The circuit will have a current flow. The amount of current depends upon the total resistance and the applied source voltage. Each component in the circuit develops a voltage drop. The voltage drops are directly related to the amount of resistance of the component. There is one compo-

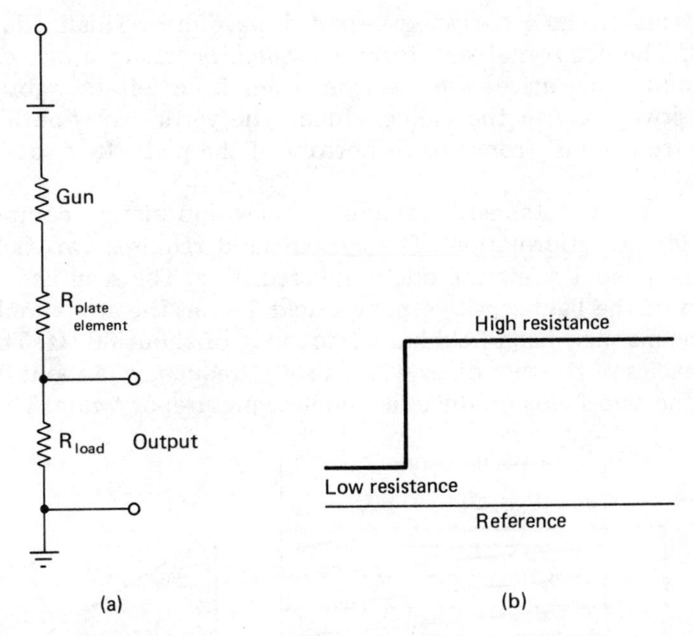

Figure 1-2 (a) Electrical equivalent of the camera and its output circuit. (b) Varying voltage that develops across the load resistor, shown in graphic form.

nent with the capability of varying its resistance. This is the light-sensitive plate in the vidicon. Each element on this plate acts as an individual resistance.

The amount of light striking the picture element determines its resistance. A high level of light, which we call *white level*, raises the resistance to a relatively high value. The voltage developed across the load resistance drops to a low level. If the light source is very low, the resistance of the element is also low. The variations in light develop a voltage across the load resistor that is related to the light intensity striking the element of the vidicon tube. This is illustrated in Figure 1-2(b). This concept is basic to the conversion of light energy into electrical energy for use in the television system.

Broadcast standards used in the television systems dictate that each element of the vidicon is transmitted in sequential order. This requires a method of positioning the electron beam on each picture element. This method is called *scanning*. Two types of movements are required in this system. One relates to the horizontal position of the electron beam. The other relates to the vertical position of the beam. The scanning position is developed by the use of two magnetic fields. A *deflection yoke* is placed around the sides of the vidicon tube. This yoke has two sets of coils. One is connected to a horizontal scanning system. The other is connected to a vertical scanning system. Each set of coils develops a magnetic field. The strength of the magnetic field is related to the electrical current flowing through the coils. The scanning systems produce a sawtooth-shaped waveform. This is illustrated in Figure 1-3. The horizontal waveform, or *signal*, occurs at a rate of 15,734 times a second. This drives the electron beam from left to right across the light-sensitive plate in the vidicon tube. The vertical sawtooth signal drives the electron beam from top to bottom of the plate. Its frequency is 59.9 hertz (Hz).

Industry standards establish a television picture composed of 525 lines of picture information. The system used requires two fields of horizontal lines. Field 1 scans the odd-numbered lines. The scan goes from top to bottom of the light-sensitive plate. Field 2 scans the even-numbered lines in the same manner. Each field has a frequency of about 60 Hz. The sum of the frequencies of the two fields will scan at a frequency of about 30 Hz. Interlacing of the two fields produces a complete picture, or *frame*. This consists of 525

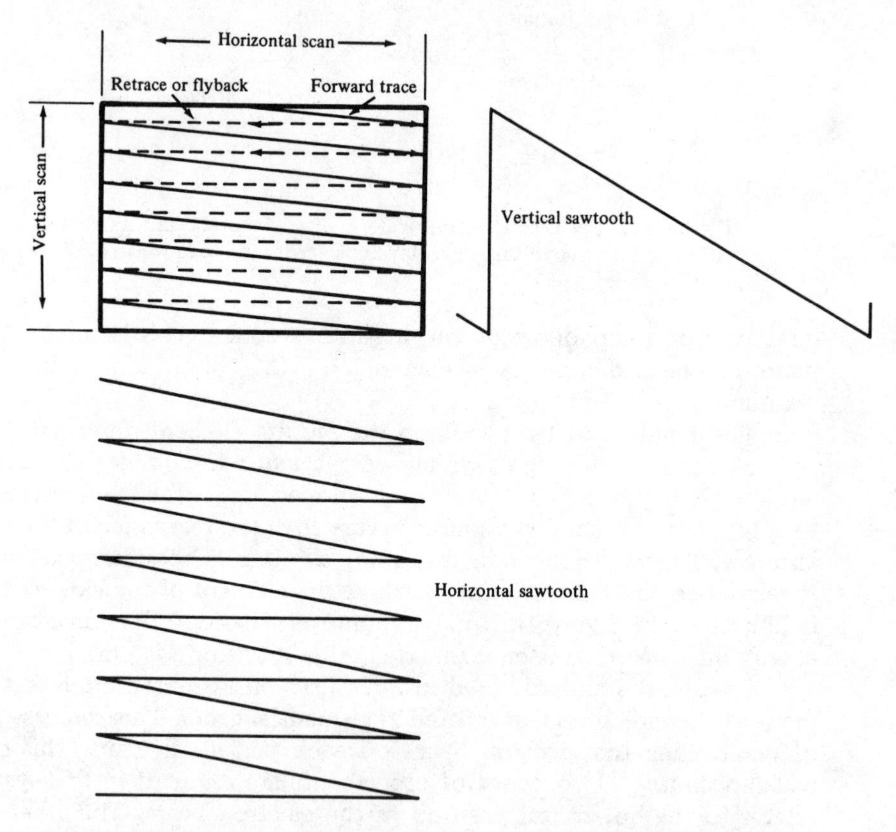

Figure 1-3 Sawtooth-shaped waveform, produced by the scanning section of the system. This waveform is used in positioning the beam in the camera and picture tubes. (Clyde N. Herrick, *Television Theory and Servicing: Black/White and Color, 2nd ed.*, 1976. Reprinted with permission of Reston Publishing Co., a Prentice-Hall Co., 11480 Sunset Hills Road, Reston, VA 22090)

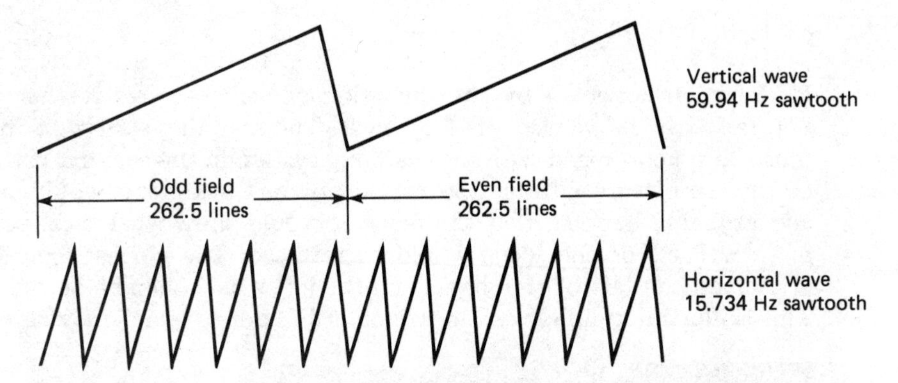

Figure 1-4 The relationship of the horizontal and vertical systems' sawtooth waves.

lines of picture information at a rate about 30 Hz. The horizontal rate of 15,734 Hz is developed from this format ($525 \times 59.9 = 15{,}734$). The waveforms and an explanation are illustrated in Figure 1-4. This process requires two scanning generators connected to the television camera vidicon-tube deflection circuits.

Each picture element is scanned in this manner. The result is several thousand individual picture elements, each with a voltage, or signal, level of its own. The information developed in this manner is further processed to produce individual lines of picture information. This is illustrated in Figure 1-5. Television standards identify a *white level* and a *black level* of picture information. White has a very low value of signal. Black, at the other extreme, develops a high signal level, or voltage. The camera signal is developed as the electron beam scans an individual line of picture information. The result is an electronic signal that corresponds to the picture being viewed on a specific line. This information is processed by the camera and the transmitter and broadcast to television receivers. This system produces the television picture information and will ultimately be returned into a viewable picture in the receiver.

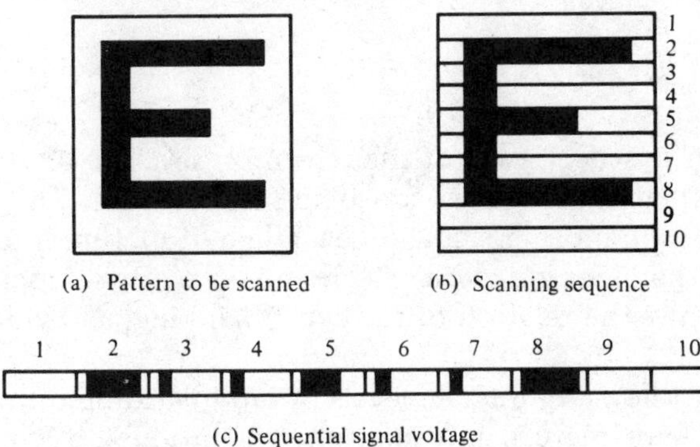

Figure 1-5 Development of a varying signal voltage as the figure is scanned by the camera tube circuits. (Clyde N. Herrick, *Television Theory and Servicing: Black/White and Color, 2nd ed.*, 1976. Reprinted with permission of Reston Publishing Co., a Prentice-Hall Co., 11480 Sunset Hills Road, Reston, VA 22090)

SYNCHRONIZING PULSES

Information collected by the television camera requires further processing before it can be viewed on a receiver. The scanning system in the receiver must be synchronized with the scanning system in the camera. If this did not occur, the picture information would not be the same. It would be possible, and probably happen, that the receiver would show what is called a *floating picture.* Parts of the scene would be reversed. The normal right side might appear on the left of the screen and the left would then be on the right side. This is illustrated in Figure 1-6, which is an "out-of-sync" picture.

Figure 1-6 Sync pulses are used to lock the picture in the receiver. A loss of sync will often produce a picture such as this.

Synchronizing pulses are developed by a *sync generator.* The sync pulses have a specific timing and shape. These pulses become a part of each line of picture information. Both vertical sync and horizontal sync pulses are produced. The vertical pulses are used to synchronize the vertical sweep in the television set with the vertical sweep at the camera. This is also true for the horizontal sync pulses. These pulses have different durations and timing. Both are required to synchronize picture information. Further details related to the function of sync pulses are presented in Chapters 16, 18, and 19.

BLANKING PULSES

The sync pulses are not seen by the viewer of the television set. The picture information is turned off when these pulses are being transmitted. This is accomplished by the addition of another signal on each line of picture information. This signal is called the *blanking pulse.* The composite video signal is shown in Figure 1-7. This shows the relation of the sync, blanking, and video information.

Broadcast standards for television include levels of modulation required for the picture. Absence of any picture information will produce a white

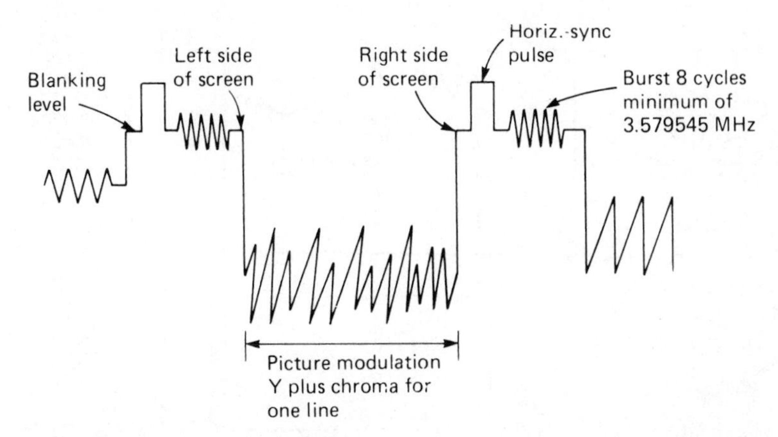

Figure 1-7 A line of composite video information. Note the placement of picture, blanking, color-burst, and sync signals. (Joel Goldberg, *Radio, Television, and Sound System Repair: An Introduction*, © 1978, p. 135. Reprinted by permission of Prentice-Hall, Inc.)

screen. The level of signal for this is about 12.5% of the total allowable signal modulation strength. The systems used to produce a television picture may be described as a *negative system*. In other words, the strongest level of signal produces a black picture. The lowest level of signal produces a white picture. The standard for a black picture is 75% of the allowable signal modulation strength. This point is the black level. The remaining portions of the allowable signal occur between the 75 and 100% level. This portion is called "blacker than black."

Signals transmitted in the blacker-than-black levels are not seen by the viewer. This portion of the picture signal contains synchronizing information. The picture tube is neither conducting nor producing a picture during this period. It is said to be "blanked." Pulses produced at the transmitter for this purpose are blanking pulses.

Figure 1-8 shows the relationship of the blanking pulses, sync pulses, and video information. This information is related to specific portions of the picture being shown on the receiver. The lighted screen of the receiver picture tube is called a *raster*. The shaded areas in the top picture are blanked. These are normally not seen by the viewer. The picture is enlarged or "overscanned" in the receiver so that the blanked areas are off to the sides of the picture tube.

Notice the relationship of the viewable picture and the trace scanning time. Every line has a period when the electron beam moves across the screen from left to right. This is called the *trace time*. The beam has to return to the left side to start the next scan. This period is called the *retrace time*. It is during the retrace time that the sync and blanking pulses are transmitted. The sync and blanking pulses transmitted during this period are used to control

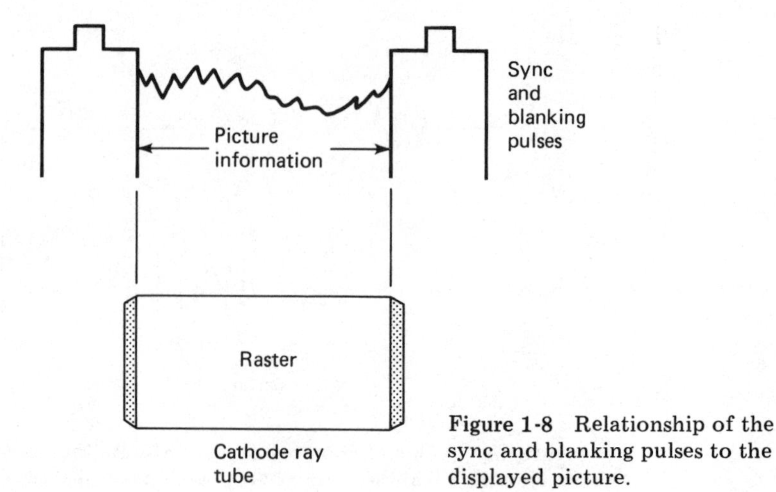

Figure 1-8 Relationship of the sync and blanking pulses to the displayed picture.

the horizontal sweep. They occur during each line of picture information. Additional pulses are required for vertical sync.

The vertical sync information is developed at the end of each picture field. Several lines of picture information are transmitted during a vertical blanking period. This period will blank between 13 and 21 lines per field,

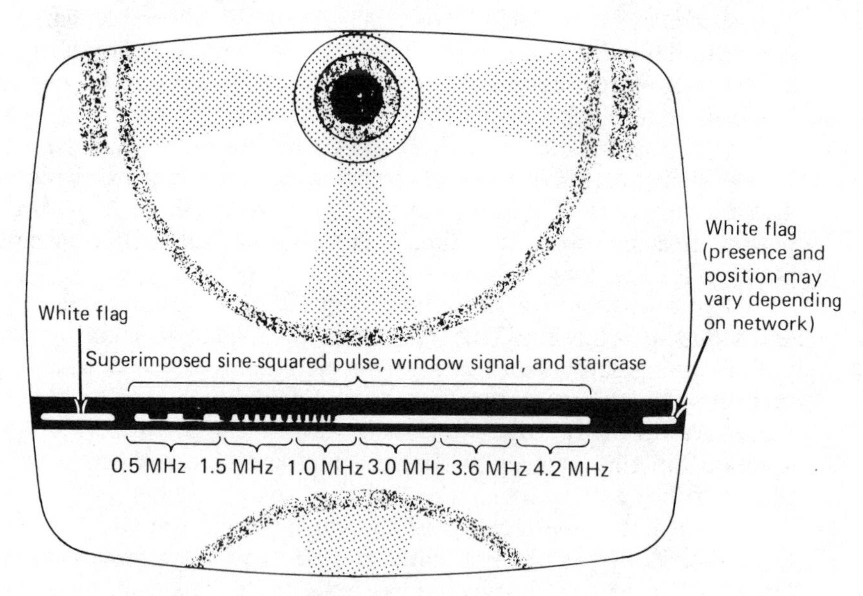

Figure 1-9 Vertical blanking interval and its relation to the observed picture. (Joel Goldberg, *Radio, Television, and Sound System Repair: An Introduction*, © 1978, p. 99. Reprinted by permission of Prentice-Hall, Inc.)

making a total of 26 to 42 lines per frame. A typical vertical blanking signal is shown in Figure 1-9. Most of this information is transmitted during the first one-half to two-thirds of the blanking interval. The equalizing and vertical sync pulses form a pattern often called a *hammerhead* due to its shape. It may be viewed by adjusting the receiver vertical hold control so that the blanking bar appears in the picture. Often, reduction of the contrast and brightness controls helps to bring this bar into view.

MODULATION

The process of transmitting radio frequency information requires a *carrier.* The carrier is necessary because the information is either too weak or on the wrong frequency to be transmitted. The human voice has a frequency range between 20 Hz and 20 kilohertz (kHz). If every person's voice could be broadcast at these frequencies, the resulting noise would be unbearable. Little communication would occur because of the large amount of interference created.

Each broadcast station has an assigned frequency, a *channel,* on which it operates. A signal, called a *carrier wave,* is created to keep the broadcast station on the proper frequency. Normally, no two stations in the same geographic area operate on the same frequency. The intelligence produced by the broadcast station is electronically added to the station carrier signal. This process is called *modulation.* The combined intelligence and carrier signals are called a *modulated carrier* signal. There are a variety of ways in which to modulate a carrier. One method is called *amplitude modulation* (AM). In this system the carrier's amplitude takes on the shape of the modulation signal. This is illustrated in Figure 1-10. The modulating process produces a carrier with the modulating signal shaping both the upper and lower halves of the carrier. If every television station were to broadcast the same information at

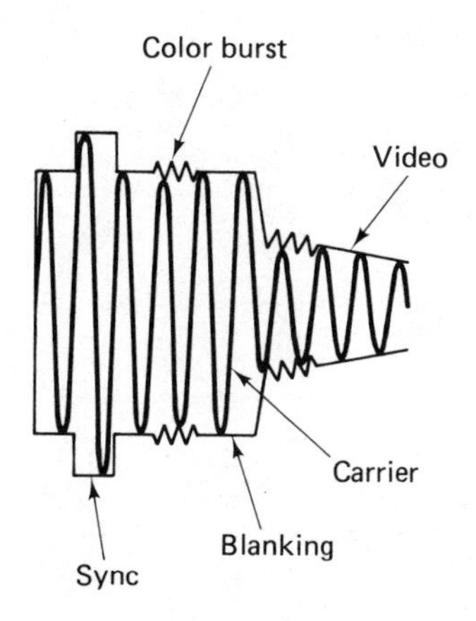

Figure 1-10 Modulated carrier wave, which assumes the shape of the modulation signal.

exactly the same time, all the signals would have exactly the same shape. The difference in the signals would be the frequency of the carrier signal.

A typical amplitude-modulated signal contains three components. The center one is the carrier. The other two are called *sidebands.* Each sideband contains intelligence and carrier information. The space occupied by the composite signal is determined by the frequencies of the modulating signal. Typically, a modulating signal with frequencies of up to 4 megahertz (MHz) will produce a carrier with a width of 8 MHz. Each of the two sidebands requires a space equal to the modulating frequencies; therefore, twice the modulating frequencies of space is required.

Broadcast standards for television signals do not permit the occupation of this much space. A system called *vestigial sideband* transmission is used instead. This system uses only a small portion of one of the sidebands. It occupies less space and allows additional broadcast stations in a given spectrum of frequencies. Figure 1-11 illustrates the shape of the television signal. The overall bandwidth of the signal is 6 MHz. The lower sideband is reduced to a very small portion of the bandwidth. Since all the information is contained in each sideband, one may be reduced and the broadcast signal will still contain all the necessary information. The station carrier signal is placed 1.25 MHz above the lower edge of the signal. The carrier frequency is different for each station. A list of assigned carrier frequencies is given in the Appendix.

Two other major frequencies are also shown. One relates to the color carrier and the other to the audio carrier. Broadcast standards place the color carrier 3.579545 MHz above the station carrier. The audio carrier is 4.5 MHz above the station carrier. Color information is not transmitted on a separate carrier. The information is transmitted using suppressed subcarrier systems.

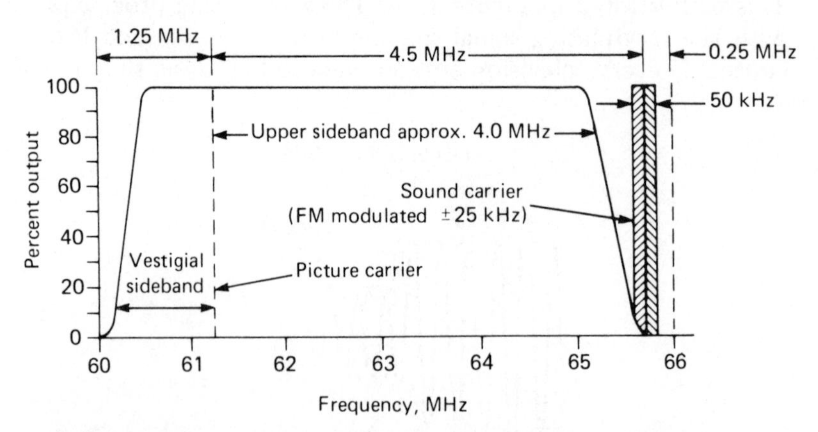

Figure 1-11 Spectrum analysis of the modulated video signal shows this waveform. (Joel Goldberg, *Radio, Television, and Sound System Repair: An Introduction,* © 1978, p. 97. Reprinted by permission of Prentice-Hall, Inc.)

The information in each sideband contains both carrier and intelligence. The color subcarrier is suppressed and only the sidebands are transmitted. The double-sideband suppressed carrier color information is modulated onto the main carrier. The process of using a subsidiary carrier to separate one channel of information from another is called *frequency multiplexing.* In this situation the color signal is separated from the black-and-white video signal. This system is suppressed carrier amplitude modulated and requires re-creation of the carrier in the receiver for detection of the intelligence. Color information consists of two signals that are 90° out of phase with each other. This important phase relationship is discussed later in the book.

COLOR SYNC PULSES

Color picture transmission requires the addition of another sync pulse. This pulse is used to synchronize the color information and to turn on the color-processing blocks in the receiver. It is called the *color burst.* Each color receiver has a color oscillator. It is required in order to re-create the suppressed carrier. Modulation systems that use a suppressed carrier require the re-creation of the carrier in the receiver so that the detection process may occur. Color hue is dependent upon the phase relation of the color information. The receiver color oscillator reestablishes the color carrier with a correct phase relation with the transmitted signal.

The color sync pulses or color burst signals consist of a minimum of eight cycles of a signal whose frequency is 3.579545 MHz. This signal rides on the "back porch" of the horizontal blanking pulse. Its position is shown in Figure 1-12. The color burst is used to turn on the color-processing blocks in addition to providing color sync for the receiver.

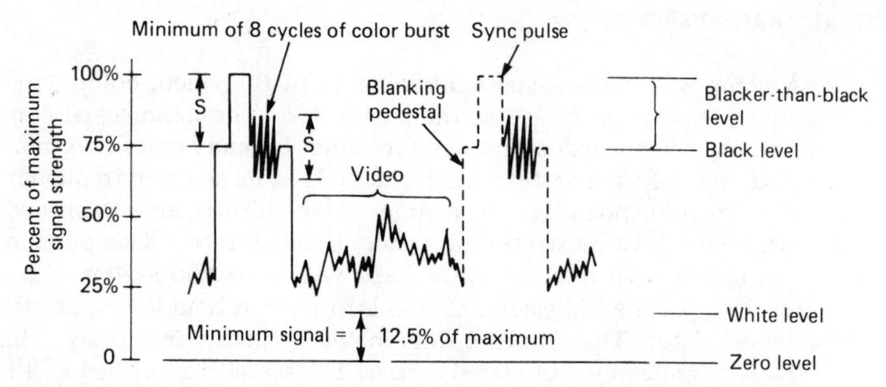

Figure 1-12 Color sync information, called burst, is placed on the "back porch" of the horizontal blanking pulse. (Joel Goldberg, *Radio, Television, and Sound System Repair: An Introduction,* © 1978, p. 101. Reprinted by permission of Prentice-Hall, Inc.)

SPECIAL SIGNALS

A portion of the vertical blanking pulse lines are used for purposes other than those for which they were originally intended. Two signals are often transmitted during this blanking period. One of these is called the *vertical interval test signal* (VITS). This information normally originates at the network. It is used to establish a reference white level at each station. This enables each broadcast transmitter to send out the same color value signal. VITS signals are transmitted using lines 17 to 20 of the vertical field. They are not used in the home receiver.

A second signal, one that is used in the receiver, is called the *vertical interval reference signal* (VIRS). This is transmitted on line 19 of each vertical field. The VIRS signal is used in the receiver. Its purpose is to establish correct color intensity and hue in the receiver. Special circuits in the receiver react to this signal in order to automatically set these controls. It is probable that other signals, such as time information, could be transmitted during the vertical blanking period in the future.

AUDIO SIGNALS

The other major signal transmitted from the television broadcast station is the audio signal. This information is transmitted as a frequency-modulated (FM) signal. Most stations use a separate transmitter for this purpose. The FM audio signal is broadcast on a frequency that is 4.5 MHz higher than the station's video carrier signal. This relationship is constant for all television broadcast stations.

SIGNAL TRANSMISSION

A composite video signal containing all of the video, color, sync and audio information is processed at the transmitter. The basic signal consists of two carriers, one for video and one for audio. In many cases two transmitters are used, one for each system. Each channel has its assigned frequency. These are given in the Appendix. The manner in which these are established is interesting. Figure 1-13 illustrates the system in block form. Each portion of the system has its own requirements. Start with the audio system. An oscillator is required for the FM carrier. Audio information is added at, or just following, the oscillator. This causes a shift in the oscillator frequency. This shift produces a frequency-modulated carrier. The signal is amplified as it is processed through additional power amplifier stages. After the final amplifier stage the signal is sent to a broadcast antenna.

The signals required for transmission of picture information require

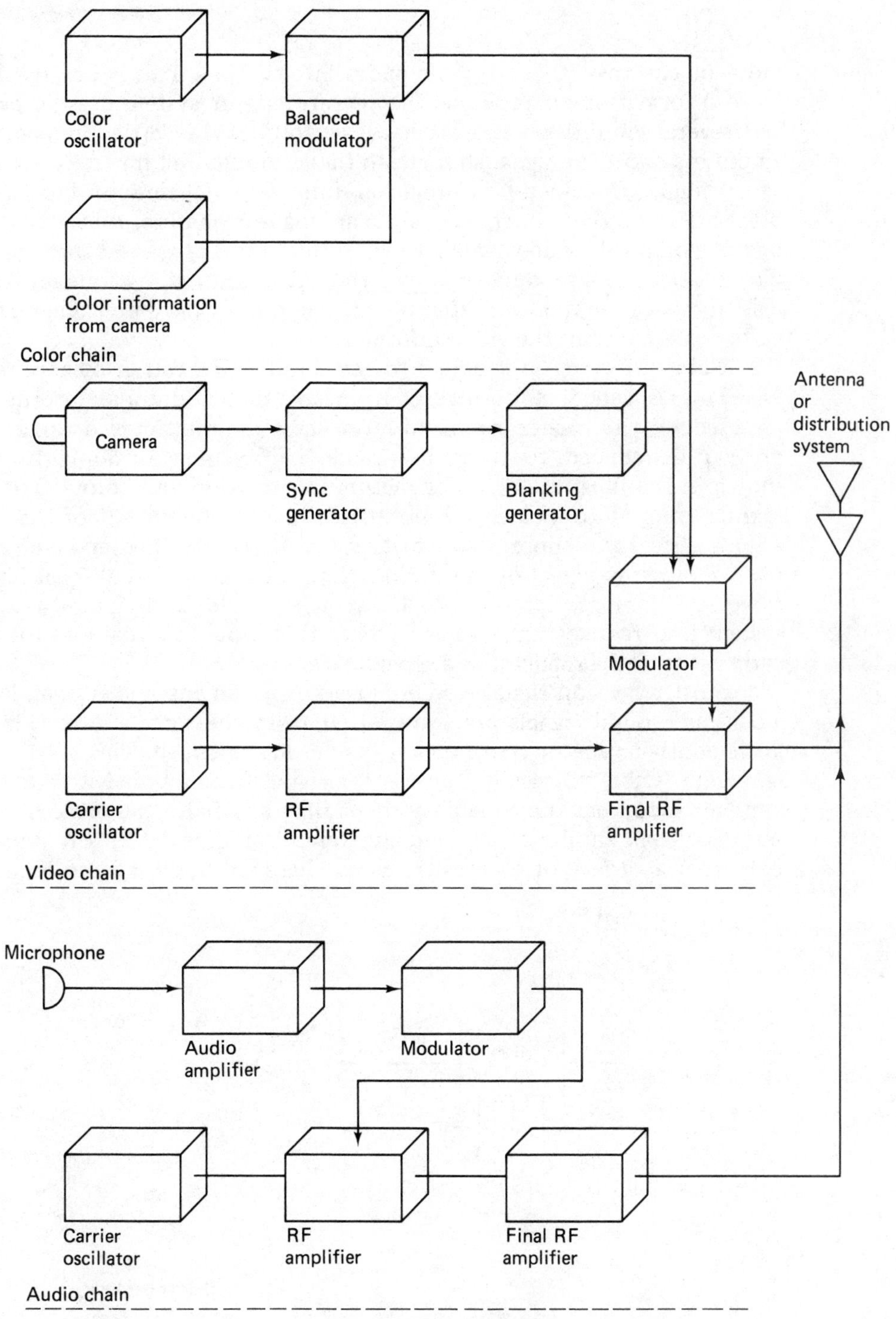

Figure 1-13 Block diagram representing signal development at a typical TV transmitter.

more blocks than those for the audio information. This is because there is more information to process. The picture carrier system uses an oscillator and several amplifiers. The oscillator establishes the carrier frequency. The amplifiers boost the signal strength to the level required for transmission.

Video, or camera, information follows a "chain" of blocks. These blocks develop the proper sync and blanking information. It is added to each line of composite video signal. This information is processed through a modulator block. The modulator allows the video and radio-frequency (RF) signals to mix in the final amplifier block. The result is an amplitude-modulated carrier that contains the video information.

Color information is added to the station's RF carrier by another process. This is called *multiplexing.* It means that additional information is imposed on the carrier signal. In most cases the multiplexed carrier is suppressed, or reduced, to a very low value. The systems for doing this are also shown in the illustration. Color picture information and color carrier information are sent to a balanced modulator block. The output of this block is a double-sideband suppressed carrier signal. It is amplified and added to the main RF carrier signal of the station. Suppression of this color carrier is necessary to eliminate interference between the color and video carriers. This system also reduces unnecessary power that would be required for a color carrier if it were broadcast as a separate signal.

Most television signals are broadcast from an antenna system. Both the video and sound signals are transmitted from the same antenna. In recent years another type of transmission system has emerged. This is called *cable television.* Cable television may use the signal from a broadcast station. The system usually has the capability of sending a signal it has created. A distribution system similar to the one illustrated in Figure 1-14 is often used. Signals from a variety of sources are used. These are amplified and sent out, via

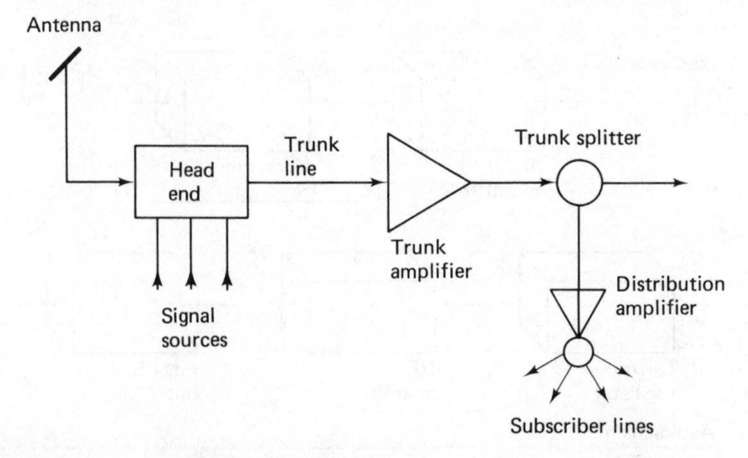

Figure 1-14 Typical block layout for a receiving system antenna distribution system.

wire cable, on several main trunk lines from a distribution center. The signals are split further and sent to individual subscribers. Systems of this type are used in homes, industry, and education. Modifications of this system are used as master antenna distribution systems in large apartments.

SIGNAL RECEPTION

The television signal is transmitted as an electromagnetic wave. Radiated signal power from the broadcast station's antenna may be several kilowatts. The signal is carried to the receiver located in the broadcast service area. This area will often vary from one geographic area to another. Typically, the signal travels on a line-of-sight basis. Television signals do not go beyond the horizon under normal propagational patterns. The signal, when it reaches the receiver, is measured in units of the microvolt (μV). Good broadcast reception standards require about 1000 μV of signal at the receiver's antenna terminals.

ANTENNAS

There is no "standard" TV antenna. There are, however, several basic types of antennas. The type of antenna used as a reference standard is called the *dipole.* It has two elements. The length of the antenna element is directly related to the electrical wavelength of a common half-wave antenna. This formula is

$$L\ (\text{ft}) = \frac{462}{f\ (\text{MHz})}$$

where L represents the length of the antenna. It is based upon the point of resonance for this type of antenna using the television channel frequencies.

Signal reception values may be increased by use of an antenna that exhibits qualities of *gain.* Gain refers to the ability of the antenna to increase the strength of the received signal to a higher level than that received with a dipole. This is accomplished by adding elements to the antenna system. Figure 1-15(a) shows a *dipole antenna;* part (b) shows a *multielement antenna.* The multielement antenna may have several reflector and/or director elements. Each additional element increases the amount of signal present on the driven element. This signal is carried on the transmission line from the antenna to the receiver.

Figure 1-16 shows the directional qualities of each antenna. The multielement antenna is called a *yagi antenna* if it is designed to operate on one channel, or frequency. Note the increase in forward reception capabilities with the yagi type of antenna.

In many instances one must use an antenna that is not specifically made for one channel. This is particularly true when there is more than one TV

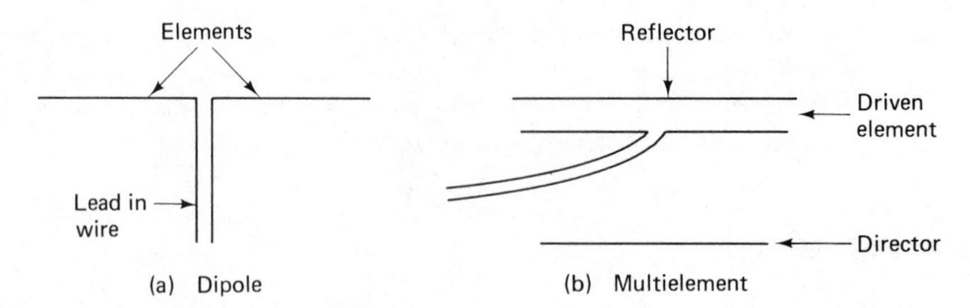

Figure 1-15 (a) A dipole antenna has one set of elements. The length is determined by the electrical wavelength of the signal. (b) A multielement antenna has director and reflector elements in addition to the element connected to the receiver.

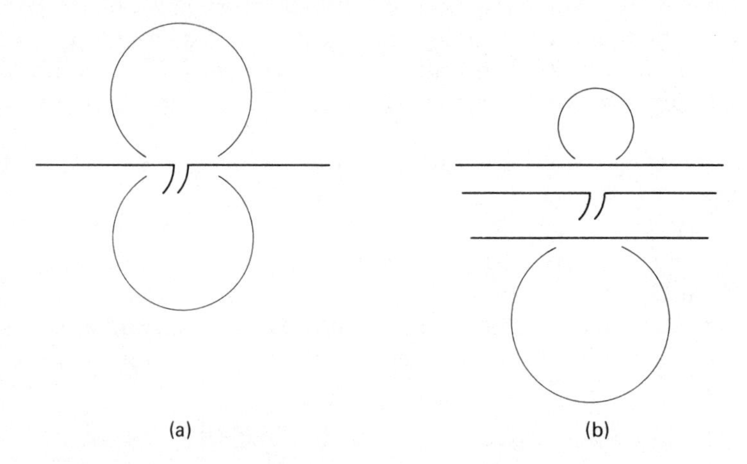

Figure 1-16 A multielement, yagi type of antenna has reception capabilities that are much better than those of the dipole antenna.

channel in the reception area. The choices available include installation of an antenna for each channel or use of a multichannel antenna. If one were to use a multichannel antenna, such as shown in Figure 1-17, the orientation of the antenna to the broadcast station may be critical. Electromagnetic waves induce the largest voltage on the antenna when the antenna is pointed directly at the broadcast station. The weakest amount of signal is received when the antenna elements are in line with the transmitter. In communities with more than one station a compromise in antenna orientation may be required. The orientation may have to be the best position for reception of all the channels. Another solution to this problem is the use of an antenna rotator. The antenna is then rotated for best reception of each channel.

One problem that occurs with reception is that related to multipath signals. The TV signal has a tendency to bounce, or reflect, off flat surfaces. This may cause poor reception. The antenna may be receiving two signals

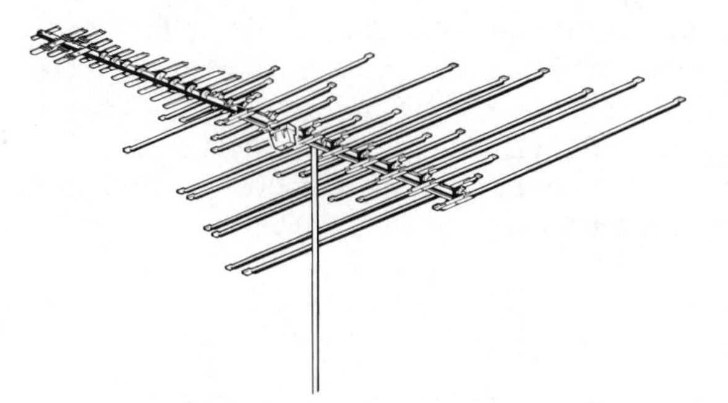

Figure 1-17 A multichannel antenna has elements of different lengths in order to properly receive different channels.

from the station. The second signal, as shown in Figure 1-18, will also influence the antenna. Its time in reaching the receiver antenna is longer than that for the direct signal. The result is a double image on the receiver called a *ghost*. The best way to eliminate this problem is to reorient the antenna for minimum ghost reception.

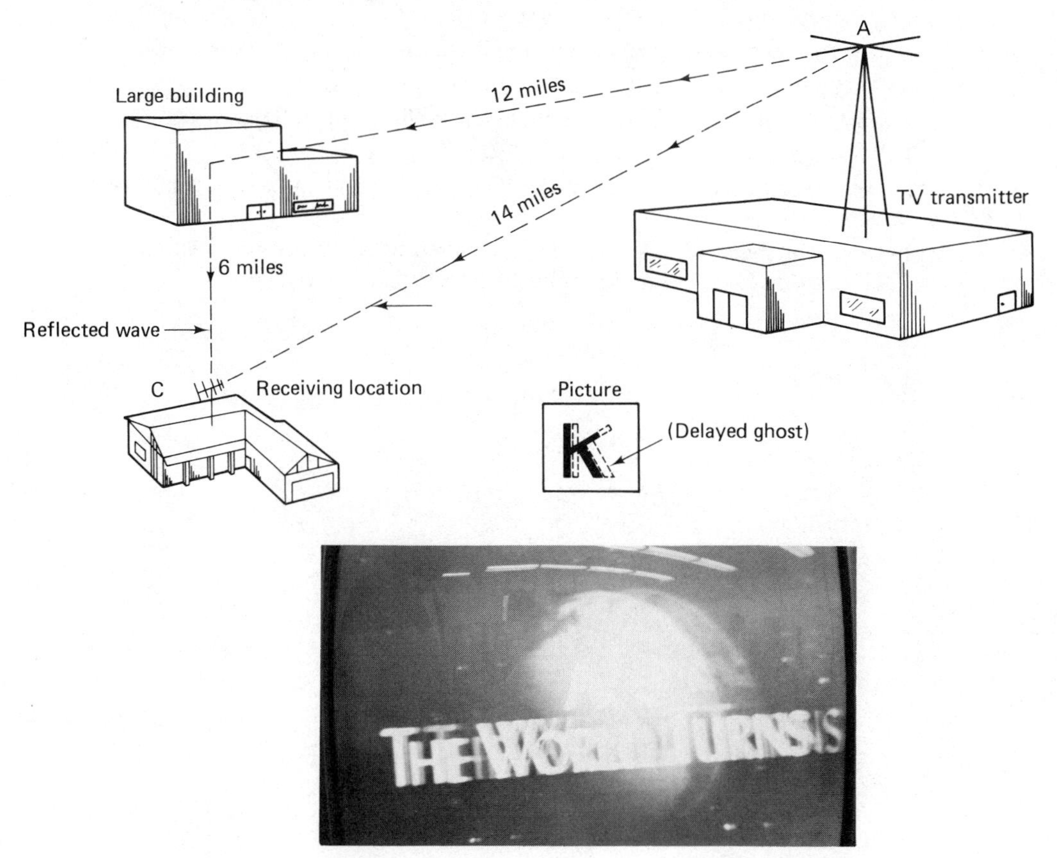

Figure 1-18 A reflected signal of the same frequency will produce this "ghost" effect on the picture.

Do not overlook the importance of a good antenna system for reception. Broadcast stations use antennas that provide gain to their signal. This may be as much as three or four times the amount of energy from the transmitter. The system increases the efficiency of the transmitter. The same is true for a receiver. A good antenna system will provide maximum signal strength reception, probably with gain, for the receiver.

QUESTIONS

1-1. Name seven major components contained on a TV station carrier.

1-2. Explain what each of the seven components does to develop a viewable signal in the receiver.

1-3. What frequency is used for horizontal signals?

1-4. What frequency is used for vertical signals?

1-5. What is the frequency of each field?

1-6. How many lines are used to develop a picture?

1-7. What percentage of signal modulation is required for a black-level signal?

1-8. What modulation signal level is used for sync pulse signals?

1-9. Name the three components found in an amplitude-modulated carrier signal.

1-10. What is meant by the term "ghost"?

1-11. Draw a picture of the modulated TV carrier signal. Identify each component in the signal.

1-12. Draw the shape of the spectrum wave used by a TV station. Identify sound and picture carriers as well as the color subcarrier signal placement.

Chapter 2

Television Receivers

Servicing of any product is usually easiest when the service technician has a good understanding of how the system functions. This is very important for today's technician. The manner in which TV sets are produced is an evolving technology. Discrete components such as tubes and transistors have been replaced by integrated circuits. The day of replacing a specific tube or transistor in order to effect a rapid repair seems to have disappeared. The service technician uses test equipment to localize a defective component. Symptoms of trouble in yesterday's TV set often relate to the same block in today's set. One can only guess how this will relate to tomorrow's set as technology keeps changing. The best way to start to service any device is to get to know how the device functions. Keep this in mind as we progress through the workings of modern TV sets.

THE BLOCK DIAGRAM

Block diagrams for electronic devices seem to be an excellent way of first describing the operation of the set. Block diagrams have not changed too much until recent years. Often, different manufacturers give somewhat different names to the blocks for their sets, but in general, the names are not too different. Learn these block names as well as what function each serves in the set. This knowledge will make servicing the set much easier. You will also have to know how to read a schematic diagram and how to use test equipment effectively to service any set successfully.

The purpose of any block diagram is to break the device into functional units. Each block in the device serves a valuable function. If it did not, it would not be included in the device. Blocks may contain one or more subunits. For example, an audio amplifier block may be constructed having two or more stages of amplification. Its purpose, however, is to amplify audio signals. All of these stages may be shown as one block. A schematic diagram would show each stage. Remember that the purpose of the block is related to *function* as opposed to specific design of the electronic circuits in the blocks. Material presented in this chapter covers block diagrams. Both color and non-color sets are discussed. Also included is the relationship between the blocks found in both types of sets.

RECEIVER BLOCKS

A block diagram for a traditional (sets produced through the late 1970 era) color TV is shown in Figure 2-1. The diagram provides the names most often used to describe each block. Each major section of the TV set is outlined with a heavy black box. These major sections are used to show the relationship between this type of set and those employing newer integrated-circuit technology. The newer style of set is described in the section of the book that follows the discussion of the traditional set.

TUNER SECTION

The tuner in any television set has the function of receiving a signal, amplifying it, and converting it into an intermediate-frequency (IF) signal. The blocks related to the tuner are shown in Figure 2-2.

Ultra High Frequency (UHF) mixer. This block is connected to the UHF antenna. It contains a mixer and RF tuning circuits. Its purpose is to receive the UHF signal and to electronically mix the received signal with a signal from the UHF oscillator. The output of the mixer block is an IF signal whose frequency is about 44 MHz.

UHF oscillator. This block creates its own signal. It receives power only when the set is in the UHF tuning position. The oscillator signal is fed to the UHF mixer to create an IF signal. The oscillator block contains tuned circuits to create the required generated frequency for the mixer. This signal's frequency varies depending upon the frequency of the received channel.

Very High Frequency (VHF) RF amplifier. This block selects the proper channel. It also amplifies the received signal. The output of the RF amplifier is sent to the VHF mixer block.

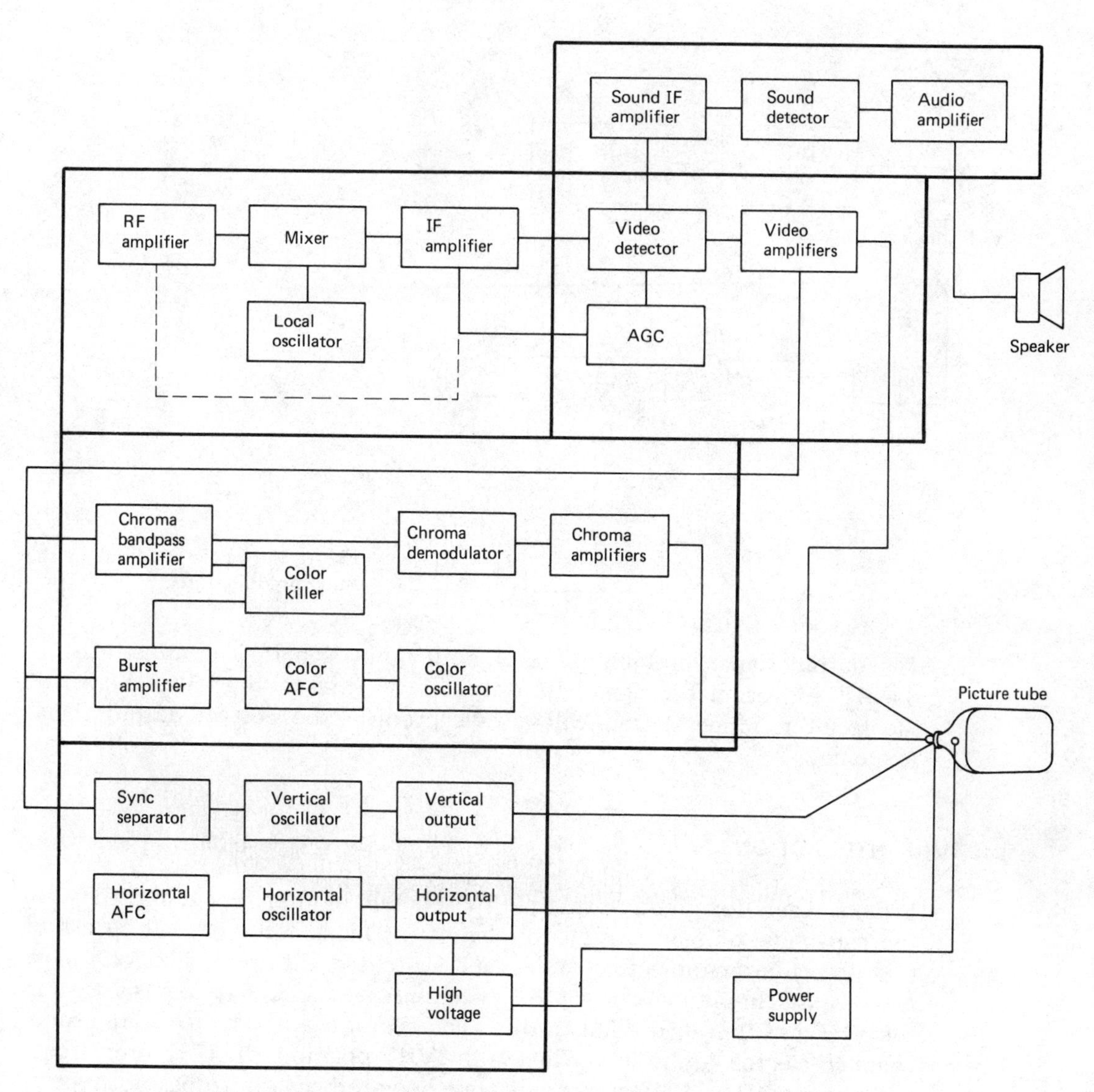

Figure 2-1 Functional block diagram of a color receiver produced in the 1970 era.

VHF oscillator. Oscillators, by definition, create an electronic signal. The purpose of this block is to create a carrier type of signal. Each channel has its own oscillator signal frequency. The oscillator block has no input. Its output is connected to the VHF mixer block.

VHF mixer. The term *mixer* is used in electronic devices to indicate a function that will accept two input signals and produce one output signal.

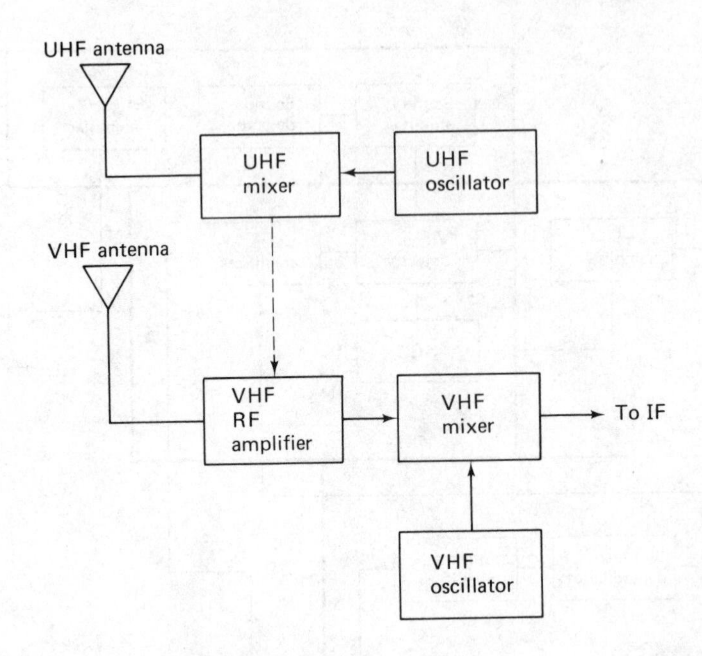

Figure 2-2 Blocks found in the tuner are illustrated.

The output signal contains parts of both input signals. The mixer accepts the RF and oscillator signal. Its output is a signal at the IF frequency of the receiver. Mixer action will always produce the correct output signal frequency.

UHF–VHF INTERACTION

Federal regulations require that all TV sets sold in interstate commerce have the capability of receiving all 82 channels. This is accomplished by use of two tuners, interconnecting cables, and switching. Figure 2-3 illustrates how this works. These tuners have a B+ switching system. This system is shown in the VHF position in the illustration. The B+ switch is attached to the tuner channel selector knob shaft. When in the VHF position, the UHF oscillator is not powered. The UHF mixer does not operate because it does not receive the oscillator signal. In effect, it is turned off. Signals present at the VHF antenna are processed by the VHF tuner blocks.

The B+ switch is thrown when the tuner is set on UHF. In this mode the VHF oscillator does not receive power. The UHF oscillator functions and the UHF mixer provides its output signal. This signal is fed to the VHF RF amplifier as an IF signal. It is amplified and sent to the VHF mixer, where it is also amplified as an IF signal.

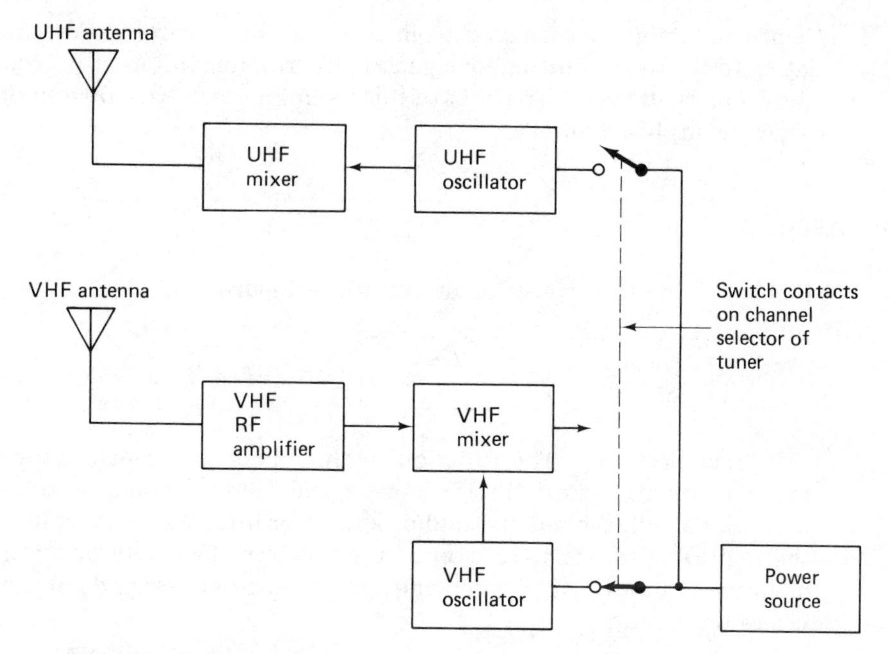

Figure 2-3 Interconnections between the UHF and VHF tuner systems.

IF SECTION

Blocks related to the IF section are shown in Figure 2-4.

IF amplifier. The signal developed by the action of the mixer is called an *intermediate frequency*. This signal receives about half of the total amplification in the set in this block. This is necessary due to the very low level of signal developed in the tuner section of the set.

Keep in mind that set design is not a constant factor. Design tends to vary from one manufacturer to another. Also, there are design differences in sets produced by the same manufacturer. This is particularily true where

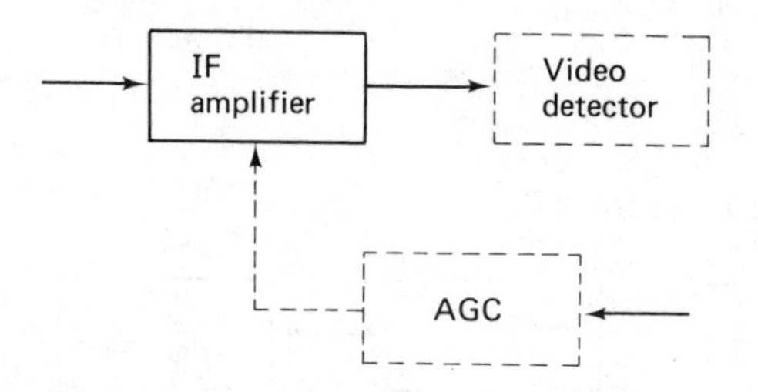

Figure 2-4 Intermediate-frequency (IF) blocks are illustrated.

signals are *split*, or branched from the main path. Information presented relating to points of splitting for signals in this section will use the block diagram shown in Figure 2-1. Variations of this technique will be shown in other block diagrams in this chapter.

VIDEO SECTION

Blocks related to this section are shown in Figure 2-5.

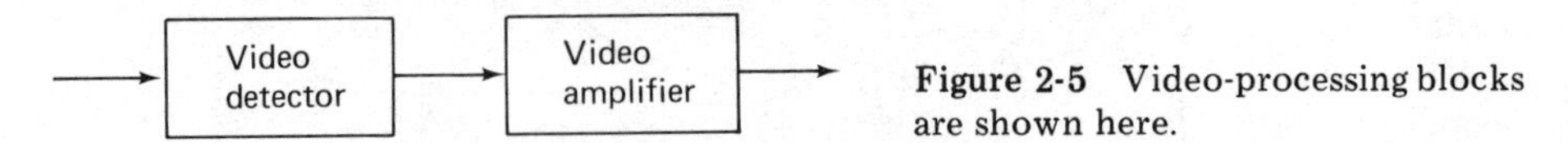

Figure 2-5 Video-processing blocks are shown here.

Video detector. The function of this block is to separate the composite video information from the IF carrier signal. The remaining signal, which contains video, sync, blanking, audio, and color information, is split and sent to the required processing section of the receiver. This may occur in the block or in many cases, in the video amplifier block. Set design determines exactly where this occurs.

Video amplifiers. Composite video information is amplified in this section and sent to the picture tube. The video amplifier block has the capability of providing almost half of the total amplification in the set. This amplifier is often called the *Y amplifier*. This term is used to identify noncolor video information in the picture, called the *luminance signal.*

AGC. The full name for this block is *automatic gain control.* This block samples some of the video signal. The sample is then filtered and converted into a direct-current (dc) control voltage. The dc control voltage is used to control bias on the IF and RF amplifier blocks. The bias sets the amount of amplification for these blocks. This amplification is dependent upon the level of received signal. The purpose of the AGC block is to set the level of amplification automatically so that a nearly constant amplitude signal is presented to the video detector.

SOUND SECTION

Sound blocks are shown in Figure 2-6.

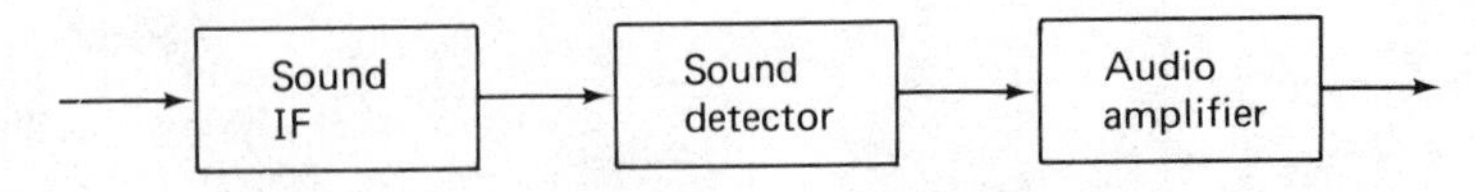

Figure 2-6 Sound section blocks in the receiver.

Sound IF amplifier. Audio information as processed by the IF amplifier is processed in this section. Audio has its own IF frequency. The FM signal is processed by all the previous blocks in the set. The video and audio carriers produced by the transmitters are separated by 4.5 MHz. A carrier IF frequency is developed when the signals are processed in the mixer and IF blocks. The 4.5-MHz sound IF frequency is amplified in this block.

Sound detector. The FM sound is detected in this block. Separation of carrier and intelligence occurs. The detected sound is then sent to the next block.

Audio amplifiers. Sound, or audio, information is amplified in this block. The signal is then sent to the speaker, where it is converted into sound waves.

COLOR SECTION

Color blocks in this section are shown in Figure 2-7.

Chroma bandpass. Color information in the receiver is tuned, amplified, and detected together with the rest of the picture information. This signal is processed as an amplitude-modulated suppressed carrier signal.

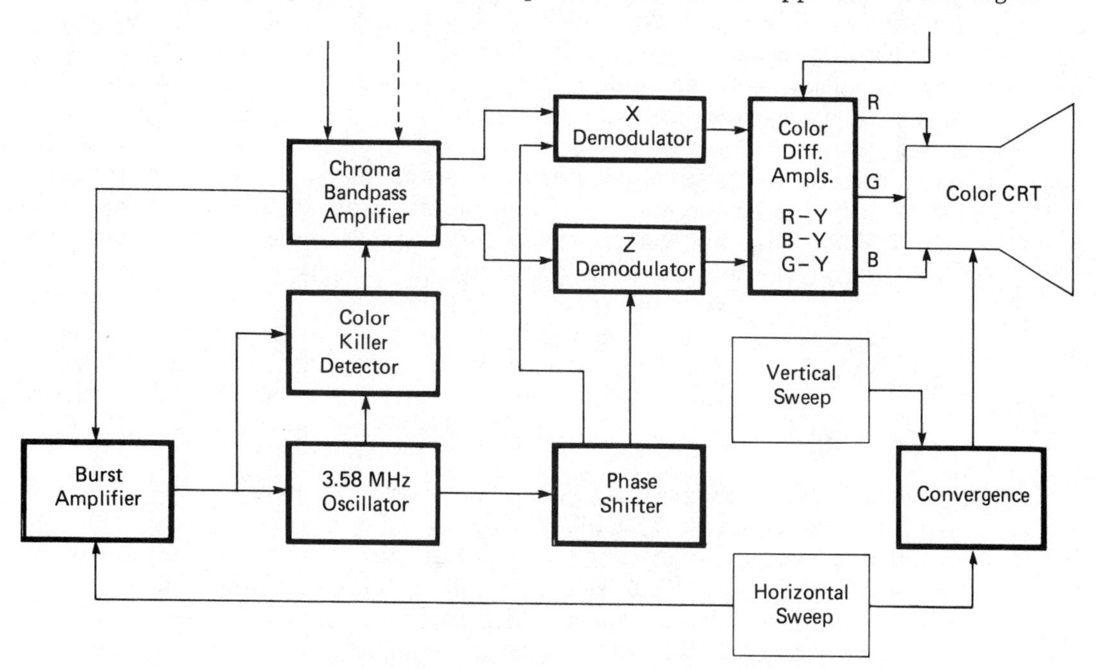

Figure 2-7 Color-processing blocks are often arranged in this manner.

It cannot be detected, or demodulated, until its carrier is reinserted in the signal. The signal is clustered around its carrier frequency of 3.579545 MHz.

A tuned amplifier circuit in the receiver will accept a signal centered at this frequency. It tends to reject signals whose frequencies are very much removed from the color carrier's frequency. The function of this block is to amplify color information.

Burst amplifier. Color burst information is also transmitted at 3.579545 MHz. It is placed on the "back porch" of the horizontal blanking pulse. This information is used to turn on the color-processing blocks and to synchronize the color oscillator in the receiver with that in the transmitter. The burst signal is amplified in this block. A feedback signal from the horizontal output transformer keeps this signal in sync.

Color oscillator. The suppressed 3.58-MHz (rounded off from 3.579545) color carrier has to be re-created in the receiver. The requirement for this signal is that is must be of the correct phase and amplitude in order to be used by the color demodulators. The output of the color oscillator is fed to the color killer and a phase shifter.

Color killer. The present method of transmitting television signals must provide a means of turning off the color circuits when a noncolor signal is transmitted. This is accomplished in the receiver. A color killer block will establish an operating bias for the bandpass amplifier. This bias cuts off the signal. This, in turn, shuts down the color-processing blocks. The color killer is on when a burst signal is not present.

Phase shifter. This block is often a part of the output circuit of the oscillator. Its purpose is to create two color oscillator signals that have a 90° phase shift. These re-create the missing carriers used when the color information was multiplexed to the station carrier signal.

Color demodulators. Color information and its missing carrier join again in these blocks. Once they are reestablished as amplitude-modulated signals, they may be demodulated. The terms "X" and "Z" are used because the two signals have different phases when compared to the basic color signal phases.

Color-difference amplifiers. The three basic color signals of red, green, and blue are created in this block. Keep in mind that there are no color electrons. The reference here is to a waveform that will be used to control the picture tube's electron gun related to one specific color. Most color picture tubes require signals that control each of the three electron guns. The three guns when given the correct percentage of signal will create a white picture.

The terms R-Y, G-Y, and B-Y refer to the R, G, and B signal information without the Y, or video information. When two color signals are demodulated, the third signal is created by mixing action. This will create the missing color signal. Color information and video information are combined either in this block or in the picture tube to create the color image.

Convergence. This term means "meeting." Each of the electron beams from the gun in the picture tube must meet at the same point on the face of the tube. Not only must they meet, they must meet at the same time. The physical layout of the electron gun often will look as shown in Figure 2-8. When the electron beam from each of the guns hits the center of the tube, they all are the same length. When these beams are deflected due to the action of the magnetic fields in the yoke, they change length. Individual lengths depend upon the beam position in the tube. Different lengths are related to different times of deflection. Compensation for the different timing of each electron beam is done in the convergence block. The shape of the sweep wave is changed so that all beams meet at almost every point on the screen of the picture tube. This problem is not as apparent with the "in-line" type of picture tube as it is with the older tri-gun tube.

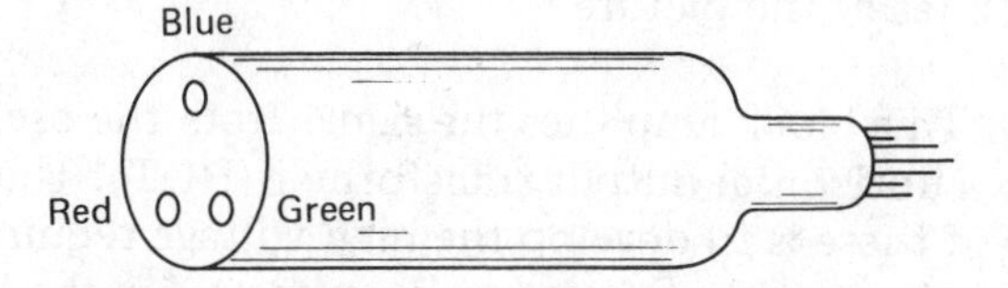

Figure 2-8 The geometry related to the three guns in a color CRT.

SWEEP SECTION

All sweep-related blocks are shown in Figure 2-9.

Sync separator. Another signal that requires processing is the sync signal. The sync signal is processed in this block. Other information that is with this signal is not processed in this block. The output of the sync separator is sent to the vertical and horizontal sections. It is used for timing of these oscillators.

Vertical oscillator. This block develops the vertical sawtooth waveform. This shape of waveform is required for proper scanning of the picture tube. Sync signals from the sync separator are used to determine the correct phase of this wave.

Vertical output. The vertical output block amplifies the signal received from the oscillator. The output of this block is used to develop a magnetic deflection field in the vertical windings of the yoke. Another output is sent to the picture tube to blank it during retrace periods.

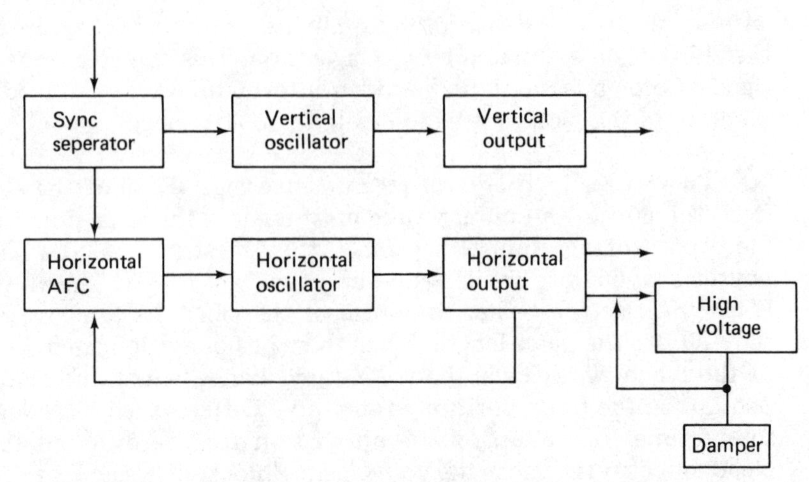

Figure 2-9 Blocks containing sweep circuits are shown.

Horizontal oscillator. The horizontal oscillator develops a sawtooth-shaped wave at the horizontal frequency of 15,734 Hz. This signal is used for scanning of horizontal lines of the picture.

Horizontal output. This block amplifies the signal from the oscillator. Its output is connected to a horizontal output transformer (HOT). The HOT has several functions. One of these is to develop the high voltage required for the picture tube. Another is to provide the proper waveform for the horizontal deflection yoke winding. A third output is used to control the frequency of the horizontal oscillator and color signals. In many sets the horizontal signal is also used to develop a low-voltage power source for other circuits.

Horizontal automatic frequency control. This block, commonly called AFC, does exactly as its name implies. It is used to control the frequency of the horizontal oscillator. This is accomplished by comparing two signals. The frequency of the horizontal sync signal is compared to the frequency of the horizontal signal. The horizontal signal is fed back from the output stage. The result of this comparison is a dc control voltage. The dc control voltage is used to establish the bias and thus the operating frequency of the horizontal oscillator. A shift in frequency causes a shift in the dc voltage. This then changes the bias and the frequency of the oscillator. Sampling is done at the horizontal rate of 15,734 Hz. Frequency correction keeps the oscillator on the proper frequency.

High voltage. Each picture tube requires a large value of high voltage. This voltage is developed during the retrace period of the horizontal waveform. The horizontal output transformer has a high-voltage winding. When the trace "flies back" to its starting point, the high voltage is developed. This transformer is also called a *flyback transformer* because of its action. The high voltage will be between 600 and 34,000 volts (v). The specific voltage is dependent upon the requirements of the picture tube. Generally, the higher voltages are required for the larger tubes and lower voltages for the smaller tubes.

Damper. The purpose of the damper block is to control undesirable oscillations. High voltage is produced as a result of horizontal oscillator action. The resulting wave is amplified and sent to the high-voltage transformer. This transformer, as all transformers, has a resonant frequency. It tends to produce a wave, or signal, at this frequency. It also tends to want to keep on producing this frequency after the input signal is stopped. The damper action will cause this oscillation to decrease. This is necessary for proper scanning. Without damper action, scanning would either stop or be highly irregular.

POWER SUPPLY SECTION

Blocks for this section are shown in Figure 2-10. Normally, one block is used for the power supply. This is shown in part (a) of the figure. The subblocks shown in part (b) are presented to provide a better understanding of this block.

Transformer. The transformer serves two basic functions. One is to provide electrical isolation between the operating circuits of the set and the power line source. The second is to provide the proper voltages and current (power) to the set.

Rectifier. Rectifiers change the alternating current from the transformer into direct current. The dc is required for necessary operation of the tubes, transistors, and integrated circuits in the set. The output of the rectifier is a pulsating dc voltage.

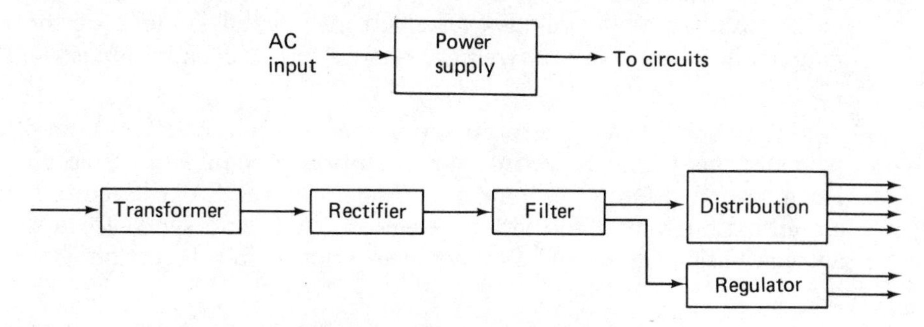

Figure 2-10 Power supply blocks.

Filters. Filters are required to smooth the varying dc output from the rectifier. Many sets have multiple rectifier and filter circuits. These are necessary when there is more than one dc voltage required for set operation.

Regulator. Newer technology, solid-state sets require a fairly constant voltage from the power supply. Voltages often have to be held at some value for a variety of output load demands. This requires a special circuit. This circuit is called a *voltage regulator.* It is designed to have a specific voltage available as load resistance varies.

Distribution. Each TV set has several blocks. Each block requires its own voltage level. Power variations due to set operation in one block must not influence any other blocks. This is accomplished by use of a distribution system.

MODULAR RECEIVER CONSTRUCTION

It seems that no two TV sets are identical. This presents some minor problems when one writes a book intended for general coverage. There are, however, some very common elements within all receivers. One way of describing these is shown in Figure 2-11. The development and extensive incorporation of integrated-circuit technology in sets of the 1980s make this concept very important. Each of the blocks shown in the illustration contains at least one of the individual blocks described earlier in the chapter. Look at each of these blocks. Try to determine the location of the blocks shown in Figure 2-1 as contained in this set.

Tuner section. Tuners used in current model sets use electronic tuning rather than mechanical switching. A complete discussion and notes on troubleshooting tuners are presented in Chapter 10. These tuners include UHF mixer and oscillator, VHF RF amplifier, mixer, and oscillator, and in many cases circuitry that provides an electronic digital readout of the received channel number. Many tuners have remote tuning capabilities as well.

IF section. The IF circuitry combines several of the blocks in older sets. In many sets there are two or three multipurpose integrated circuits that replace several discrete components. Functions of the IF circuits include IF amplification, sound and video detection, AGC, and sync separation. Tuner automatic fine tuning (AFT) is another function in this section.

Sound section. This section contains the blocks required for sound detection and amplification. The output of this block operates the speaker.

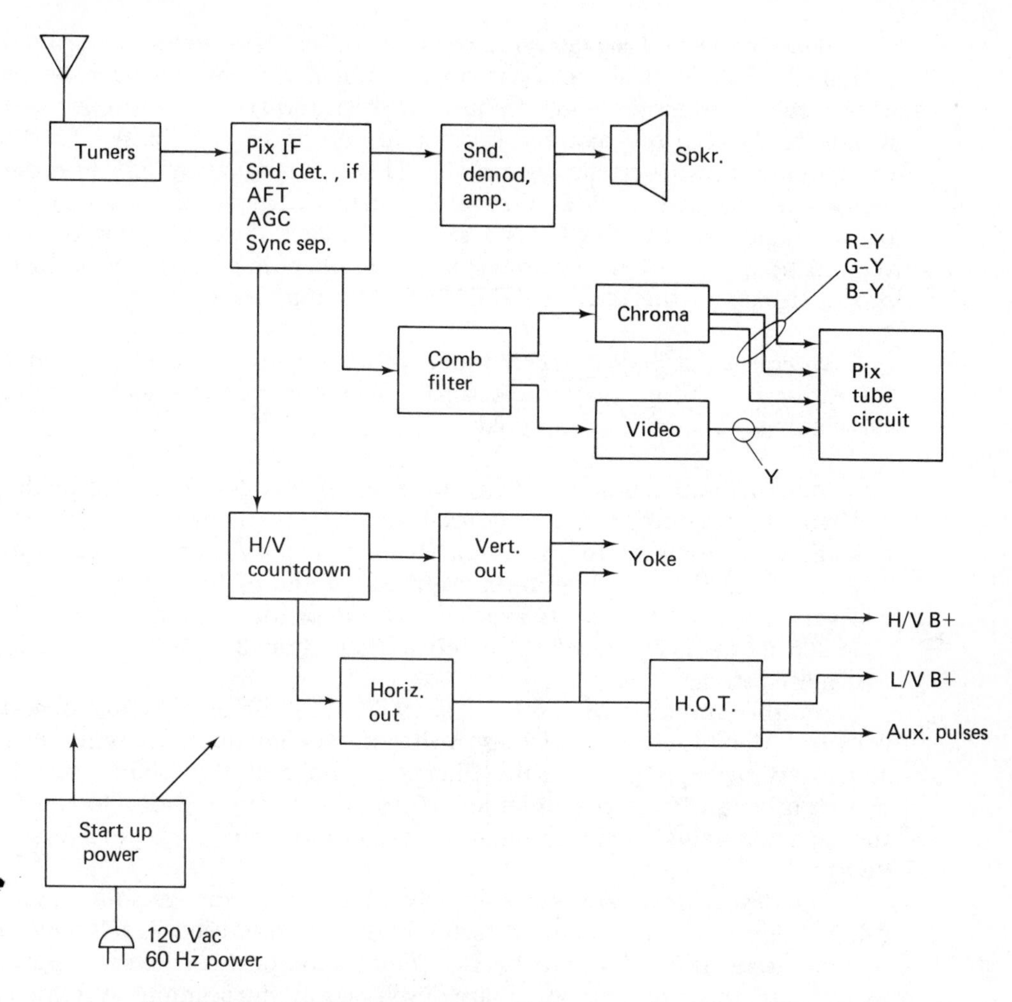

Figure 2-11 Block diagram of a 1980 era receiver. Several blocks are combined in addition to some new ones.

Chroma section. This section contains all the color-processing blocks. A composite color signal is found at the input to the section. The output consists of three separate color signals. These signals are sent to the picture tube.

Comb filter circuitry. One of the problems requiring correction is the interaction of the color and video information at unwanted places in the set. The result of this interaction is a herringbone-like effect in the picture. This is particularily noticeable when viewing plaid woven patterns. This section uses integrated-circuit technology to successfully separate these two signals and eliminate this problem. The output of this section feeds the correct signals to the video and color-processing sections.

Horizontal/vertical countdown section. Another new concept is found in this block. Poor vertical sync has been a problem in many sets over the years. This system uses an oscillator frequency of 31,468 Hz. A countdown system is used to develop the required horizontal rate of 15,734 Hz. A second part of the countdown system divides the 31,468-Hz signal by 525 in order to create the required 59.9-Hz vertical frequency. The result is a very stable picture. Another feature of this system is the lack of either a horizontal or a vertical hold control for customer use. The outputs of this section provide proper signals for the vertical and horizontal amplifiers.

Vertical output section. This block is similar in action to that found in the earlier discussion. Its purpose is to provide the correct signal to the vertical windings of the deflection yoke.

Horizontal output section. One function of this device is to provide the high voltage required for the picture-tube circuits. Large-screen TV sets require at least two levels of high voltage. One is the 5 kV used as a focus voltage. The second is the very high level required for the picture-tube anode. This will vary from set to set. Its exact value is dependent upon the size of the tube and its operating needs. It could be as high as 30 to 32 kV in a large-screen TV set.

Another function of the HOT is to develop one or more low operating voltages. The exact value of these voltages depends upon set requirements. Many sets develop picture-tube filament voltage at this point. Additional operating voltages on the order of 12 to 250 V are also developed. Here, too, specific values and the quantity of output voltages depend upon set design.

Set design also determines exactly where these voltage sources are developed. Early integrated-circuit technology sets often used signals developed before the horizontal output block. These accomplished the same purpose. In any case, these power sources are developed by the scanning system. They are often called *scan-derived* or *derived* power supplies. This is because they are developed from the scanning circuits. Older, more traditional sets developed all operating voltages from one common power supply section.

Another output of this circuit is in the form of pulses. These pulses are at the horizontal rate of 15,734 Hz. They are used to provide blanking and AGC timing information. Some sets also use these pulses for color sync and sync separation.

Power supply section. This block provides startup power for the set. This power is required to turn on the scanning circuits. Once the scanning circuits are functional, power for the balance of the set is available. Some sets also have shut down circuits in their power supplies. These turn off operating power in case of overcurrent or too-high operating voltages.

BLACK-AND-WHITE RECEIVERS

Blocks found in noncolor receivers have names and functions to simulate those found in color receivers. The major difference is the lack of color-

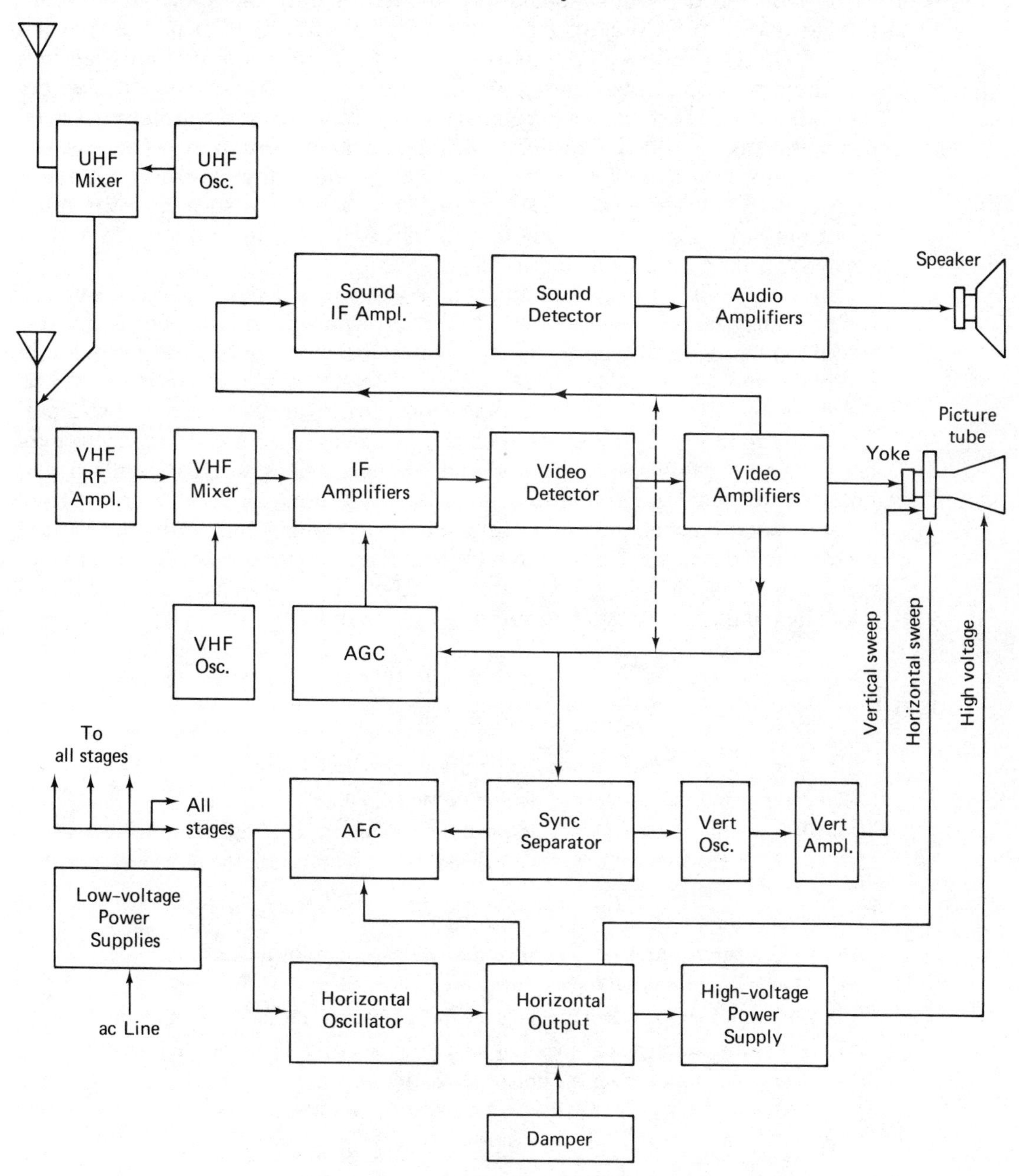

Figure 2-12 Black-and-white receiver block diagram.

processing blocks. A diagram for a typical black-and-white TV set is shown in Figure 2-12. Note the similarity of the names for the blocks in this type of set. The lack of color-processing blocks leads to simplification of the set. There is no need for convergence when there is only one beam of electrons in the picture tube. Traditionally the term "Y amplifier" is not used when describing video circuits in this set. Almost all noncolor sets are sold in a very highly competitive market. Prices and related cost of production are relatively low. Most of these sets use integrated-circuit technology to keep production costs down. The technological advances found in color sets will tend to be "spun off" for noncolor sets. Some manufacturers will adapt quickly to the new integrated circuits. Others tend to stay with the more traditional transistor components. In either case, the blocks are almost always like those shown in Figure 2-12.

Functionally, the blocks in a black-and-white TV perform the same as their counterparts in the color set. Color information will be received by this set. It will not be processed in the receiver because there is no circuitry in the black-and-white set for processing it. The signals related to color will be found in the tuner, IF, and video sections, however. The present broadcast standards for television require that the broadcast signal and the receiver are both capable of processing color and noncolor signals. This standard was initiated so that the transition to color reception could be accomplished without scrapping all noncolor sets. There are many homes in the United States that do not have a color TV receiver. They, too, must be able to receive a TV signal and have it reproduced on their set. This is accomplished with the compatability system presently being used.

QUESTIONS

2-1. Draw a block diagram of a color receiver. Label each block.

2-2. Describe in brief terms the function of each block.

2-3. Describe the action of the VHF mixer block when a UHF signal is being received.

2-4. What do the terms "R-Y," "G-Y," and "B-Y" represent in the receiver signal processing system?

2-5. What is the function of the comb filter?

2-6. What is the function of the horizontal/vertical countdown block?

2-7. What is a scan-derived power source?

2-8. What subblocks are used in a power supply? Name each and explain its function.

2-9. Why is the convergence block not used in a black-and-white color receiver?

2-10. Name the outputs from the horizontal output block.

Chapter 3

Signals and Signal Processing

There are several methods of troubleshooting any electronic device. One way is to follow the path(s) of the electronic signals. To do this one must know both the signal shape and the path(s) it takes. It is almost impossible to memorize the size, shape, and location of each signal for every TV set. It is much better to have a general knowledge and to use service literature available for the specific set as the reference source. This is particularly true as the technology undergoes change. In the days of vacuum-tube radios it was possible to memorize a schematic diagram. This was because almost all sets used the same circuits and tubes. This is not true for modern TV sets. Each set is generally similar, but it seems that almost all are different. Servicing from memory appears to retard, rather than increase, service efficiency.

THE ELECTRONIC SIGNAL

In understanding how a TV set functions, it is necessary to know what the electronic signal represents. Several factors are included in the makeup of the signal. These factors, shown in Figure 3-1, include shape, frequency, duration, position in time, and amplitude.

Shape. The signal contains some form of information. This information may not be in a form for human use. It may, instead, be used to control a circuit rather than be used to create a picture or make sounds. Each signal used in a TV set has a classic form. The *exact* shape may vary between sets, but the general form remains constant.

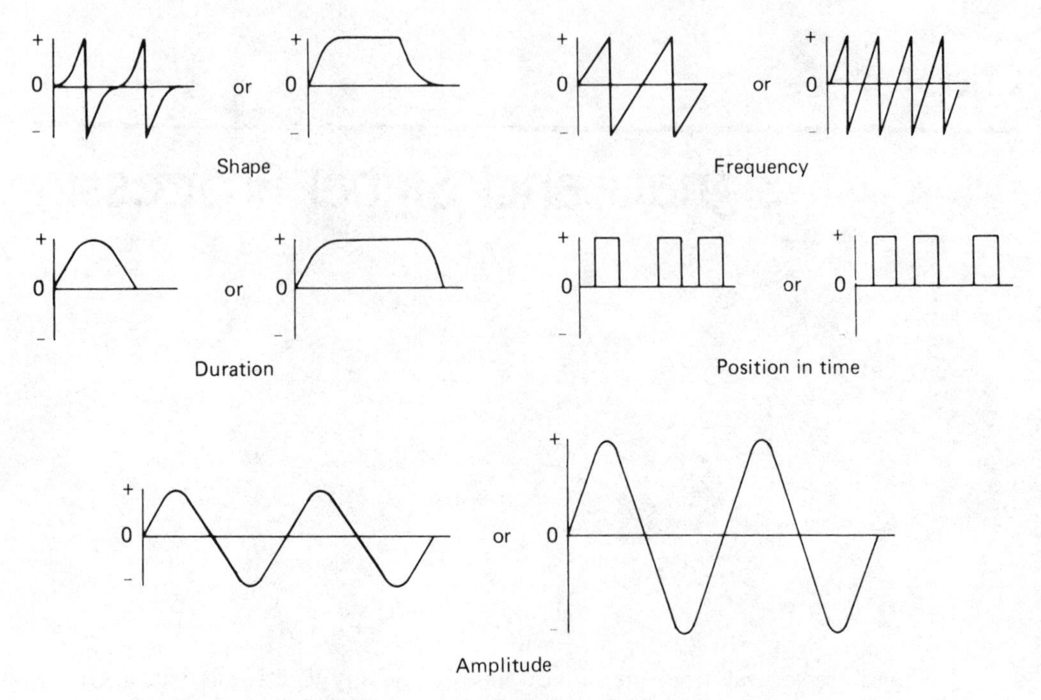

Figure 3-1 Each electronic signal will be developed using the factors illustrated here. (Joel Goldberg, *Radio, Television, and Sound System Repair: An Introduction*, © 1978, p. 87. Reprinted by permission of Prentice-Hall, Inc.)

Frequency. This is the repetition rate of the signal. Vertical frequencies for a TV set are near 60 Hz. Horizontal frequencies are 15,734 Hz. Audio frequencies range between 20 Hz and 15 kHz.

Duration. This reference is to the length of time the signal is "on" during one complete time unit. (Time units in electronics are based upon the second.) This information is usually found in service literature for the set.

Position in time. This reference is for the starting time of the signal. It is compared to a reference time. Often, the position in time is called the *phase relation* or *phase* of the signal. Time position is discussed in degrees, using the full circle value of 360° as a reference. The full-circle value refers to one complete cycle of the wave as created by a mechanical generator. A signal that starts 90° later than a reference is said to be "90° out of phase" with the reference signal.

Amplitude. This reference is to the height of the signal. It is usually measured in units of the volt. Signals, being a varying value, are usually given ac voltage values. These signals are shown on a schematic diagram as peak-to-peak (p-p) voltage values.

With these terms in mind, let us look at a specific signal and break it down into its component parts. The signal used for discussion purposes is shown in Figure 3-2. It is the signal that is received at the antenna terminals of a TV set. The instrument used to observe this signal is an oscilloscope. This signal is an amplitude-modulated carrier. The high-frequency carrier, a sine-wave form, is shown in Figure 3-2(a). Its frequency is the assigned station's frequency. The carrier is created in the transmitter as a constant-amplitude continuous wave. The shape illustrated in part (b) is the composite video signal. It contains sync, blanking, color, and video information. The process of amplitude modulation produces the composite signal shown in part (c). The modulating signal changes the shape of the *amplitude* portion of the carrier to that of the modulating signal. The process develops signals that both add to and subtract from the carrier frequency. The result is a signal that has both a positive (upper) and a negative (lower) shape. The shapes are exactly alike in form, but one is the mirror image of the other.

The signal developed in the transmitter is radiated from the antenna as electromagnetic waves. These waves travel from the transmitting antenna to the receiving antenna. The waves move across the metal parts of the receiving antenna. They create a voltage on the receiving antenna. This works in the principle of electromagnetic induction in a manner similar to the action in a transformer. The electromagnetic signal has all the properties discussed earlier in this chapter.

It is very important to recognize the signal as a voltage value in the receiver. Too often people fail to associate the signal as a varying voltage.

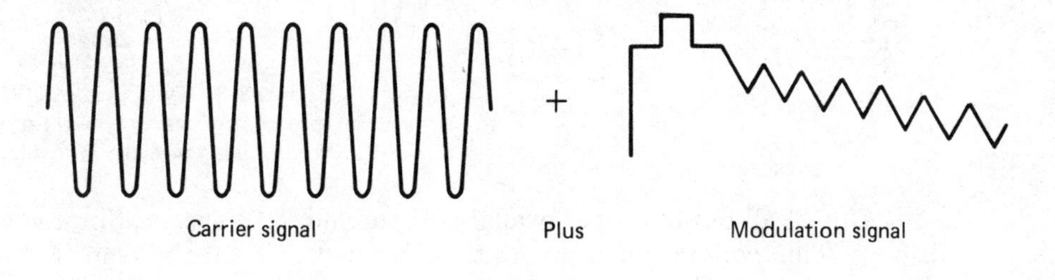

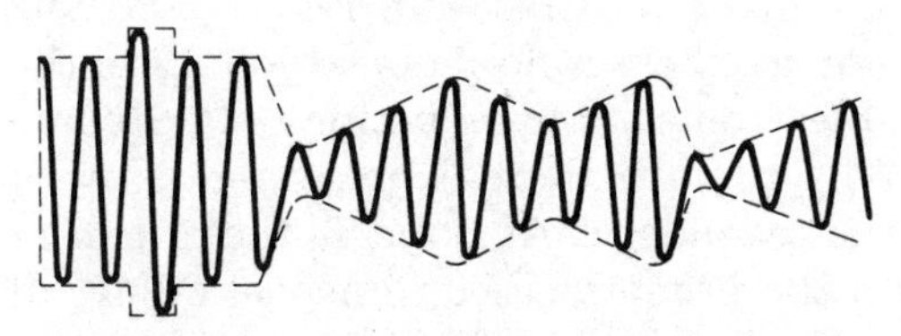

Figure 3-2 Shape of a TV signal observed at the antenna terminals of the receiver.

Having a separate name, it becomes something different and separate from the set. This is a false assumption. The signal is a working voltage in the set. It has a very important function. The signal is processed through the set from one stage to another. It normally moves, or "flows," from the input of the set to the output of the set. In some instances certain signals are processed within the set. An example of this is the horizontal AFC signal. It is used only inside the set. It never appears at any of the output devices of the set.

If we accept the statement that an electronic signal is a varying voltage, we can move on to a detailed discussion of how it works.

The oscilloscope will display a picture of an electronic signal. The oscilloscope display has two dimensions. One dimension is time. The second dimension is amplitude. Time is normally displayed on the horizontal axis. Amplitude is shown as the vertical axis. Figure 3-3 illustrates this point. Two lines of picture information are shown. All the information is displayed above the horizontal line. This line is a zero voltage reference in this display. All voltages are above the line, so are considered positive in nature.

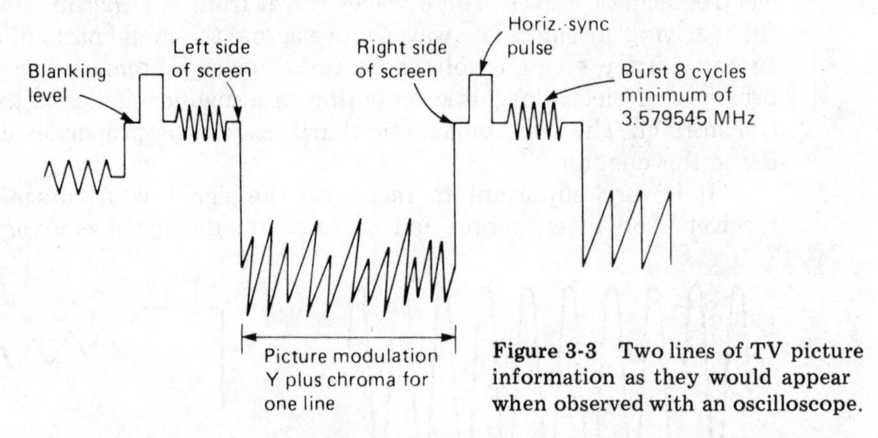

Figure 3-3 Two lines of TV picture information as they would appear when observed with an oscilloscope.

Almost all oscilloscopes available in the market have a calibrated vertical display. This calibration is in units of the volt. The technician is therefore able to measure the amplitude of the signal at any point in the set.

Let us now relate signal processing and its reaction in a circuit with a transformer. This is illustrated in Figure 3-4. With no signal applied to the primary windings, there is no induced voltage in the secondary. This is shown at point A. Electromagnetic induction only occurs when the magnetic field is changing. At zero volts there is no current flow and, therefore, no induced voltage in the secondary. At point B, the primary voltage is rising. A current is flowing in the primary. Electromagnetic lines of force are radiating from the primary winding wires. These lines of force induce a voltage on the secondary winding wires. The voltage is of opposite polarity to the applied primary voltage. This is shown at point B in the secondary.

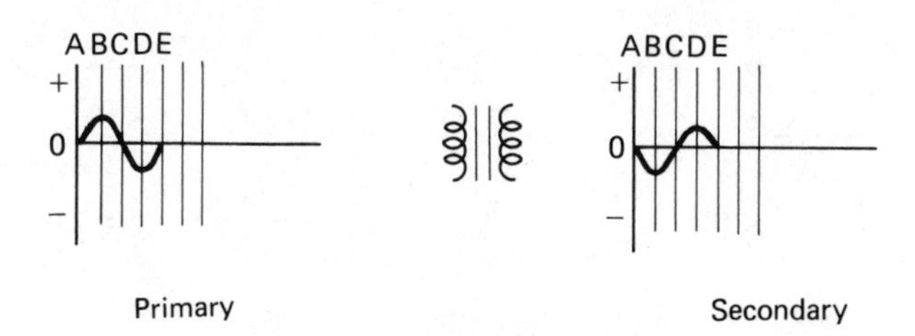

Figure 3-4 Transformer action and the resulting induced secondary voltage.

Increasing primary voltage develops an increasing opposite polarity secondary voltage. The opposite is also true. A decreasing primary voltage will produce a decreasing opposite polarity secondary voltage. Compare signals at points B, C, D, and E in the illustration. The exact value of secondary voltage depends upon the specifications of the transformer.

A similar type of action occurs with a capacitor. This is illustrated in Figure 3-5. The difference is that the capacitor transfers a charge instead of a voltage. Current flowing in the primary circuit develops a voltage drop across resistor R_1. Capacitor action creates a charge that is of opposite polarity at the output circuit plate. Changing charges on the input side of the capacitor develop changing charges at the output plate. This creates a voltage that has values similar to those at the input. Only those values that are changing are affected in this manner.

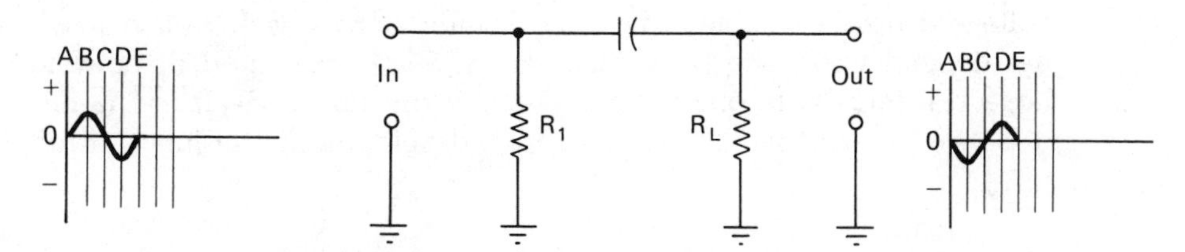

Figure 3-5 Charge transfer as it occurs in a capacitor.

The next phase of this discussion relates to the action of a transistor in a circuit. The sample circuit is shown in Figure 3-6. Operating conditions for the transistor are established by R_B and R_L. This is done with no signal applied. Kirchhoff's voltage law states that voltages in a closed-loop, or series-circuit, add. The input signal is a voltage. It is applied between the base and common in this circuit. The signal voltage adds to the operating voltage established by the transistor emitter-base junction, R_E and R_B. When the input signal is a sine wave, as illustrated, the positive and negative values are added to the base-common connection. This changes the voltage at the base to a range from +3.0 to +1.0 V (+2.0 + 1.0 = +3.0 and –1.0 + 2.0 = +1.0). These voltages affect the operation of the transistor. Increasing base voltage increases emitter–collector conduction in the transistor. Decreasing emitter-base voltage decreases conduction in this type of transistor.

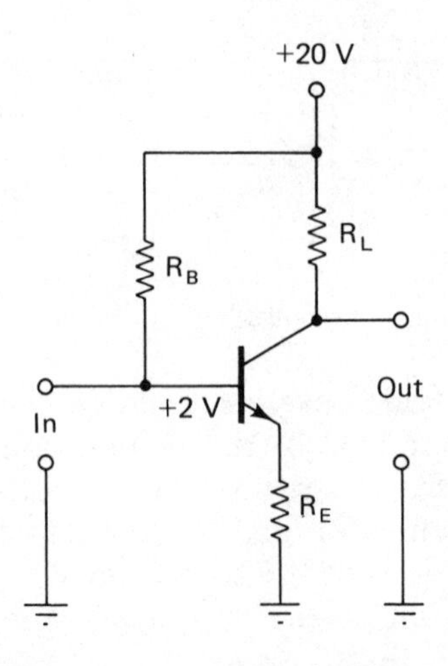

Figure 3-6 Schematic diagram of the transistor amplifier circuit described in the text.

The output circuit consists of three resistances. These are R_E, R_L, and the emitter-collector circuit of the transistor. Two of these are fixed values. The emitter-collector value changes. Its resistive value is determined by the amount of emitter-base voltage and conduction. Therefore, we have a series circuit consisting of a power source, two fixed resistor values, and one variable resistor. This is shown in Figure 3-7.

The rule developed by Kirchhoff is also applied here. There are three voltage drops in this circuit. They develop across each resistor. Specific values depend upon the resistance values. Large voltage drops develop across large resistances. If one were to assume the values given in the illustration, it is easy to see that a voltage drop is developed at the junction of R_L and

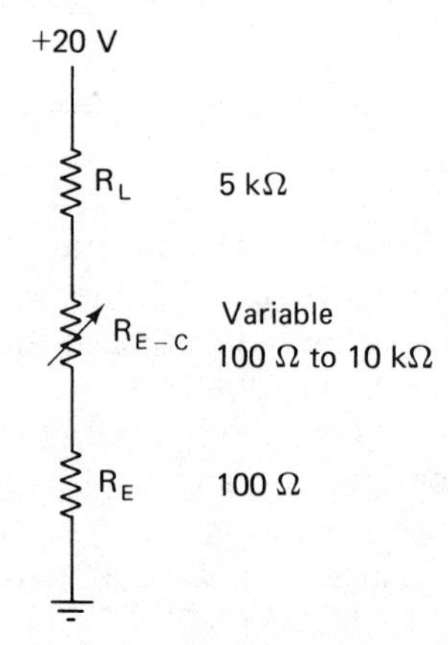

R_{EC}	R_T	I_T	E_{E-C}
5 kΩ	10100	0.00198	9.9
10 kΩ	15100	0.00132	13.2
1 kΩ	6100	0.0032	3.2

Figure 3-7 Equivalent circuit of the transistor amplifier shown in Figure 3-6. Voltage values for different operating conditions are shown in the box on the right.

$R_{E\text{-}C}$. This drop is measured from that point to circuit common. Varying the value of $R_{E\text{-}C}$ will change the voltage at the point. A high resistance at $R_{E\text{-}C}$ will produce a high-voltage drop. A low resistance at $R_{E\text{-}C}$ will produce a low-voltage drop. This principle may be applied to the action of a transistor or tube.

The component $R_{E\text{-}C}$ is in reality the emitter–collector connection of a transistor. The amount of resistance present at these points is determined by the base voltage action of the transistor. Figure 3-8 illustrates this point. An input signal with a value of 2 V p-p is applied to the base circuit of the transistor. This causes a swing in base voltage from a high of +3 V to a low of +1 V. This change in voltage affects the resistance of the emitter–collector circuit. Increasing base voltage (and its related current) produces a decrease in the resistance between emitter and collector. This, in time, permits more current to flow in this secondary circuit. It also produces a lower voltage drop across the output terminals. *Remember, this is an operating voltage value!* A decrease in the amplitude of the input signal produces a higher resistance in the transistor output circuit. This raises the voltage between this point and circuit common.

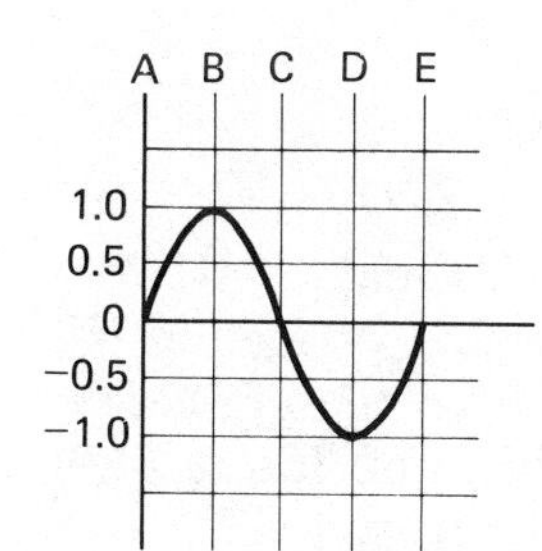

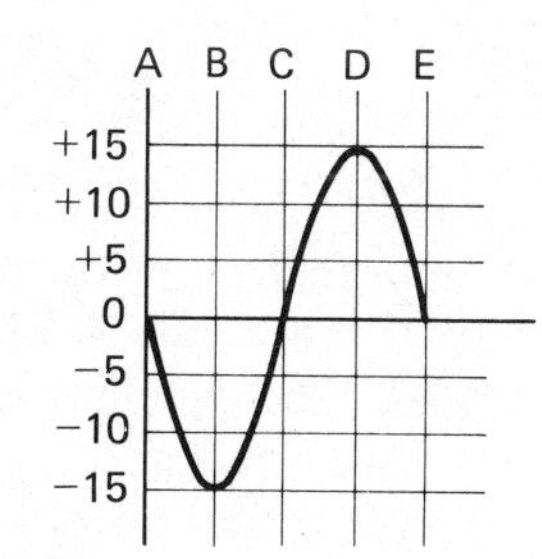

Figure 3-8 Signal transfer in a common-emitter amplifier. A small change in input signal level controls a larger change in output level.

The result is a dc operating voltage that has taken on the shape of the input signal. The signal is inverted due to the action of the transistor in this circuit. Inversion does not always occur. This effect will depend upon the type of circuit used. The most important point to remember is that this so-called "signal" at the output circuit is actually a dc voltage that has taken on the form of the signal. These variations in voltage (and related current) may be transferred to another stage by either transformer or capacitor action. Only those values that are evidenced in the change are transferred, or coupled, to the next stage. Assume the following. A 100-V source is required in the circuit. The input signal produces an output signal voltage of 20 V p-p as illustrated in Figure 3-9. The collector-common dc voltage in the circuit is 60 V. The collector-common signal swing is from 70 to 50 V. Only the 20-V p-p charge is coupled to the next stage. There is no influence by the operating voltage when either transformer or capacitor coupling is used.

There are certain circuits that are directly wired, or coupled. In these

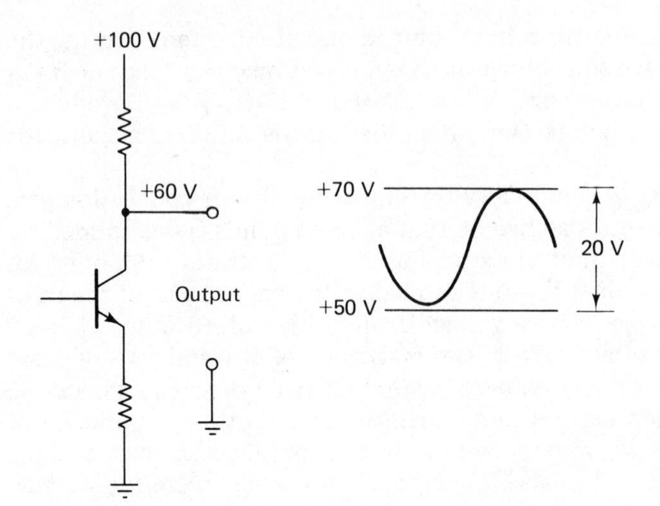

Figure 3-9 The output signal is actually a changing level of dc voltage measured between the output point and common.

circuits a wire connects the output of one circuit to the input of the next circuit. In these circuits the entire dc operating voltage is applied to the following stage.

SIGNALS IN THE RECEIVER

Now that you have a better understanding of the term "signal," let us see how this signal is processed in the receiver. A block diagram of a color TV receiver is shown in Figure 3-10. The amplitude-modulated picture carrier and the frequency-modulated sound carrier induce voltages on the receiver antenna. This voltage is tuned by the RF amplifier in the receiver. The outputs of this block are amplified versions of the input signal. They have the same form, but they are larger in amplitude. The design of the tuned circuits on the receiver permits all signals within a 6-MHz bandwidth to be processed at the same time. The receiver oscillator generates a carrier wave. This wave is unmodulated. Its frequency is always 45.75 MHz higher than the broadcast station picture carrier. This signal is fed to the mixer.

Mixer action produces many signals. There are four basic signals produced in the mixer as the result of the injection of the oscillator and one carrier. These signals are: the oscillator frequency, the carrier frequency, the sum of the two frequencies, and the difference between the two frequencies. In the TV receiver, only the one frequency that is the difference between the two signal frequencies is utilized. The TV receiver processes both video carrier and sound carrier signals in the RF amplifier and mixer. Mixer action develops two output signals on all channels when a signal is processed. One of these is a frequency-modulated carrier that contains sound information. Its frequency is 41.25 MHz. The other frequency is an amplitude-modulated picture signal. Its frequency is 45.75 MHz.

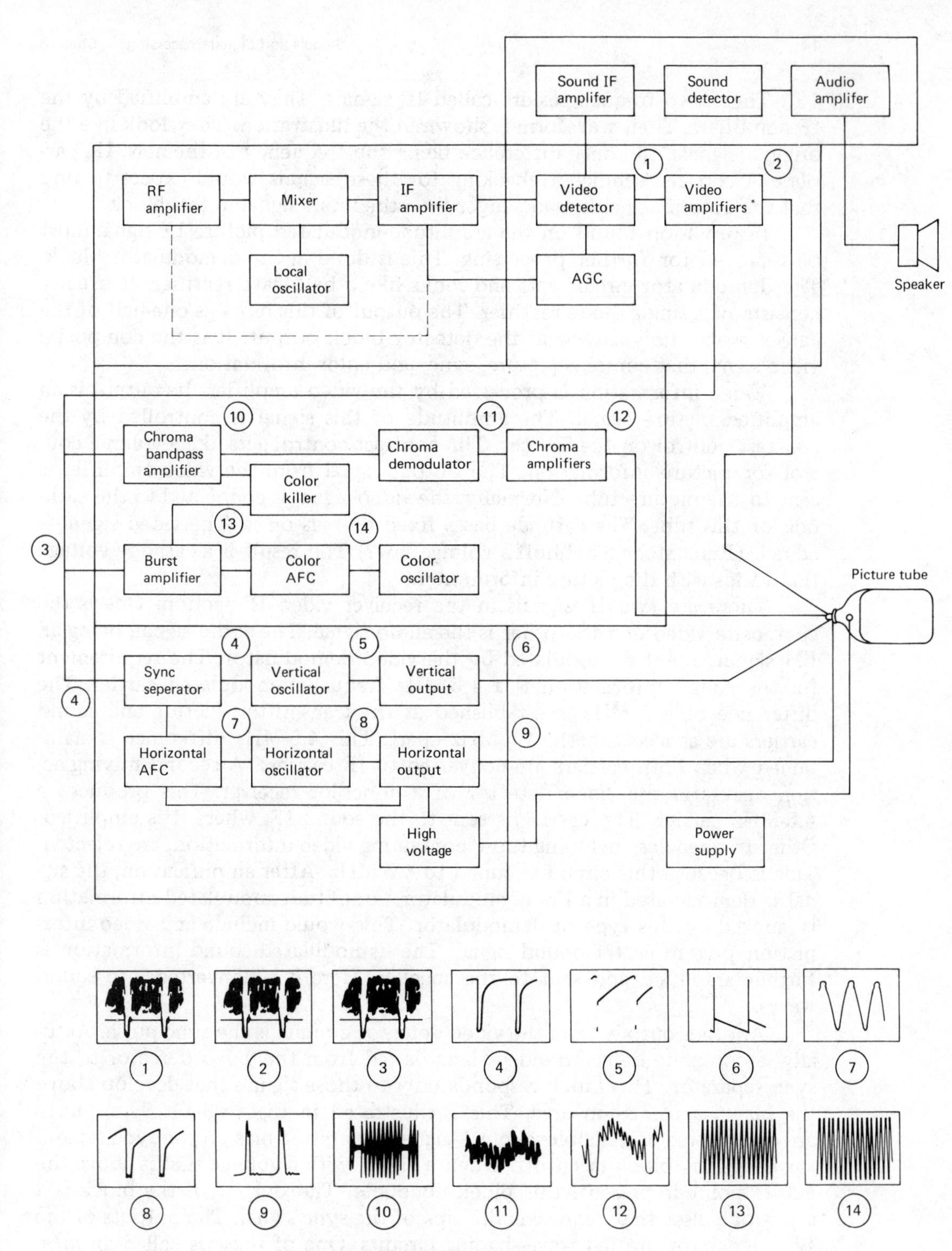

Figure 3-10 Block diagram of a color TV receiver.

These two frequencies are called IF signals. They are amplified by the IF amplifiers. Their waveform is shown in the illustration. They look like the original signals, the only difference being the frequency of the new, IF, carriers. A service technician checking for these signals would expect to find that the output signal is made larger than the input signal in this block.

Information found on the amplitude-modulated picture IF signal must be removed for further processing. This is done in the demodulator block. The demodulator circuit acts and looks like a half-wave rectifier. It usually consists of a single diode rectifier. The output of this block is one-half of the carrier wave. It is shown at the detector block output. It is the composite video signal that contains picture, sync, and color information.

Video information is processed by the video amplifier. Its output is an amplified picture signal. The amplitude of this signal is controlled by the contrast control on the TV set. The contrast control acts like a volume control for picture information. The output signal from the video amplifier is sent to the picture tube. Normally, the video signal is connected to the cathode of this tube. The cathode has a fixed dc bias on it. The video signal is added. (Remember Kirchhoff's voltage law?) The result is a cathode voltage that varies with the picture information.

There are two IF signals in the receiver video IF section. One is the composite video and the other is the audio signal. The audio signal, being an FM signal, is not demodulated by the video demodulator. The requirement for the sound information is a 4.5-MHz frequency-modulated carrier. The difference of 4.5 MHz is established at the transmitter. Picture and sound carriers are spaced exactly 4.5 MHz apart. This 4.5-MHz difference is maintained when both carriers are converted to IF carriers. A second mixing action occurs in the video detector of a noncolor receiver. This produces a 4.5-MHz carrier. The carrier is sent to the sound IF, where it is amplified. Other frequencies, including those containing video information, are rejected. This is because this circuit is tuned to 4.5 MHz. After amplification, the signal is demodulated in a FM demodulator. Amplitude-modulated information is ignored by this type of demodulator. This would include any video information present in the sound signal. The demodulated sound information is further amplified and sent to the speaker. Here it is converted into sound waves.

Another output from the video detector circuits is the sync pulse. Actually, a complete detected video signal is fed from the video detector to the sync separator. This block responds only to those signals that develop above the black transmission area. This is illustrated in Figure 3-11. Sync pulses are transmitted as a high-level black signal. Operating bias on the sync separator keeps the block in cutoff. When a signal with a voltage that is above the cutoff point is present, this block conducts. The output of the block is a series of pulses that represent the tips of the sync signal. The outputs of the sync separator are fed wave-shaping circuits. One of these is called an *inte-*

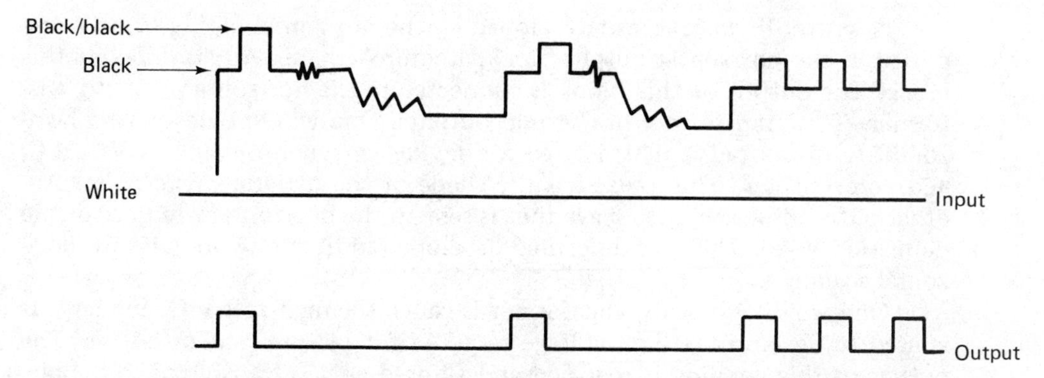

Figure 3-11 The upper diagram shows the output of the video detector. The lower diagram shows the output of the sync separator.

grator. Its output is shown in Figure 3-12. The sync pulse is changed into a sawtooth waveform. It is fed to the vertical oscillator. The second pulse is shaped by a differentiator circuit. Its pulse shape is very sharp and short in duration. It is fed to the horizontal oscillator. Both sets of pulses are used to establish operating frequencies for their respective oscillators.

Pulses developed in the vertical sweep oscillator block have a sawtooth form. The pulse is created in the oscillator block. The action of the oscillator is almost independent of the rest of the set. It is a free-running oscillator. It does not require an input signal in order to operate. The only purpose of the sync pulse is to time the oscillator with the sweep in the camera.

The oscillator's output is fed to the vertical amplifier. Here it is amplified. Its form is usually changed to compensate for the inductance in the defection yoke. The basic form of the sawtooth wave is maintained, however. A pulse called a *vertical blanking pulse* is developed in this block. It is used to cut off electron-beam current in the picture tube during the vertical blanking interval.

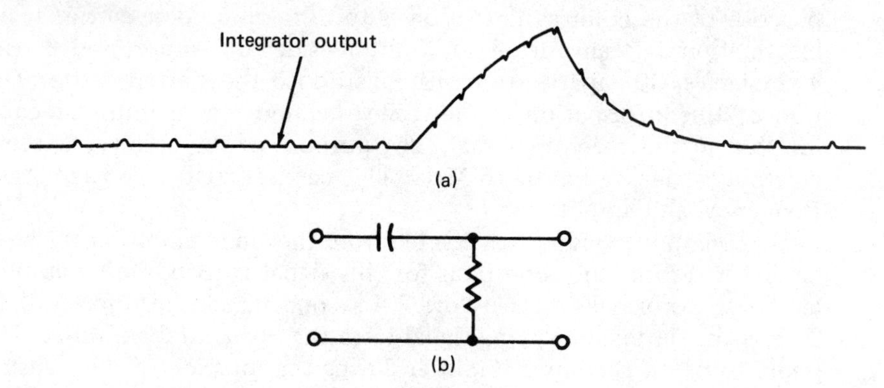

Figure 3-12 Integrator output waveform and its vertical circuit.

A sawtooth wave is also developed by the horizontal oscillator. It is amplified in the horizontal output block. Required shaping is also done in this block. The output of this block is connected to the horizontal output transformer. This transformer has several outputs. One winding develops a horizontal feedback pulse. This is used for *keying*, or synchronizing, AGC, AFC, and color signals. This pulse has the shape of the blanking/sync pulse. Another output is a sawtooth wave that is sent to the horizontal windings of the deflection yoke. The magnetic field developed in the yoke provides for horizontal scanning.

One winding of this transformer is called the high-voltage secondary. It is used to develop the high voltage required for picture-tube operation. The output of this winding is rectified and filtered as it is transferred to the picture tube. The signal is a high-frequency sine wave before rectification. After it is rectified and filtered, it appears as dc. Most scopes do not have the capability of displaying this high voltage. Schematic diagrams often state "do not measure" at this point. There is a high-voltage meter available that may be used to measure the dc voltage at the picture tube.

Part of the secondary of the horizontal output transformer is connected to the damper. The operation of this section of the receiver, which develops a dc voltage, is described in Chapter 19.

Many TV sets use the high-frequency horizontal output transformer to develop secondary operating voltage sources in the set. The requirements for this type of circuit are very similar to those for a traditional 60-Hz system. The differences are in the requirements for diode action and filter capacitor resistance. Both must be capable of using the 15,734-Hz horizontal frequency. Use of this transformer's frequency reduces size and weight of power supply components.

Color-processing blocks found in a TV receiver are shown in Figure 3-13. The color-processing system shown here is subdivided into two major areas. One of these is called color sync. The other is the chroma section. The purpose of the color sync section is to develop a color carrier signal. Color information is transmitted as a double-sideband suppressed carrier signal. Two signals, 90° apart, are multiplexed onto the station carrier. Demodulation of this information in the receiver cannot occur until the carrier is reinserted in the received signal. The purpose of the color sync section is to create a carrier at 3.579545 MHz. This carrier must have the correct phase, frequency, and amplitude.

A composite video signal is fed from the video amplifier to the bandpass amplifier. From here, one path for this signal is to the burst amplifier. This amplifier is normally biased "off." A second signal is also fed to this section. This is the timing, or keying signal from the horizontal amplifier. The keying signal turns on the burst amplifier during that portion of the transmitted signal which contains the color burst. The output of the burst amplifier is the

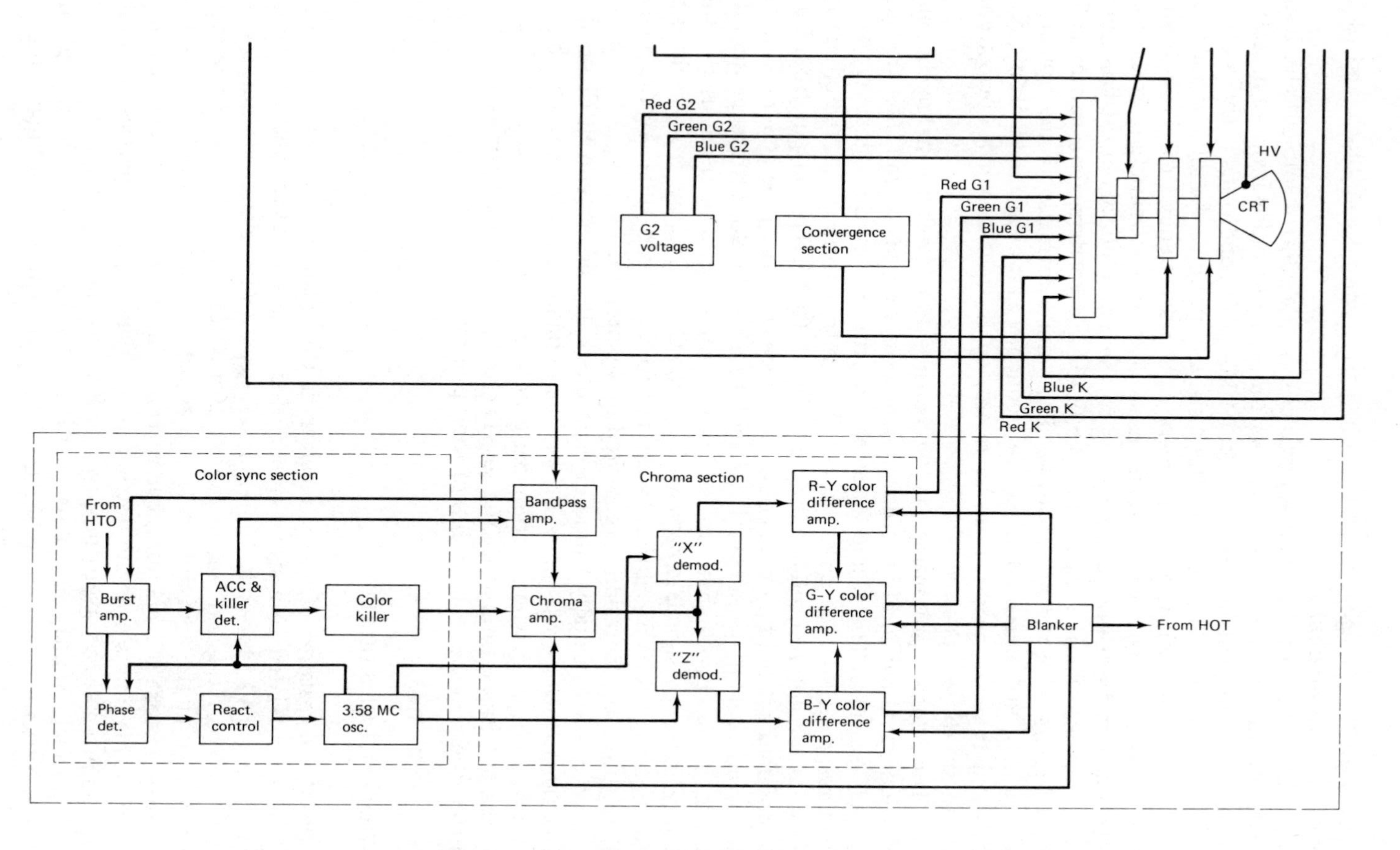

Figure 3-13 Color-processing blocks in the receiver.

burst signal. This signal is then applied to a color killer block and to the color oscillator block.

The output of the color killer block is a dc control voltage. This bias is used to turn on the chroma amplifier block in the color processing section. The color killer acts as a switch. Under no-color conditions it shuts off the chroma amplifier. When a color burst signal is present, the killer signal is off.

The color oscillator block operates at a frequency of 3.579545 MHz. A phase-shifting network is usually used to control the timing of this circuit. Two outputs from this block are fed to the demodulators in the color processing blocks. These outputs are 3.57945-MHz carriers. One carrier is 90° out of phase with the other carrier. The color information is multiplexed as two signals, 90° apart, on the same frequency carrier at the transmitter. An identical type of carrier must be re-created in the receiver if correct color signals are to be processed.

The signals representing the two carriers do not exactly match the phase of the original red and blue color information. For this reason they are called X and Z signals. The X and Z signals and the two carrier signals are fed to their respective demodulator blocks. The output from the X demodulator is fed to a red color-difference amplifier block (R-Y). The output from the Z demodulator is fed to a blue color-difference amplifier block (B-Y). One output from each of these blocks is sent to a third color-difference amplifier block. These two signals are added electronically. The result is the third color-difference signal, representing green (G-Y). Outputs from each of the three color-difference amplifiers are fed to the picture tube. An explanation of picture-tube operation is presented in Chapter 13.

Waveforms usually found at each block are also shown in Figure 3-12. The signal waveforms shown for the demodulator signals represent the type of signal observed when a color bar generator is connected as a signal source at the antenna terminals of the set. These signals have the same shape. Their difference is in the relationship of their phases.

AUTOMATIC CONTROL SIGNALS

One of the features of newer receivers is the automatic color circuitry. This circuitry is related to a signal transmitted on line 19 of the vertical blanking bar. It is called the vertical interval reference (VIR) signal. Its purpose is to adjust color level and tint in the receiver automatically. It uses a reference established at the transmitter. If one adjusts the vertical hold control of the receiver so that the blanking bar is seen, a line containing colors may be observed. This line is near the bottom of the blanking bar. Analysis of its information produces a waveform shown in Figure 3-14.

Special circuitry in the receiver reacts to this signal. A block diagram of this subsystem is shown in Figure 3-15. Sync pulses, blanking pulses, and

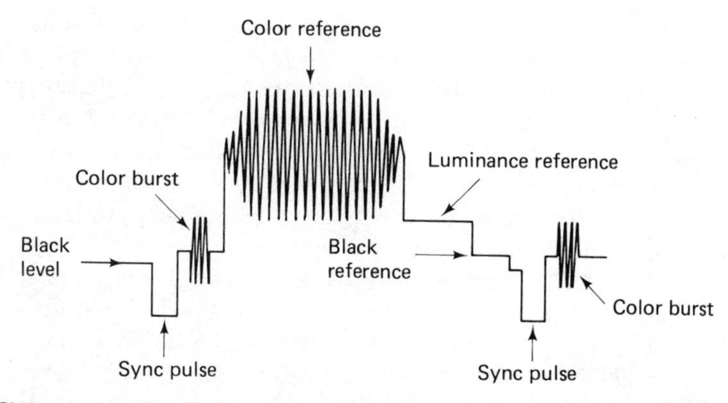

Figure 3-14 The vertical interval reference signal (VIRS) on line 19 of the picture signal.

video information are processed by a system. This system reacts only to information transmitted in line 19. Unless all three signals are present at the proper moment in time, this block is turned off. When these three signals are present, the block develops two output signals. One of these is sent to a tint control block.

The tint control block requires two input signals. One is a keying pulse from the line 19 recognizer block. The second is a signal from the R-Y circuit. These two signals produce a bias voltage. The bias voltage is used to control oscillator phase in the color demodulator.

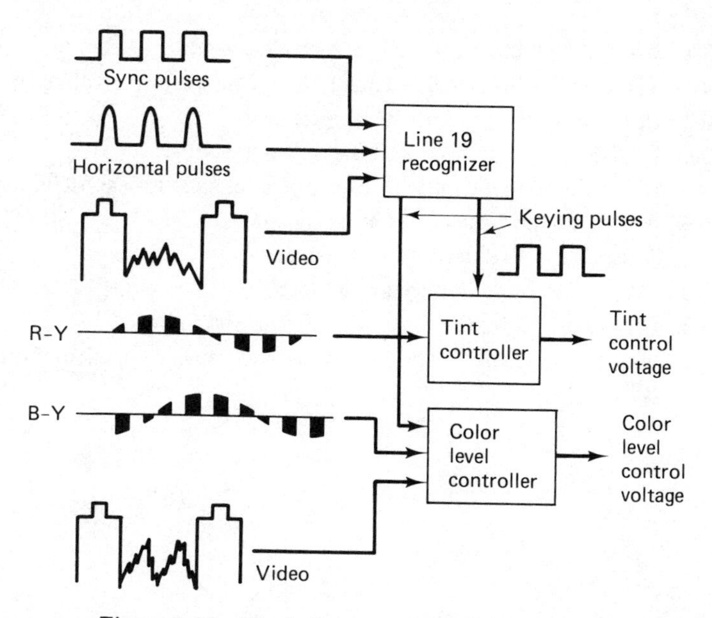

Figure 3-15 Block diagram of the VIRS decoder.

A second signal from the line 19 recognizer is sent to a chroma level block. This block requires three input signals to obtain one output signal. Video information, color information, and line 19 information are all required at the input. The output is a control voltage that is used to adjust the amplitude, or level, of the color signal. Since this system is transmitted as a part of the vertical signal, it has a frequency of 59.9 Hz. This means, among other things, that a receiver with this system automatically adjusts color tint and amplitude almost 60 times a second.

Each of the blocks described in this chapter is discussed later in individual chapters. Specific operating circuits and troubleshooting approaches are also discussed. The intent at this point is to provide an overview of the blocks and the signals processed in the various blocks. Typical waveforms are offered so that one will begin to associate these "signature" waveforms with specific blocks in the receiver.

Coverage of a noncolor receiver is basically the same as far as signals are concerned. The major difference is the lack of color-processing blocks. Color signals are not processed by a noncolor receiver. The signals are present in the receiver, but there are no blocks that will process them to reproduce a color picture. This compatible color and noncolor system was one requirement for color broadcasting. The development of the system makes our present broadcast system useful for both color and black-and-white receivers.

QUESTIONS

3-1. What three factors are used to describe a signal?

3-2. Explain why the signal is actually a varying operating voltage in the receiver.

3-3. How many IF signals are in the receiver?

3-4. What is a trap and how is it used in the receiver?

3-5. How does the sync pulse affect horizontal and vertical oscillators?

3-6. What is the purpose of the burst signal?

3-7. Where is the VIR used in the receiver?

3-8. How is the sound IF signal developed?

3-9. What is meant by the term "free-running oscillator"?

3-10. Why is the shape of the horizontal and vertical signals important to the scanning systems?

Chapter 4

Introduction to Troubleshooting

Each person has his or her own method of starting a project. This is also true in the field of television servicing. When a set requires repair, the customer usually does not know exactly what is wrong with it. The service technician is usually asked to estimate the cost of the repair and the length of time required to fix the set. Realistic estimates can be made only after determining the location of the problem. To do this, the technician requires information. Successful technicians seek information and try to localize the problem in an orderly manner. The best way for a service technician to handle any set that requires repair is to establish a working system for approaching the repair problem.

Often, a repairer wants to open up the set as a first step. The next step this person takes is usually to start to poke around in the back of the set to see what can be found that looks like a trouble area. Many hours of time are wasted with this approach. The goal of a high-quality television repair person is rapid and correct repair. In many service organizations the technician receives a base salary and a commission. Take-home pay is based upon successful production or repair of sets. Efficiency is important to the servicer if paychecks are to contain a large amount of dollars.

The highly successful technician does some work before opening up the set. This work is required to increase efficiency and productivity. The work plan is firmly established in the technician's mind. Often, a checklist is used to help novices become experts in this field.

The first step in any repair is to talk to the customer. If a customer brings a set to the shop, a technician should greet the customer. Questions related to the sounds, sights, and smells associated with the malfunctioning set are asked. The customer will often help the technician in localizing a service problem in this way. Careful listening to the answers given by the customer is very important. The attitude of the technician toward the customer is also very important at this contact. Very often the customer is certain that he will be cheated in some way by the servicer. Too many stories are circulated that show the technician to be dishonest. A major role for the technician is to gain the confidence of the customer. This is done when the technician is neatly dressed and clean. Information is obtained from the customer by asking questions related to how the set operated as it failed. The customer's description is heard. Questions again are asked as more information is collected. The technician should also attempt to obtain a service-related history of the set from the customer.

Once the technician has listened to what the customer has to say about the operation of the set, the set is plugged into a power source and turned on. At this point the back has not been removed. The technician listens for abnormal sounds in the set. These include frying, popping, arcing, and of course, a lack of any noises. The technician also carefully looks for signs of smoke and attempts to smell for overheated components.

At this point a first major decision is made. There are two directions in which to go. One is determined by a "dead" set. It has the appearance of not receiving any power. None of the usual operating sounds, sights, or smells are present. The decision to make is: Should the back of the set be removed at this time? The answer is *no!* There are still more checks to make before removing the back. These checks include the following:

1. Is there a circuit breaker or fuse on the back of the set? Often, the circuit breaker opens and all that has to be done is to reset it. In some cases a fuse will fatigue and fail. This, too, may be checked from the back panel of the set before the back is removed.
2. Are the controls in their proper position? Misadjusted customer controls, such as volume and brightness, may give the appearance of a nonoperating set. Adjusting the controls will often resolve what was thought by the customer to be an operational problem. There have been many instances where the service switch on the back panel of a color set was accidentally moved during house cleaning. Not realizing that this was done, the customer requires the services of a technician. When this is found, the technician should politely explain the problem and show the customer what had happened.

If after performing these steps, the set still does not operate properly, it is time to remove the back. Very often this is done by the technician after the customer leaves. It may sound foolish, but the person receiving the set for repair should note any missing knobs, scratches, and other details regarding the cabinet. These are called cosmetic defects. Where possible, these should be mentioned to the customer when the set is received for repair. This will save arguments and bad feelings from developing later.

All parts removed from the set should be placed in a bag. This procedure minimizes the problem of lost parts or fasteners. Some repair facilities will wash knobs and clean cabinets as a regular part of the repair process. One may purchase cloth bags that have a drawstring on them. These are ideal for holding knobs and fasteners. The bag string is tied to the set to keep all the parts together.

Another point to remember is the cosmetic appearance of the set. Cabinets and picture tubes may become scratched if one does not take care. A piece of clean carpeting is placed on the work surface before the set is placed there. The carpeting protects the exterior of the set from damage.

When the set back is removed, a replacement power line cord is used to provide power for the set. The wise technician has an isolation transformer on the workbench. This transformer provides electrical isolation between the set and other electrical equipment. In many TV sets one side of the power line is connected to the chassis. It is possible to receive an electrical shock unless the chassis is isolated in this manner. Connections are made to the antenna as well as the power source. The set is turned on again. If the set works at this point, the power cord is probably defective. Replacing it will repair the set. If this is not the problem, a further repair system is needed.

At this point the technician has to make another major decision. This is based upon the information collected so far. The question to be answered is: Is the set totally dead? If the answer is "yes," certain key parts of the set are located and checks are made. If the answer is "no," other types of checks are made. Each of these checks is made based on the knowledge of how any TV set functions. Service information must be available for use by the technician. Information related to block diagrams and signal paths is also important required knowledge for the technician.

DEAD SET

A dead set indicates a power supply problem. If there are any tubes in the set, their filaments are not glowing. No sound is heard in the speaker. This also includes a low-level 60- or 120-Hz hum. The presence of hum usually indicates that the power supply section is working. If the determination of a dead set is made, the technician follows repair techniques described in Chapter 9.

If the tube filaments light and one can hear some sound in the speaker, other types of decisions have to be made. At this point the technician looks for broken wires, charred parts, broken tubes or components, or any other visual signs of defects. Some people prefer to do this as a first step before the set is turned on. In either situation if smoke signals appear, one will usually find burned or charred components or wiring. Smoke signals often help localize a problem. Usually, the identification of the component will relate it to a specific circuit in the set.

If the set is not completely dead, it is possible to localize problems by observing and listening. This process will help localize the trouble area to one major section of the receiver. Figure 4-1 identifies major areas of the receiver and relates these areas to the appropriate blocks. Use this information as you read the following section.

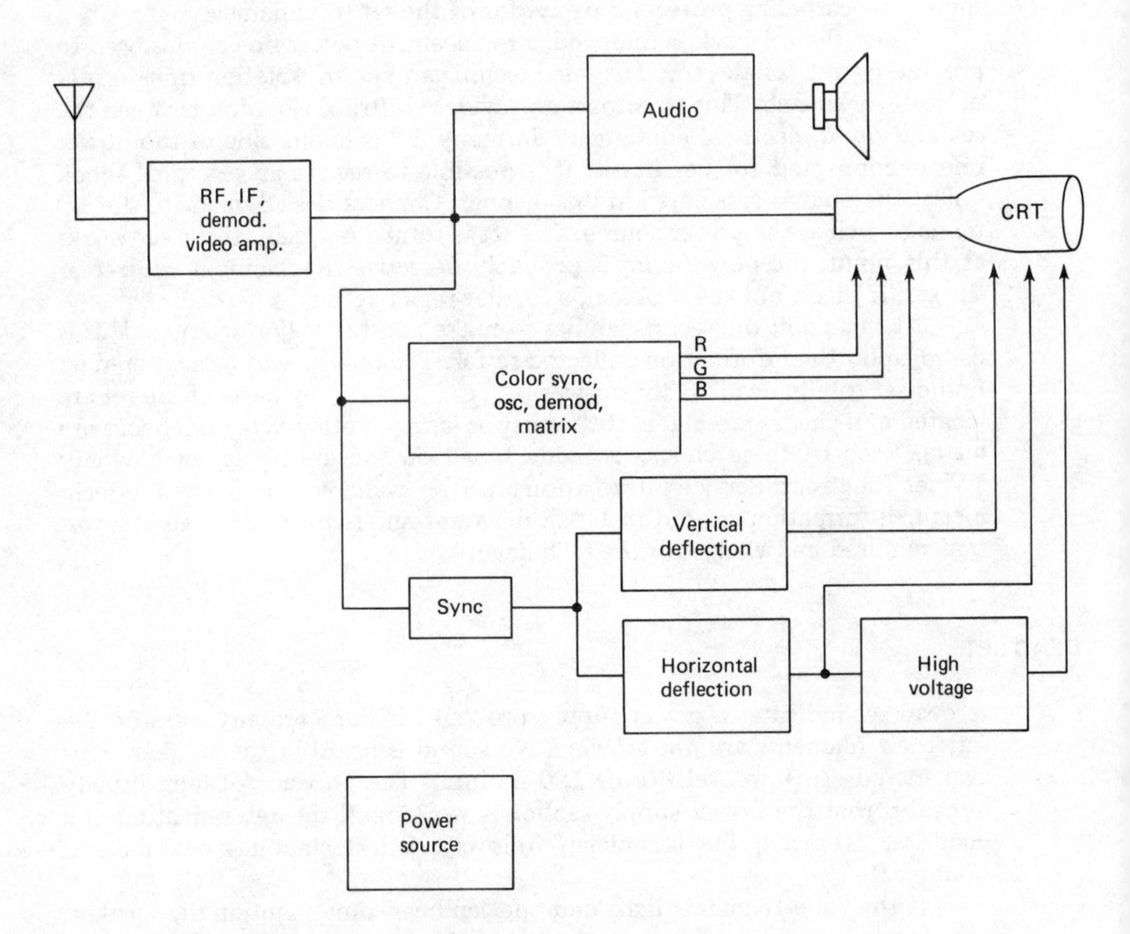

Figure 4-1 Major areas of the television receiver as they relate to functional blocks.

SYSTEM DIAGNOSIS

The first step in system diagnosis is to observe. There are four major events that occur in a television receiver. These are sound, raster, video, and color. The service technician uses the set controls to check for each of these. Value judgments related to the presence of each, the proper amount, and the correct phasing must be made.

Sound. Is sound present? Is there sufficient sound from the set? Is the sound clear or distorted? Set controls related to sound are the volume control and the fine tuning control.

Raster. Raster is the presence of a white screen in the picture tube. The raster is produced by the sweep section of the receiver. Both horizontal and vertical sweep and high voltage are required. Controls found in this section include brightness, horizontal hold, and vertical hold.

Video. Video describes picture information. It is possible to have a raster and not have picture information. It is not possible, however, to have video information displayed without the presence of a raster. The amount of video information is another value judgment. Controls in this area are contrast and AGC. In addition, sync pulses for both oscillators are needed.

Color. Color information, when transmitted, is added to video information. Questions to be answered cover the quantity, the quality, and the presence of color information. Be certain that a color program is being broadcast. People have made judgments assuming that color broadcasts were being transmitted only to discover that this assumption was incorrect. Controls in this area are color, tint, and color killer.

After these four factors are evaluated, the technician is able to relate observations to major sections of the receiver. The identification of a major section reduces the area of the problem from the entire receiver to one major area. The next step is to reduce the problem area to one block. This is accomplished by use of the controls on the set. In most sets controls may be classified as either consumer-operated or service types. Service-type controls are usually located at the rear of the set or inside, on the chassis.

Figure 4-2 shows the relationship between receiver controls and related blocks. Use this information and the following discussion to relate control function to specific blocks.

Channel selector. This control provides information related to tuner operation. Wiggling the knob shaft or rotating it slightly will indicate if the

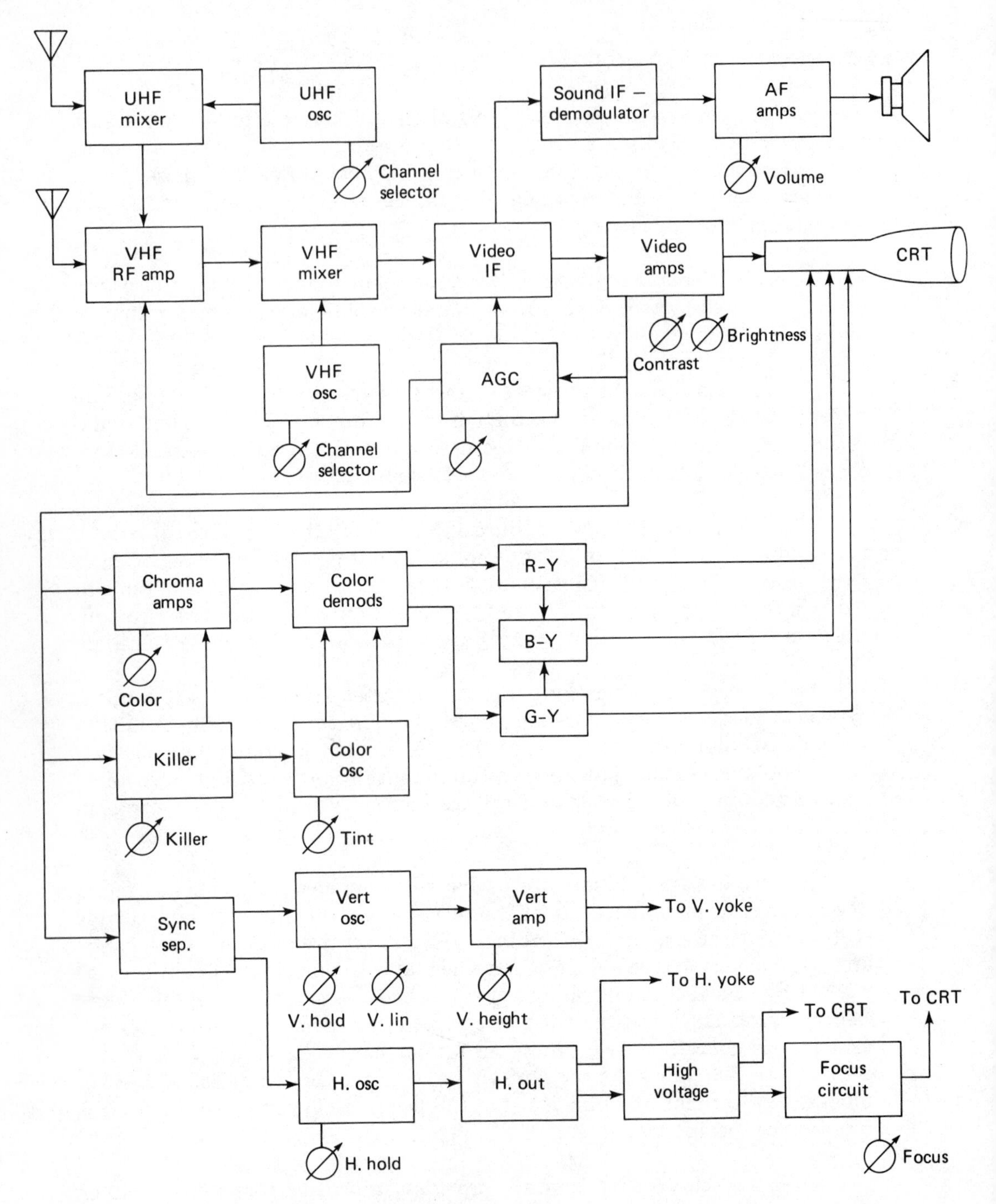

Figure 4-2 Receiver controls can be used to determine if specific blocks are working properly.

tuner is dirty and requires cleaning. Changing from UHF to VHF will establish the correct operation of the two oscillators in the receiver. Flashes of white on a dark screen as the channel selector is rotated often indicates an AGC problem.

Volume control. This control is usually found at the midpoint of the audio system. Rotating the control from minimum to maximum will tell where the problem is in the audio section. If one hears hissing noises that increase in volume, the problem is before the control. It could be that the IF or the detector blocks are not receiving a signal. If no noise is heard but there is a low level of hum that increases in loudness, the problem is also before the control. No signal is being received at the control. Both of these tests indicate that the audio amplifier and speaker are working. A completely dead (no noise or hum) output indicates that the problem is after the control.

Contrast. This control adjusts the amplitude of the video signal. It is located in the video amplifier block. Rotation of this control should vary the amount of picture information displayed on the picture tube.

Brightness. This control is also located in the video amplifier block. Its purpose is to adjust the operating bias on the picture tube. Rotation of this control will vary raster brightness. The range is from total black to full white.

Drive controls. These controls are associated with the color picture tube. They are used to establish a balance between the three electron guns.

AGC. This is an amplitude control for the AGC circuits. At minimum it will reduce IF and RF amplifier gain. The result is a very low video level. At maximum the result is a reverse, or negative, picture, usually accompanied with a buzzing sound in the audio.

Color. This controls the amplitude of the color signal. At low level it shuts off the color processing, resulting in a black-and-white picture. At high level there should be too much color in the picture. This is a color-intensity control.

Vertical height. This adjusts the fullness of the picture. This could be considered a vertical volume control. The range of this control is from no height, or a horizontal line on the middle of the screen, to an overscanned picture. The overscanned picture is too large for the screen. This control is usually in the vertical output block of the receiver.

Vertical hold. This is related to the vertical oscillator. It is used to adjust the frequency of the oscillator. Misadjusting this control makes the picture roll vertically.

Vertical linearity. This control adjusts the shape of the vertical signal in the vertical amplifier. The shape of the signal determines the spacing between the vertical lines on the screen. A nonlinear picture will either stretch the height of the display or shorten it. The result is a distorted picture.

Color killer. This control is used to adjust the threshold of color. When properly adjusted, it shuts off the color-processing blocks when a noncolor signal is present. If it is misadjusted, it will allow these blocks to operate without any color signal. The result is flecks of colors on the picture tube.

Tint. This is a phasing control. It adjusts the phase of the color oscillator signal. Proper adjustment produces normal percentages of color on the screen. Usually, it can be adjusted so that flesh colors are correct. Misadjusting produces too much green or blue in the picture.

Horizontal hold. This control regulates the frequency of the horizontal oscillator. When misadjusted the picture-tube raster appears as diagonal lines.

High voltage. This control is normally found on a color receiver. It adjusts the amplitude of the high voltage required for picture-tube operation. It should not be adjusted unless a meter is used to monitor the voltage level.

Focus. This is a picture-tube control. It is used to provide the proper operating voltage at the focus element of the tube. It should be adjusted so that a clear sharp picture is obtained.

Screen controls. These controls are found only in a color receiver. There may be either two or three of them. They are used to set the dc bias on the screen elements of the picture tube. When properly adjusted the color picture will appear as white on the picture tube. This is true only when setting up the set and using test procedures established by the manufacturer.

The relationship of each of these controls to a specific block should now be more meaningful. Keep in mind that all sets do not have every one of these controls. As sets have become more "customer proof," some of these controls have been eliminated. This would include vertical and horizontal hold controls as well as some of the color and tint adjustments. One will also find controls with names other than those given in the text. Sometimes the function is the same, but the name has been changed.

SIGNAL PATHS

Effective troubleshooting and repair is very dependent upon knowledge of how the receiver is supposed to work. This knowledge aids the productive technician. The presence or absence of signals at key points in the receiver aids in the decision-making process. The effective technician learns to eliminate quickly those blocks that are working correctly. The effect is to reduce the suspected area of trouble to a very small section of the receiver. Using information obtained by operating the receiver's controls assists in this process. Knowledge of the signals that are present in each block or major area also aids in this process. Figure 4-3 shows the major areas in a color receiver. Each area is identified. The layout is common for most receivers. Also included in each area are the types of signals normally found there.

Let us see how this system works. The picture tube requires several signals to produce a picture. These include video as well as red, green, and blue color information. A correct display on the picture tube indicates that all systems between it and the input to the receiver are working properly. There

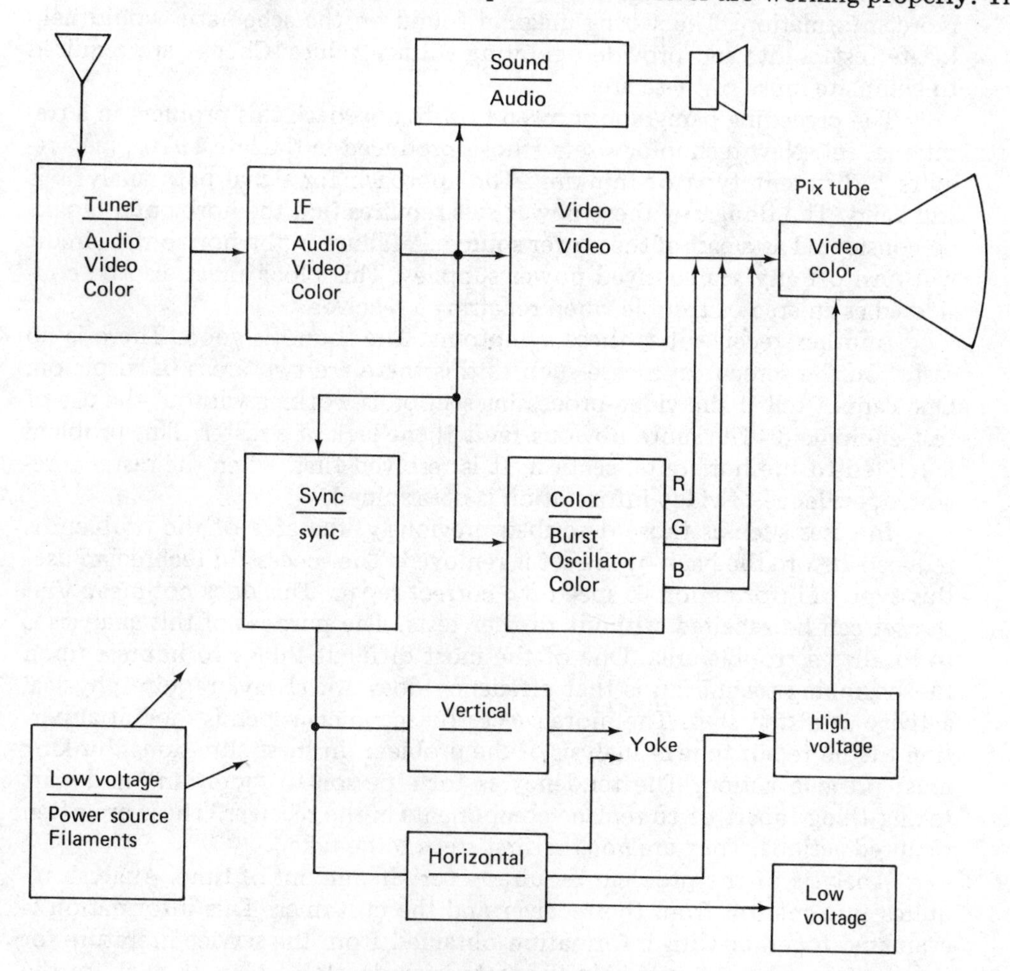

Figure 4-3 Major areas of the color television receiver.

is no need for the technician to check back from the tube to the antenna. This effort is wasted time and energy. It is an inefficient work habit. Unless there was a lack of sound, or low sound, the set would be in working condition.

In another receiver one finds these conditions. There is a raster displayed on the screen of the picture tube. No evidence of sound or a picture appears when the controls are operated. This type of problem requires the use of a schematic diagram. In a traditional type of receiver a problem such as this would probably be found in an area that processed both video and sound signals. This would be somewhere before the sound and video signals separated. The probable area of trouble would be in the IF strip or in the video detector block. There is also a possibility that the trouble is in the low-voltage power supply distribution system. Location of the specific trouble area requires more information. The wiring diagram found on the schematic would help locate test points and provide operating voltage values. Checks are required to eliminate more of these areas.

The preceding paragraph showed how to approach this problem in a traditional set. New-technology sets (those produced in the late 1970s, may require a different type of thinking. The approach for signal path analysis is still valid. The design of these newer sets requires that the horizontal blocks be considered as a part of the power source. A failure in the horizontal circuit will turn off any scan-derived power supplies. This block must also be considered as an area of trouble when repairing a receiver.

Another receiver has these symptoms. The sound is good. There is no raster on the screen. In a case such as this there are two areas of suspicion. One cannot tell if the video-processing section is working without the use of test equipment. The more obvious fault is the lack of a raster. This problem is related to the horizontal section. It is resolved first. When the raster is restored, evidence of video information is determined.

In cases such as those described previously, the area of the problem is reduced before the back of the set is removed. The successful technician uses this type of information to speed the correct repair. This does not mean that the set can be repaired without further tests. The purpose of this analysis is to localize a trouble area. One of the most difficult things to impress upon the beginning technician is that efficiency does not always require physical activity as a first step. The more successful technician spends the initial portion of the repair time in analysis of the problem. In these situations thinking must precede action. The tendency is for a person to "jump in" and start taking things apart or to replace components in the receiver. These are often required actions. They are *not* the first steps to be taken.

Analysis of the problem requires a certain amount of time. Analysis requires information from the receiver and the customer. This information is evaluated together with information obtained from the service literature for the receiver. These inputs help the technician develop a plan of work for the repair. The plan of work requires knowledge of signals, signal paths, and expected waveforms, as well as how operating power is obtained and distributed in the set. This approach is valid for other appliances, for automobiles, or for any other type of device. The technique of planning based upon analysis will help develop proper repair procedures.

OPERATING POWER CIRCUITS

Problems created by operating power circuits must be considered during the repair process. There are two types of problems that may occur. One of these is related to the failure of a filter capacitor. This may cause a low operating voltage. It may also produce a 60- or 120-Hz wave in a section of the set. If this occurs in the audio section, the result is a low-frequency hum that is reproduced and heard at the speaker. If the problem relates to other blocks, it may often be observed on the picture tube as a slowly moving bar of information. It often is darker than the picture information because of its low frequency.

Another type of filter capacitor failure allows signals from one circuit to enter a second circuit by way of the power supply. Filter capacitors are actually "decoupling" capacitors. Their purpose is to remove any variations in operating voltage. Signals are variations in operating voltage in many circuits. In the collector circuit of a transistor this is desirable. The variations are then coupled to the next stage. This variation in operational voltage in one stage must not be transmitted to any other stage through the power source. Capacitors C_1 and C_2 in Figure 4-4 are used as decoupling capacitors.

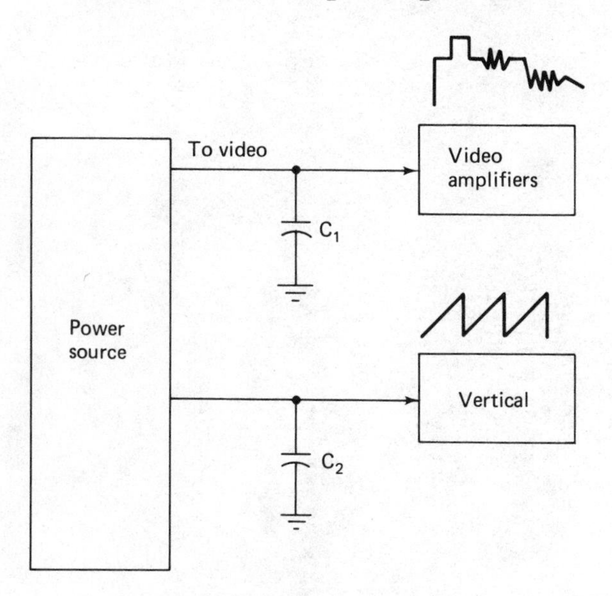

Figure 4-4 Decoupling capacitors keep signals from traveling through the power source to other areas in the receiver.

Any variations in operating voltage caused by the signal are decoupled, or bypassed, to common by capacitor action.

A second type of operating power failure is caused by a component failure. This failure requires dc circuit analysis and test equipment. A full discussion of this problem and how to analyze it is presented in Chapter 7.

At this point you will realize that the successful technician requires much information. This information is needed throughout the repair process. Decisions must be made at each step of the repair. Information is required for valid decision making. The next three chapters provide further types of information that is required for analysis of the problem.

QUESTIONS

4-1. What steps should be performed before the back of the set is removed for servicing?
Which blocks of the receiver are adjusted by the following controls?

4-2. Volume

4-3. Color

4-4. Brightness

4-5. Contrast

4-6. Vertical hold

4-7. Horizontal hold

4-8. Tint

4-9. What major events should be observed when evaluating receiver performance?

4-10. What is the importance of discussing receiver problems with a customer?

Chapter 5

General Troubleshooting Procedures

Procedures outlined in Chapter 4 provide a broad background for troubleshooting. Once these procedures are developed the technician is ready to start the repair. In this process the technician makes tests and measurements in order to locate a specific problem or defective component. The technician starts by suspecting that the whole set is bad. Using a well-developed system reduces the suspected trouble area. This may require several steps and measurements. Each step further reduces the area by eliminating properly working sections until only a small section of the receiver is left. If this procedure is correctly followed, the problem is then isolated.

There are several procedures that can be used when locating trouble areas. One method starts at the signal input of the receiver. Signals are traced through the set until they disappear. When this occurs the trouble area is located. It is between the last two test points. A second method is the reverse of the one just described. Tests are started at the output of the set. The procedure then is to work back toward the input point. When the signal is located, the area of trouble is also located.

Both of these methods are slow. They are time consuming and inefficient. There is a much better way to troubleshoot any receiver. Information in Chapter 4 covers techniques for analysis by sight, sound, and smell. This procedure should be developed until the technician is expert at it. It may be refined further as knowledge of set operation increases. The next phase of troubleshooting relates to a system for analysis. This system may be applied to signal paths and to operational voltages.

The analysis procedure uses five basic systems. These are the only systems that are used in electronic devices. They should be learned. Recognition of each is important to the analyst. Each of the five systems is discussed in this chapter. The information is then related to typical receiver blocks and circuits. The use of the proper system at the appropriate time will raise the effectiveness of the technician.

LINEAR PATHS

A linear device is one that is in one line. Such a system is shown in Figure 5-1. Input to the system is on the left side. There is only one output, on the right side of the diagram. Signals are processed through each block from the left to the right. The signals may change in size or shape as they are processed. Service literature for a specific receiver will provide specific information related to size and shape. The basic fact is that there is only one path for movement in a linear system. Signal flow should not be interrupted.

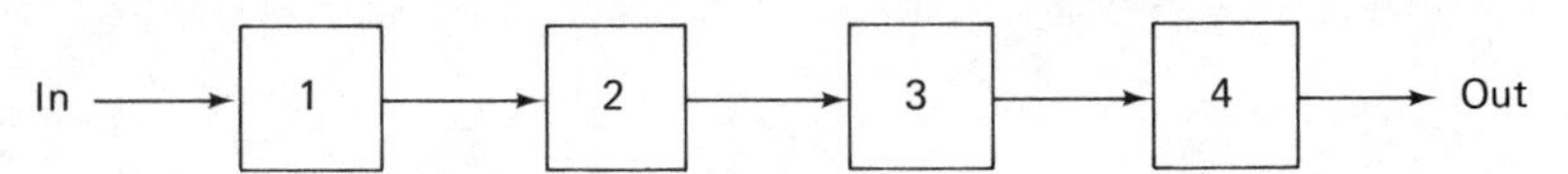

Figure 5-1 Linear signal flow path system. Signal flow is from left to right.

The procedure for troubleshooting the linear system is:

1. Be sure that the input signal is present.
2. Check the output. If the proper signal is present, the whole section is working properly. If there is a wrong output, or no output, the system is not working properly.
3. The next check is to be made at or near the middle of the path. This point is located between blocks 2 and 3 in the illustration. This procedure is shown in Figure 5-2. The results of the test will provide information for the next step.
4. Evaluate the test results. Figure 5-2(a) shows a set of brackets at each end of the blocks. This is the original area of trouble. It has to be reduced to a very small portion of the system. One of two results will be found at the midpoint of the system. Either signal will be present or it will not be present. One of the brackets is moved, depending upon the test results. If signal is present at the midpoint, blocks A and B are working properly. If no signal is present, blocks C and D are working properly.
5. Move one of the two brackets around the system to the midpoint. In the example the right-hand bracket is moved because no signal was found at the midpoint. This is shown in part (b) of the illustration.

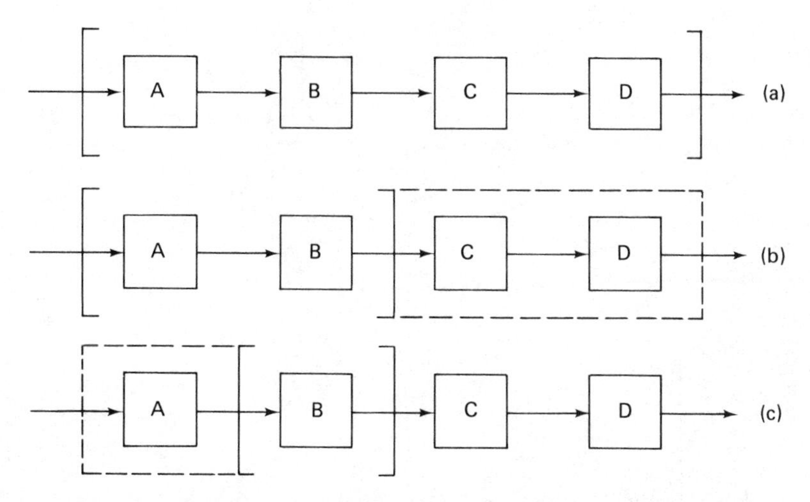

Figure 5-2 The moving-bracket system is used to reduce the area of trouble to one block.

6. Make another test at or near the midpoint of the remaining blocks. The evaluation at this point is the same as performed in step 4. In the sample a signal is found at the test point. The area of trouble is then considered to be in block B. The left-hand bracket is moved to this block. The brackets now surround the one block in which trouble is located. The next step is to locate a specific component that has failed. Procedures for doing this are discussed in Chapter 7.

This procedure is easily related to a system in a TV receiver. Figure 5-3 shows the video processing portion of a solid-state receiver. Each of the triangular blocks represents a stage of amplification. The input to this system is an IF signal from the tuner. The output is a video IF signal. This signal is fed to the comb filter. The comb filter, which is not shown, separates the video and color components of the signal.

There are signal takeoff points in this system. We will ignore them at this time. The basic IF signal processing occurs as a linear path form. This system consists of three integrated-circuit amplifiers and two transistor amplifiers. These components are identified as $U_{301\text{-}A}$, $U_{301\text{-}B}$, $U_{302\text{-}A}$, Q_{303}, and Q_{305}. The midpoint of this system is $U_{302\text{-}A}$. An initial check is made at pin 5 of $U_{302\text{-}A}$.

The presence of a signal at this point indicates that all is functioning properly between this checkpoint and the input from the tuner. The trouble area is between this point and the output. The next check is made at the emitter of Q_{303}. One of the brackets is moved, depending on the test results. If no signal is present, the problem area is between this test point and the previous test point. Another test is then made at pin 8 of $U_{302\text{-}A}$. It could

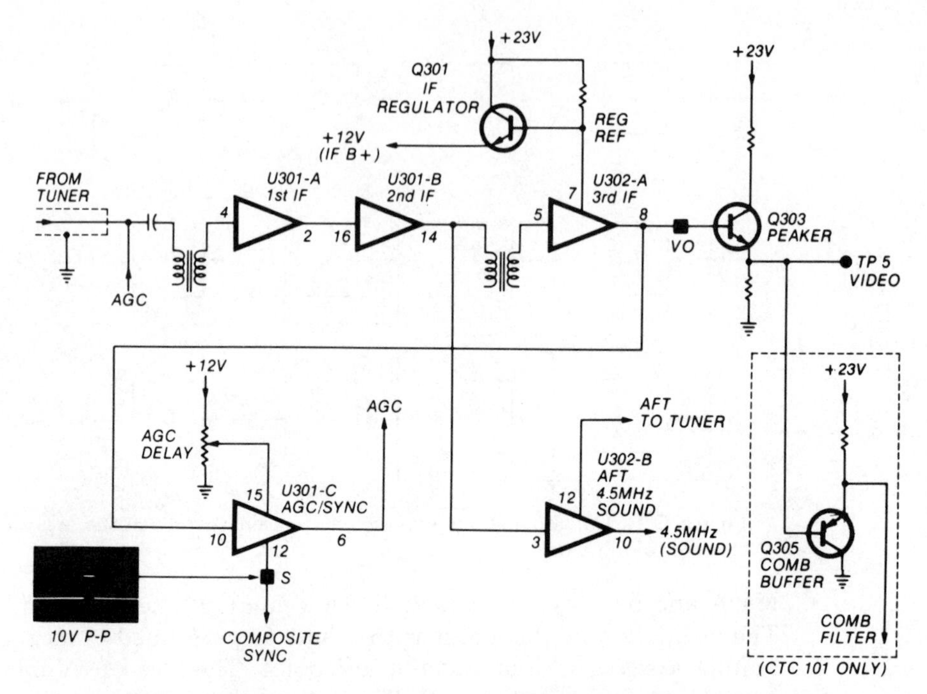

Figure 5-3 Video-processing section of a receiver. Each triangle represents an amplifier stage. (Courtesy of RCA Consumer Electronics.)

also be made at the base of Q_{303}, since both points are electrically the same. If signal is present at this point, the problem area is Q_{303}.

This system eliminates working sections by constantly reducing the suspected trouble area. Eliminating the good areas increases the efficiency of the technician. The purpose of repair work is to diagnose the problem and locate the defective component. After this the technician makes the repair and finally checks to see that the repair is correct. An efficient system of troubleshooting eliminates wasted effort. It allows the technician to concentrate on the problem without bringing in useless information to confuse the problem.

One function of the technician using this system is to make a circuit analysis. Operating voltages must be present. They must be of the proper value and polarity. Current flow, and its resultant power value, must occur if work is to be performed. Amplification is a form of work. Therefore, current flow must occur in the system. A malfunctioning system might be caused by a power problem.

The technician needs to know how the devices in the receiver operate. The system described as a linear flow system also applies to operating power circuits. The circuit shown in Figure 5-4 is used to illustrate this principle. The circuit shown is a common-collector (or emitter-follower) amplifier. The input is to the base and the output is at the emitter. Resistor R_E establishes the emitter voltage and is the load for the output signal.

When this circuit is functional, the collector voltage is 23.0 V. Emitter voltage is about 2.0 V. This information is found on the schematic diagram included with service literature. The current-carrying components in the output circuit form a linear circuit. These components are R_E and the transistor emitter–collector leads. Operating current flows in this linear path. Voltage drops develop across each component. The action of the current flow produces the operational voltages shown at each point on the diagram.

Tests performed by the technician show that the input signal at the transistor base is good. There is no output signal. The immediate thought would be to replace the transistor. This may cure the problem, *but* it may not cure the problem. Other tests should be made before any components are changed.

The first step is to check the input of the operating voltage. This system is also a linear system. The steps used for testing a signal processing system also apply for this circuit. If input voltage (+23.0 V) is correct, a second test is made at or near the midpoint. This is at the emitter of the transistor. Information given on the schematic shows 2.0 V at this point. If this voltage is incorrect, the reason has to be determined.

A measurement of 0.0 V at this point usually indicates an open circuit between the test point and the power source. The transistor is open. A measurement of 23.0 V at this point indicates a no-current condition. Voltage drops only occur when current flows. No voltage drop indicates that no current is flowing. This is an open circuit. This opening is between the test points. One meter lead is at circuit common. The second lead is at the emitter of the transistor. The opening is between these points. A resistance measurement will help to determine which component has failed.

Once the specific defective component is identified, it is replaced. This will restore the circuit to its operational condition. The procedure for troubleshooting a linear current path is the same as for troubleshooting a signal path. The difference is in the measuring equipment. In either type of circuit a schematic diagram is used to supply normal operating circuit values.

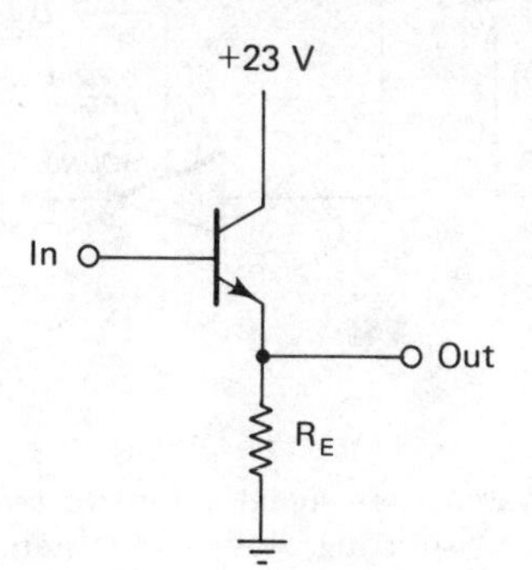

Figure 5-4 Emitter-follower amplifier stage is a good example of a linear flow path system.

SPLITTING PATHS

The splitting path system is used to separate two or more signals. The diagram shown in Figure 5-5 is used for reference in this type of signal path system. An IF signal is fed to the input of $U_{301\text{-}B}$. This is a composite signal. It has both sound and video components. The amplifier has one input and two outputs. One of these is fed to $U_{302\text{-}A}$. The other is fed to $U_{302\text{-}B}$.

This system is checked at the point, or component, where the split occurs. The procedure for troubleshooting a splitting path system is:

1. Check to be sure that the input is present.
2. Check both outputs.
3. Evaluate the test results. If both outputs are correct, the unit is good. If only one output is working, the problem is in $U_{302\text{-}B}$. If neither output is working, the problem is also in $U_{302\text{-}B}$.

The effectiveness of this approach is its adaptability. The technician is able to apply this approach when first evaluating the receiver. The presence of both audio and video indicates that the system (which is a splitting signal system) is functioning. If the set has no audio but has video, the trouble is *after* the split. The linear path portion of the audio section is checked. If

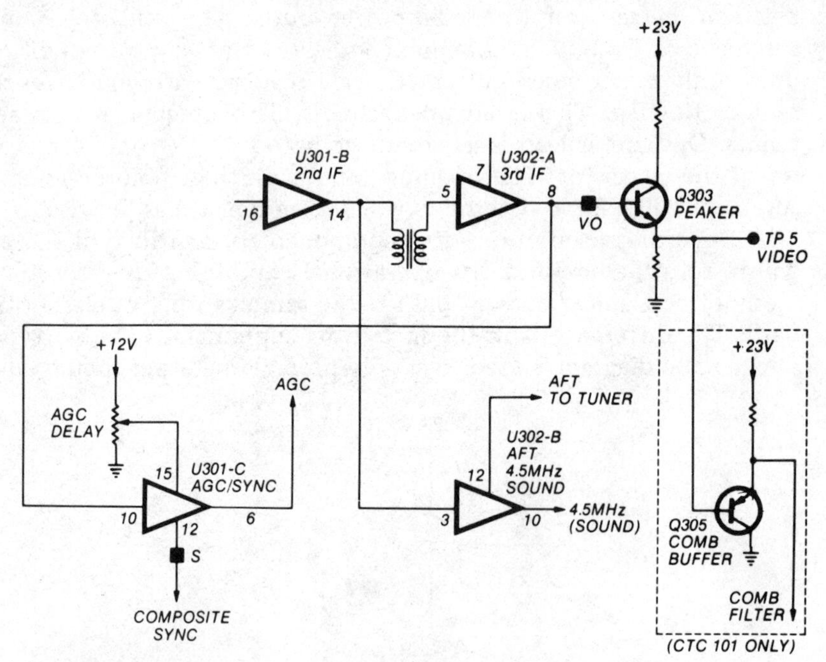

Figure 5-5 In this system one signal is sent to two separate sections. This is the separating, or splitting, flow path system. (Courtesy of RCA Consumer Electronics.)

neither audio nor video is present, the problem is located in some common point. This point could be anywhere from the antenna to the point where the separation of signals occurs. The first check the technician makes is at the point of separation. This is the output point of the linear system and the input point for the separation system. A block diagram for this combined system is shown in Figure 5-6. Results obtained from this test determine which system is checked for malfunction.

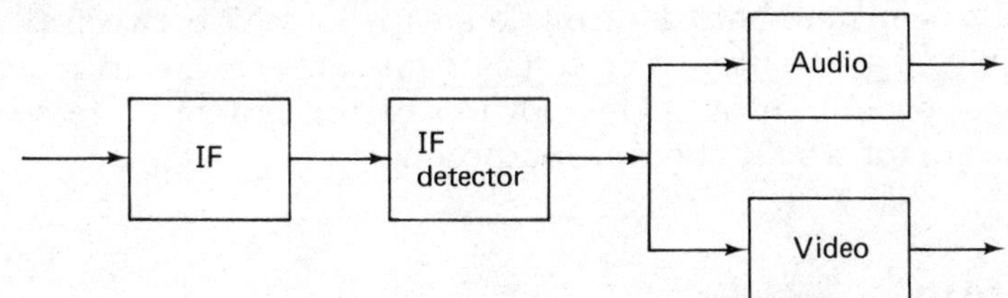

Figure 5-6 Block diagram of the signal separating system.

The approach used to check signal path systems can also be used for power source systems. Many of the power supplies found in TV receivers use a separating current flow path system. One such system is shown in Figure 5-7. There is one output from the power supply block. Current paths are

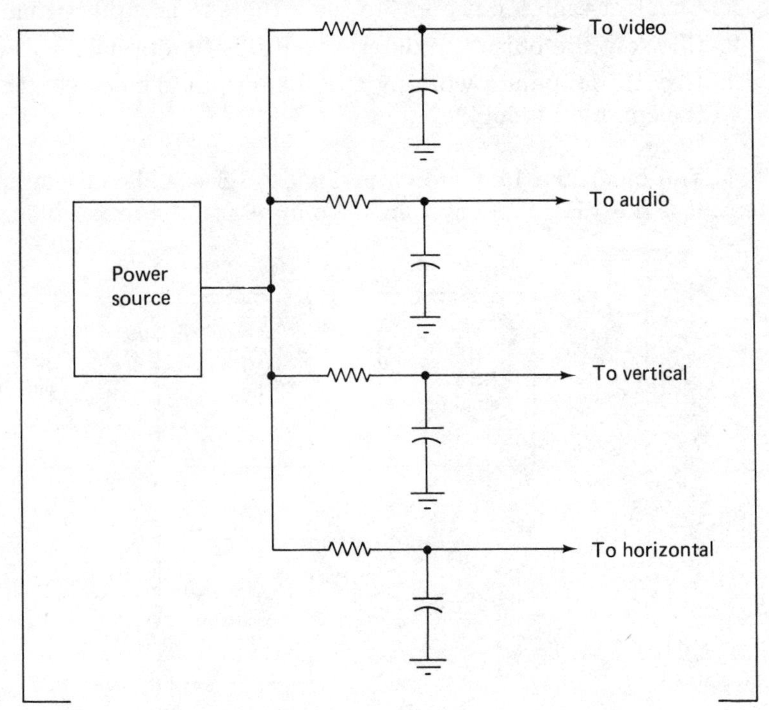

Figure 5-7 Separating system used in the power supply section of a receiver.

separated after this output in order to supply four systems. If all systems were not functioning, the most logical test point is at the point of separation. A measurement at this point moves one of the brackets. If voltage is present at this point the bracket on the left side is moved to the test point. The power source block is good. If there is no output at this test point, the power source block is not functional and the right-hand bracket is moved.

If the input point is good, the outputs are checked. If one of the outputs is not present, the trouble area is located between the point of separation and that output point. The right-hand bracket is moved to the right of that specific output. This portion of the system is treated as a linear path system for troubleshooting purposes.

MEETING PATHS

This system is the opposite of the separating path system. Two or more signals enter a junction. Some sort of electronic interaction occurs. There is one output from the junction or meeting block. The procedure used to troubleshoot a system such as this is as follows:

1. Check at each input point to see if there is an input signal.
2. Check at the output to determine if it is functional.
3. If both inputs are working and the output is incorrect, the trouble is in the common block.

The block diagram shown in Figure 5-8 is a classic example of this system. It is the VHF tuner system. Two inputs to the mixer block are required.

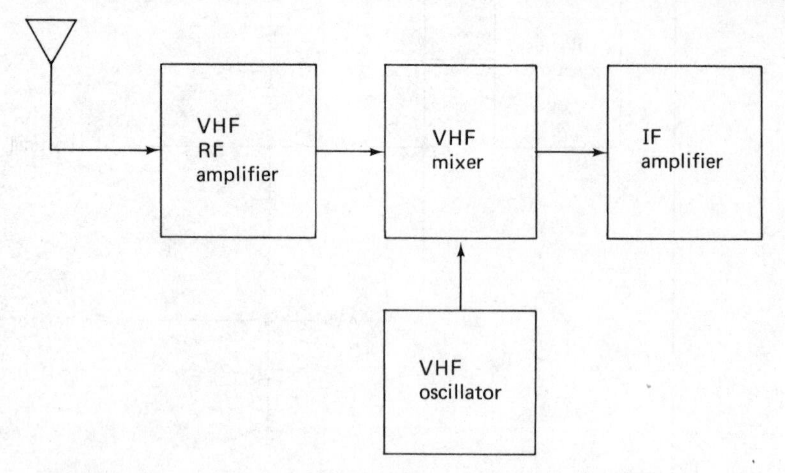

Figure 5-8 The tuner section of the receiver is a good example of the meeting signal path system.

One of these is the modulated RF carrier signal from the transmitting station. The second is the unmodulated carrier signal from the local oscillator. These two signals are fed to the mixer block. The output of the mixer contains four basic signals. These include both of the original signals, a modulated carrier whose frequency is the sum of the two carrier frequencies and another modulated carrier whose frequency is the difference between the two original carrier frequencies. The latter signal is usually the IF amplifier frequency.

A second circuit using the signal meeting principle is shown in Figure 5-9. This is the horizontal AFC circuit used in many receivers. Two signals are required for an output. One signal is fed from the sync separator. The second signal is a pulse from the horizontal amplifier circuit. These two signals are fed into the horizontal AFC block. The output of this block is a dc control voltage. The control voltage is used to regulate the frequency of the horizontal oscillator. If either of the two signals is incorrect, the output control voltage is not present. The result is an off-frequency horizontal oscillator.

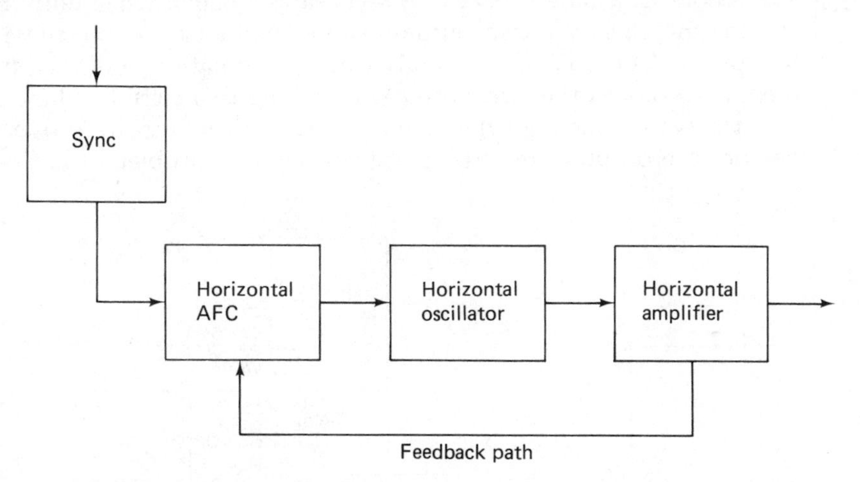

Figure 5-9 The horizontal AFC circuit is another example of a signal meeting system.

Meeting paths are also found in electronic power circuits. One such circuit is shown in Figure 5-10. A common-emitter amplifier is illustrated. The meeting path is located at the input to the transistor. This is the base circuit. Resistor R_B, the base-emitter junction of the transistor, and resistor R_E form a current path. This establishes a bias of 4.0 V at the base. A signal source of 2.0 V p-p is also connected to the base. When both bias and the signal voltage are present, the base voltage will swing from 3.0 to 5.0 V. This produces an output voltage swing of 10 V p-p. If both input voltages are present and there is no output voltage, the transistor circuit is malfunctioning.

A combination meeting-separating circuit is illustrated in Figure 5-11. The comb filter circuit requires two inputs. One of these is a video signal.

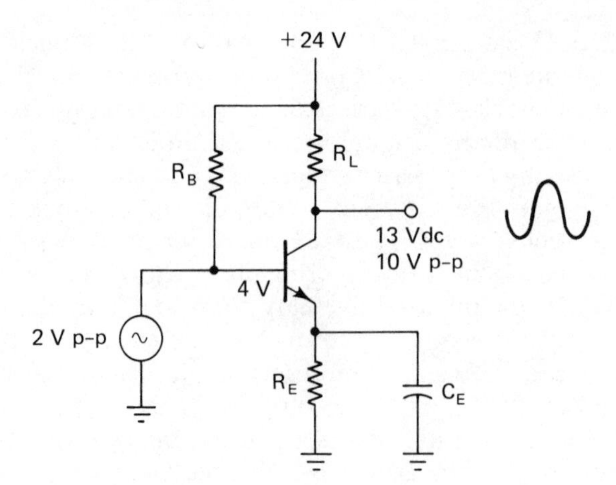

Figure 5-10 An amplifier also uses the meeting principle for its operation.

The second is a 3.58 (3.579545)-MHz carrier signal. When both input signals are present, there are two outputs from this block. One of these is color information. The second output is video information. Both inputs must be present to obtain the two outputs. Checking this circuit is done in the same manner as the meeting type of circuit. If both inputs are proper and either one or both outputs are missing or incorrect, the problem is in the filter unit.

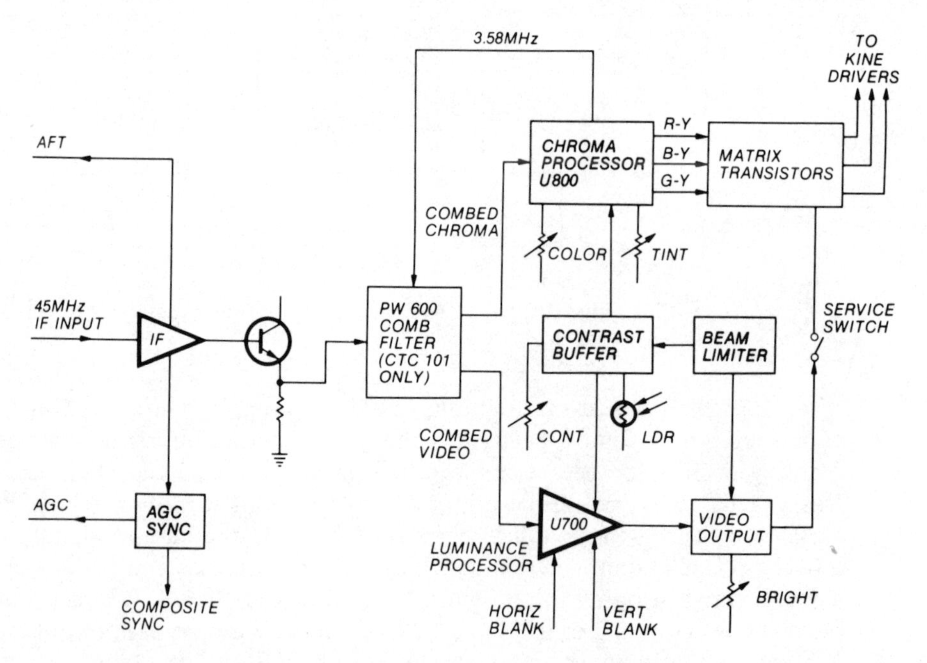

Figure 5-11 The comb filter found in new receivers uses the meeting principle. (Courtesy of RCA Consumer Electronics.)

FEEDBACK PATHS

Automatic control circuits have been used in receivers for many years. One of the first types of control circuits is the automatic gain control used in the IF section. The term "gain" is used to show that signal amplification occurs. Most of the automatic circuit controls use a sample of the output signal to control the amount of signal gain that occurs. The AGC block functions in this manner. A portion of the output signal is fed back to the input. This sample signal is used to control the bias on a stage of the IF amplifier. Figure 5-12 shows this.

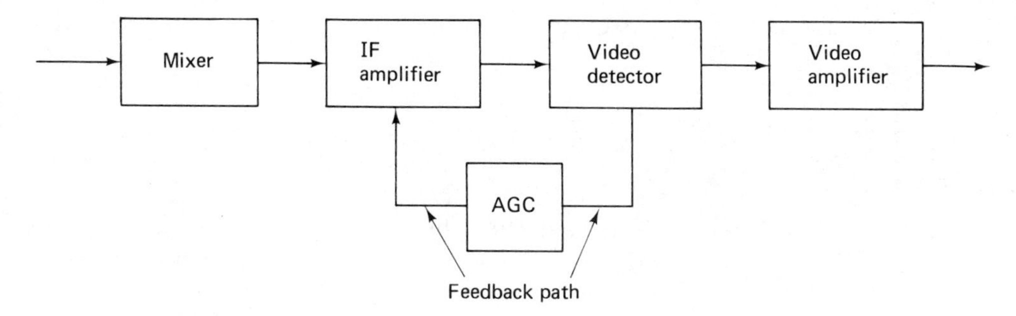

Figure 5-12 AGC systems use a feedback principle for operation.

The rule for troubleshooting any feedback type of system is:

1. Open the feedback path. If normal levels of signal are restored at the output, the problem is in the feedback circuit. Often, a technician will apply a dc voltage from another power source to the AGC circuit instead of disconnecting components in the set.
2. If the circuit is still not operable, the problem is not in the feedback path. It is probably in the forward path. In this case the forward path is the IF amplifier and detector blocks.

Many receivers use a pulse obtained from the horizontal output circuit to time, or key, another circuit. Typically, these keying pulses are used for AFC, color processing, and AGC. This is a feedback type of circuit. It uses a signal pulse rather than a dc voltage to turn on the blocks it controls. Each of these blocks then requires two signals in order to produce the proper output. The keying pulse portion of the receiver is considered a feedback path. Processing that occurs at the blocks is most likely to be part of a meeting type of system.

SWITCHING PATHS

The final signal path system is called a *switching system*. In this type of circuit signals are directed by a mechanical or electrical switch. Some circuits, such as the one shown in Figure 5-13, are used to select one input to an audio amplifier. The rules for this type of circuit are:

1. Use the switch to check each output.
2. If all outputs are functioning, there is no problem. If all outputs are not working, the problem is a block that is common. The only block that is common is the audio amplifier block.
3. If one output is not proper, the problem is in that circuit.

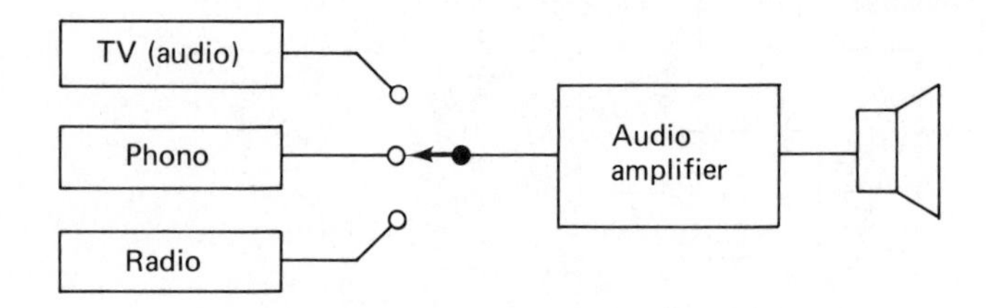

Figure 5-13 Many multipurpose audio systems use a switching system to select one output.

This approach may also be used for power circuits. A switchable power supply circuit is shown in Figure 5-14. If all three of the output blocks do not work, the problem is in the power supply or its switch. If one of the three output blocks does not work properly, the problem is in that block.

Each of the five signal path systems are used extensively in television receivers. The technician's role is to be able to identify the system. In most instances this may easily be done by use of a schematic diagram. Once the type of system is identified, the technician follows the procedures explained in this chapter. This will seem slow at first, but do not give up or try to bypass part of the system. Diagnostic time will decrease as experience is gained.

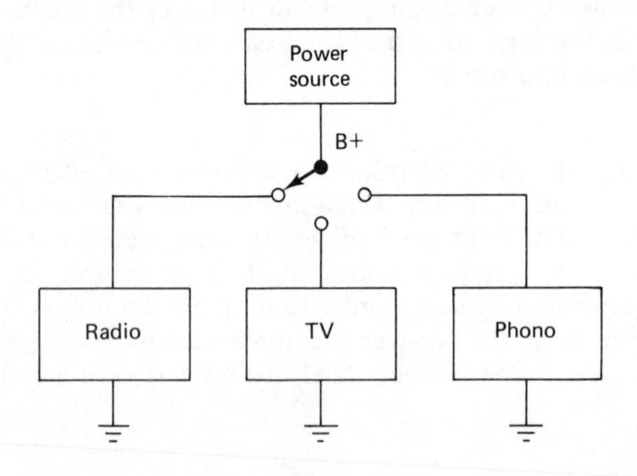

Figure 5-14 Power sources often use a switching path for direction of current flow.

After a while, use of this system will increase effectiveness. It also has the advantage of allowing a person to keep up with technological changes. Repair procedures change as components change. The diagnostic system does not change.

INTEGRATED CIRCUITS

One of the major advancements in electronics was the introduction of the transistor and its solid-state technology. A second advancement was the utilization of the transistor in the integrated circuit (IC). These components have simplified set construction. Reliability of set operation has also increased. Production costs are greatly reduced. More complex circuitry is used in today's receiver than was found in early production receivers.

This has great significance for the technician. It means that troubleshooting and repair may be faster and easier. This is only true when the technician is willing to think with an "IC" mind. Servicing techniques have changed. The successful technician changes thinking to keep up with these changes. It is important to know how to use test equipment. In addition, one must know which piece of test equipment to use for a specific job. Signal path analysis is easy when one is dealing with integrated circuits. Signal(s) at the input to the IC are checked. The signal(s) at the output are also checked. The procedures for each of the five basic systems are applied as required. Operating voltages are also checked. Decisions based upon this analysis are made. In many instances the result is the replacement of the IC. It is difficult to accept that this is really faster and easier than attempting to repair a set that uses discrete components.

A commonly used IC is illustrated in Figure 5-15. Most manufacturers show a block diagram of the circuits in the chip rather than the complete circuit. Both are shown here so that you can see the complexity of the circuit. The procedure for checking this circuit is to treat it as a linear circuit. The input is checked and then the system is split. The only coupling device that is not in the IC is one coupling capacitor. This point is checked to determine if the capacitor failed or if the failure is in the IC. Repair is based on the results of the tests.

MODULAR SYSTEMS

In many ways modules may be considered to be overgrown integrated circuits. The difference is that the modules are physically larger and contain parts in addition to diodes, resistors, and transistors. Modular servicing is similar to that used when servicing an integrated circuit. Most receiver manufacturers provide service information for their sets. This information usually

schematic and connection diagrams

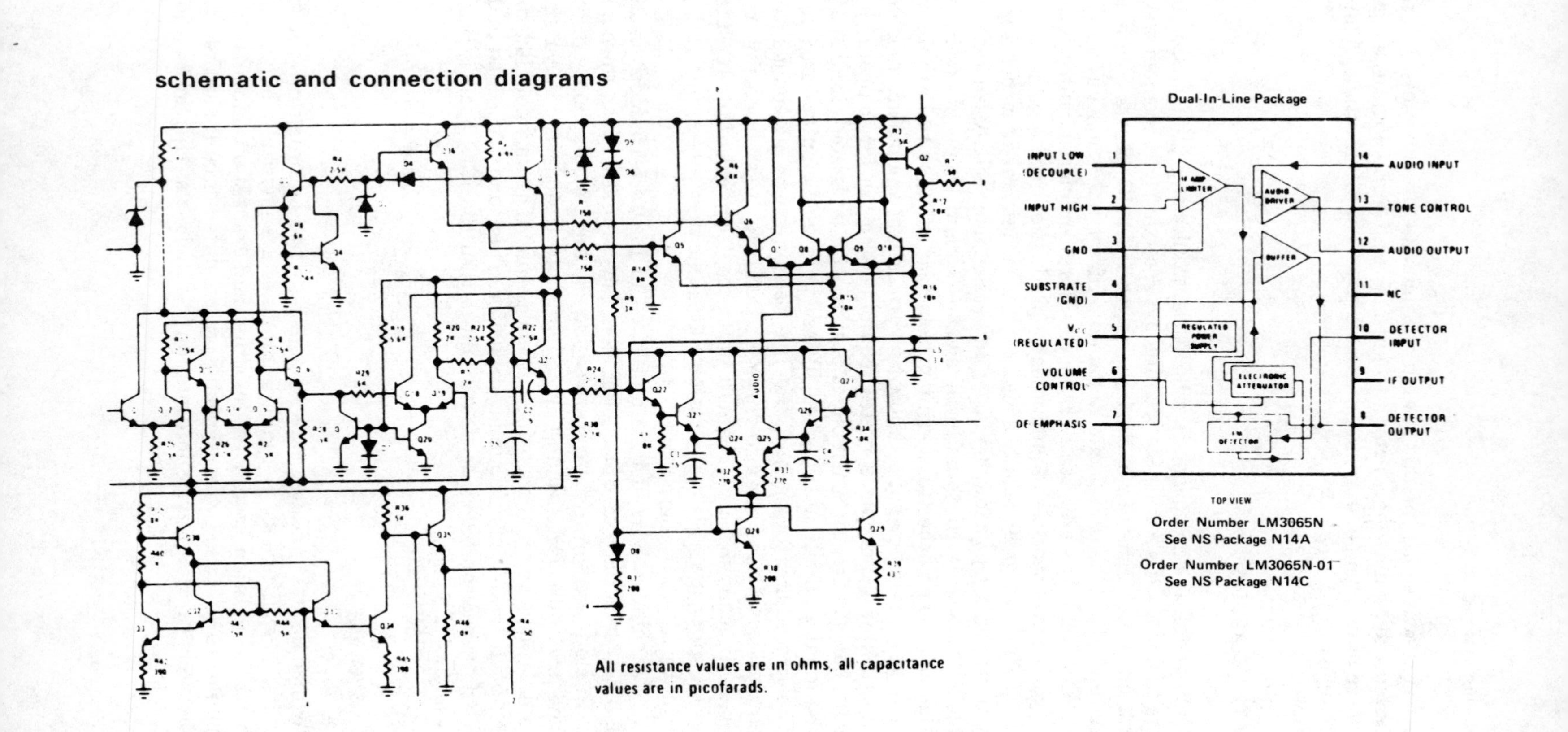

Order Number LM3065N
See NS Package N14A

Order Number LM3065N-01
See NS Package N14C

All resistance values are in ohms, all capacitance values are in picofarads.

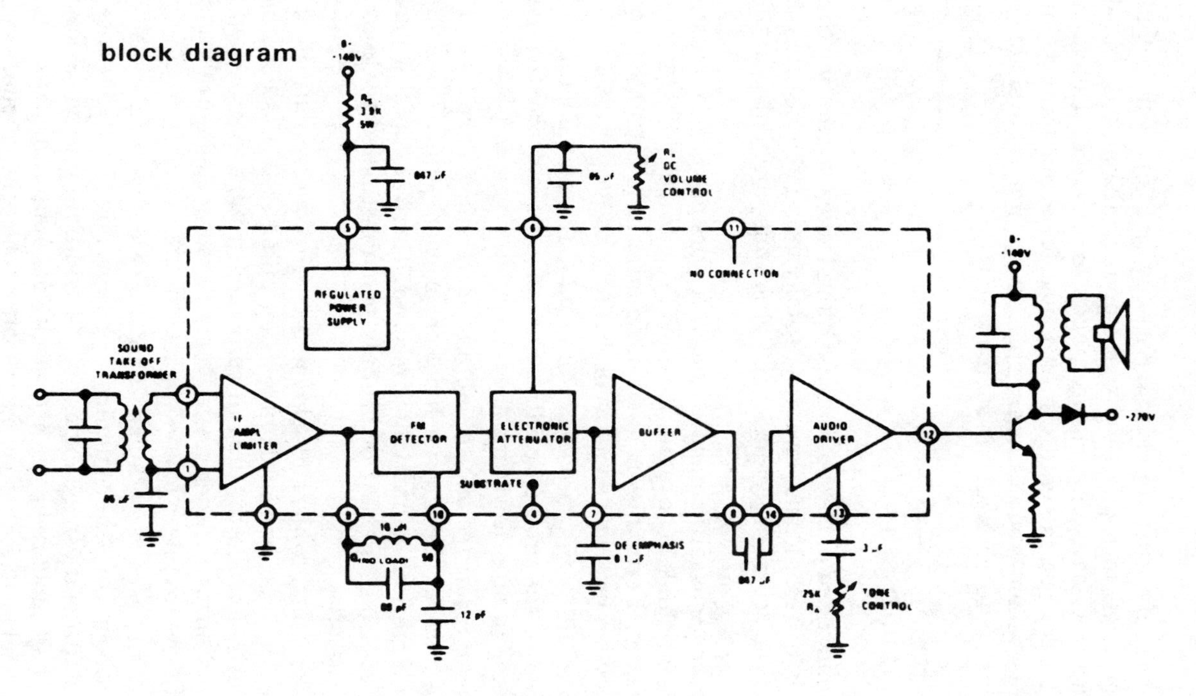

Figure 5-15 Diagnosis of IC systems uses flow path analysis to determine if the IC has failed. (Copyright 1981 National Semiconductor Corp.)

includes a diagram of the set. One such diagram is shown in Figure 5-16. Each module is shown in this drawing. Blocks found in the module are also shown. Test points (TPs) are identified to make the servicing of the receiver easier.

Each manufacturer provides service information. This includes suggestions for efficient methods for troubleshooting the module. Most modular set manufacturers have an exchange policy for modules. Defective units are returned to distribution centers in exchange for a functioning unit. Technicians have to repair modules when replacements are not available. The signal path systems approach is used in such instances.

TROUBLESHOOTING PHILOSOPHY

A certain amount of understanding is required when attempting to repair any receiver. Many technicians have been totally frustrated because they forget to follow basic repair procedures. The philosophy promoted in this book is:

1. *The receiver is an inanimate object.* It does not have a mind of its own. It cannot think. Above all, it is not "out to get you" by being devious or sneaky. Keep this in mind, no matter how frustrated you may become when resolving a circuit problem.
2. *The set did work properly at one time.* The manufacturer tested it at the end of the production line. Something occurred either when it was delivered or after it was installed.
3. *You, the repair technician, are not being asked to redesign the circuits in the receiver.* Design is a function of engineering. Your purpose is to repair a previously functioning receiver. There are times when field modification is necessary. Do this only when instructed to do so by the set manufacturer.
4. *Diagnose the problem by eliminating working sections of the receiver.* Suspect a large area of trouble as an initial step. Use controls to aid in diagnosis. Reduce the area of suspected trouble in an orderly and professional manner. Keep on reducing the suspected area until one circuit is left. The trouble is located in this circuit.
5. *Repair the receiver.* Replace only those components that have failed. The concept of "shot gunning," or replacing all components in a stage, is wasteful and can be expensive. In addition, one will occasionally replace parts incorrectly when using this approach. This introduces an additional problem rather that repairing the original one.
6. *Assume nothing.* Assumptions are often based on inadequate information. They also are based on a previous experience. It is true that some defects repeat in the same model of a set. Analyze data and use the analysis to locate an area of trouble.

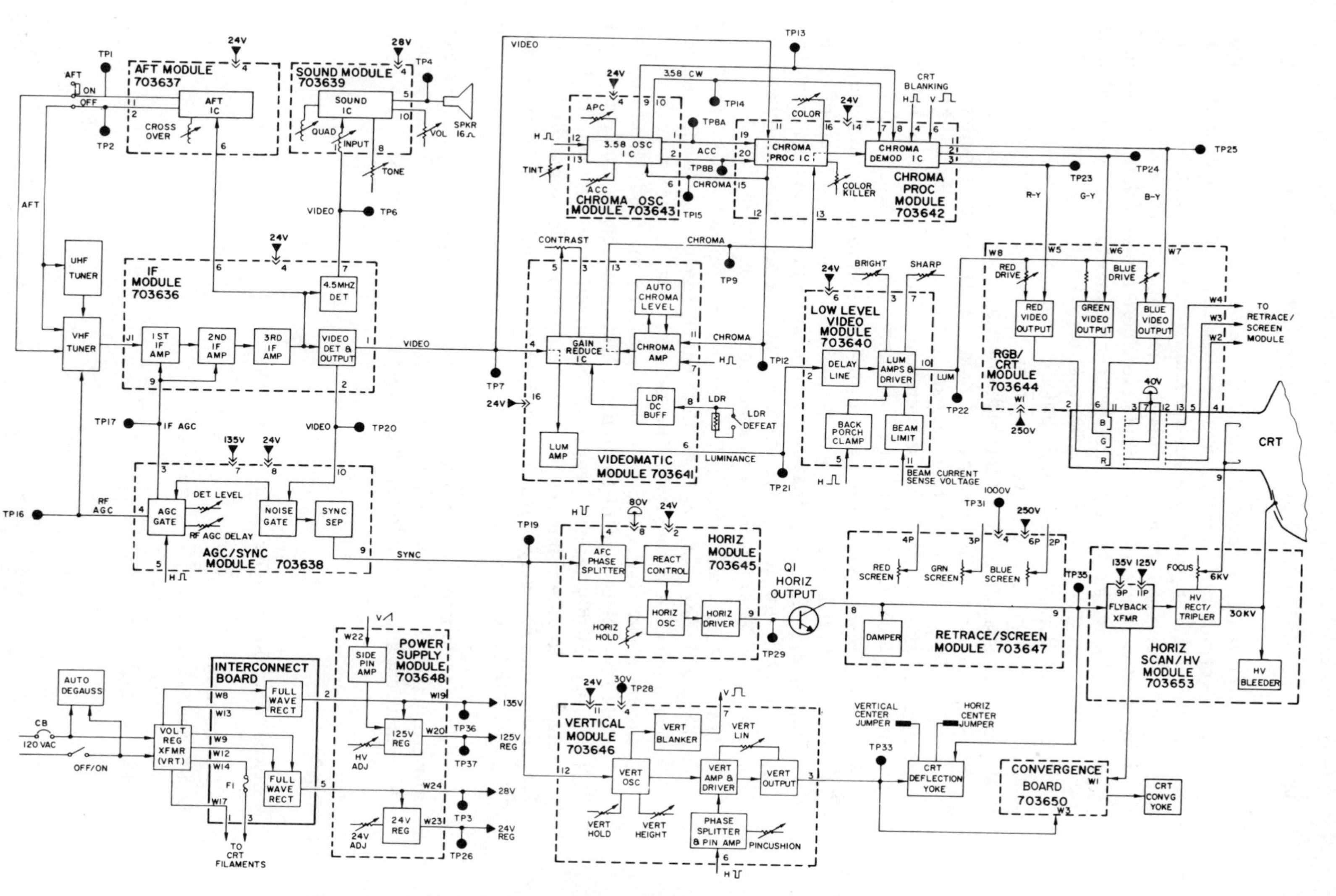

Figure 5-16 Modular block diagram used for a TV receiver. (Courtesy of Magnavox Consumer Electronics Company.)

7. *Play the odds.* In most instances the odds are correct. For example, hard-working components fail more often than components that do not work hard. Output transistors are hard-working devices. They will fail more often than transistors used as preamplifiers. Therefore, the odds are that output transistor circuits are where the set failure is found. A study of service calls as they relate to component failure was conducted by General Electric for their own receivers. The findings are that output devices fail more often than do other components in a receiver.

   This philosophy is true for other components. Low-wattage resistors fail less often than their higher-wattage counterparts. As you develop skill in troubleshooting and repair, a very definite ranking of the types of parts that fail will be apparent. Do not misunderstand this section. All types of parts fail at some time. The idea presented here is to help you look for the most obvious failures first.
8. *Test the repair.* No receiver should be returned to the customer until it is bench-tested. A good bench test includes at least two hours of operational time. Make sure that all troubles are repaired and that new ones do not turn up. Recalls are expensive in time and in lost income. They will occur. Your position is to keep them to a minimum.
9. *Last, but not least, charge for your labors.* Do not sell yourself too cheaply. Labor costs are high. Equipment, advertising, telephones, and rent are also at a high level. The technician has to earn enough to pay the bills, pay wages, and to make a profit. Some funds should be put aside for replacement of equipment. Electronic service is a professional activity. Highly trained people are employed. Their training has to be kept up to date. Time spent at a training session is nonproductive service time. It is very necessary. Service income must be adjusted to compensate for this.

The development of a good set of repair techniques will go a long way toward making one successful. The procedures discussed in this chapter provide excellent guidelines. Follow them down the road to success.

QUESTIONS

Describe signal flow paths for each of the following:

5-1. Linear system
5-2. Splitting system
5-3. Meeting system
5-4. Feedback system
5-5. Switching system

Describe the troubleshooting steps for each of the following:

5-6. Linear system
5-7. Splitting system
5-8. Meeting system
5-9. Feedback system
5-10. Switching system

Chapter 6

Television Test Equipment

One of the questions most often asked by students enrolled in a television servicing program is: What kind of test equipment do I need? This is not an easy question to answer. Several factors influence the answer. Among them is the amount of money available and the type of service that will be performed. A person who intends to perform all work in the customer's home will have different needs than one who wishes to establish a test bench in the shop. Certain basic items of test equipment are required in either case.

Test equipment may be categorized in several ways. One of these is used to establish some order to the many types of equipment available. The author does not wish to imply that one make of equipment is better than any other make. Each individual should do some research before making a purchase. The research should include price comparison, feature comparison, local availability, and convenience of obtaining repairs when required. Often, a discussion with other technicians will help bring these points into focus.

The classification of test equipment in this section is:

1. *Measuring devices.* Multimeters are used by most technicians. Often a high-voltage voltmeter is used for color TV. Another group of measuring equipment (that is not used as often as it should be used) is the oscilloscopes.
2. *Signal-producing devices.* Signal generators are used to produce a reference test signal that has constant form. These include RF, IF, color, video, and audio signal-producing devices.

3. *Testers.* This group of instruments is used to evaluate the operation of specific components. Included in this group are tube, transistor, picture tube, and resistor–capacitor analyzers.
4. *Probes.* Perhaps the least understood item available for use with other test equipment is the probe. A variety of these are used to measure RF, high voltage, and to match circuit impedance with tester input impedance.
5. *Miscellaneous items.* Some other items usually used by a television service technician include a degaussing coil, a test jig, and a power supply. Other items of interest are selected and used by individual technicians. These are too numerous to describe. A visit to a local electronic parts supply store will identify a good many of these items.

MEASURING DEVICES

Multimeters. The one test instrument found in every service organization is the multimeter. These devices measure voltage, resistance, and current. Multimeters are so named because of their multipurpose function. Figure 6-1 illustrates some basic units. The multimeter has two types of readouts. One of these is an electric meter. This is called an *analog readout device.* A scale that has numerical values is printed and placed on a metal sheet. The metal sheet fits behind the indicator pointer of the meter. An electric current flowing through an armature converts the electron motion into rotary motion.

Figure 6-1 The analog-style multimeter measures voltage, current, and resistance values. (Courtesy of Dynascan Corporation.)

The exact value is read from the printed scale. Multimeters have the capability of measuring a variety of values of voltage, current, and resistance. The range switch is used to select a specific value range for the meter.

A second type of measuring meter is shown in Figure 6-2. This meter is an advancement of the analog type. It uses solid-state technology to present a digital readout of the measured value. The electronics in this type of instrument require the addition of a power source. This may be either a battery or a source of ac. Meters of this type are called digital multimeters (DMMs).

One consideration for the new technician to keep in mind is the input impedance of any measuring device. A low input impedance will alternate, or load, the circuit being measured. A quality meter will have a constant input impedance of at least 1 megohm (MΩ) on all ranges. The importance of this cannot be overemphasized. This feature is particularly true when one measures the low values of voltage found in solid-state circuits.

Figure 6-2 Digital readout multimeters are becoming very common in the service shop. (Courtesy of VIZ Test Instruments.)

Oscilloscopes. This instrument is one that many technicians seem to be afraid to use. The oscilloscope, as shown in Figure 6-3, is one of the most versatile instruments available today. It provides a visual representation of the electrical waveform. The oscilloscope is able to measure both ac and dc voltage values. It will also display electronic signal waveforms. The oscilloscope has a time-base generator built into it. This enables the operator to synchronize the timing of the measured waveform. The result is a display of one or two of the waves.

Oscilloscopes are available with a variety of frequency responses. This means that waveforms at specific frequencies may be observed. A 10-MHz signal observed on an oscilloscope that had a 1-MHz frequency response would probably be distorted. Use of a oscilloscope that had a 10-MHz or higher frequency response will display the correct waveform. The technician needs to consider the frequencies to be measured when selecting an oscilloscope. Most

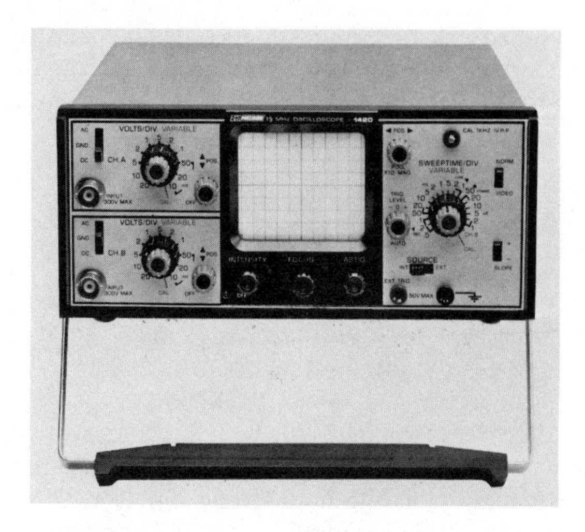

Figure 6-3 The oscilloscope, when used intelligently, will often make the service technician's work easier. (Courtesy of Dynascan Corporation.)

instruments in the 10- to 15-MHz range are adequate for the types of readings required.

Counters. An instrument that has gained popularity with service technicians in recent years is the frequency counter. One such instrument is shown in Figure 6-4. The oscilloscope has the capability of measuring frequency. The method used requires some mathematical manipulation. It is not the most accurate method known. A much better way is to use the frequency counter. This device provides a digital display of the measured frequency. It is of great importance when checking frequencies of the color, vertical, and horizontal oscillators in a receiver. Each of these counters has a frequency range. The range may be extended by use of an adapter. The adapter divides the measured signal by some known factor. The divided signal is within the

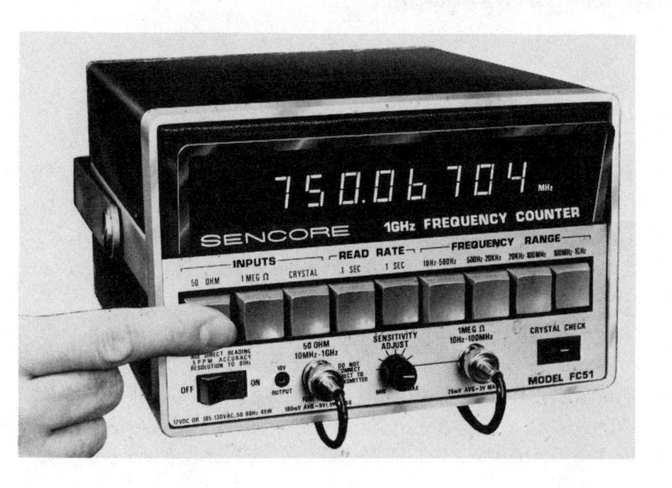

Figure 6-4 Newer-technology circuits require exact frequency measurement. The frequency counter is used for this purpose. (Courtesy of Sencore, Inc.)

capabilities of the counter. The displayed frequency is then multiplied by the technician the correct number of times to determine the measured frequency. Assume that a counter is able to measure frequencies of up to 20 MHz. One wishes to measure a 150-MHz signal. A countdown-by-10 adapter is used as follows:

measured frequency = 150 MHz
adapter divide by 10 = 15 MHz
counter display = 15 MHz
technician multiply by 10 = 150 MHz

The use of an adapter of this type permits the measurement of frequencies that are higher than the upper-frequency limit of the counter.

Field strength meters. It is important that the technician know the amount of signal strength at the antenna terminals of the receiver. An instrument that is designed to measure this quantity is the field strength meter. One such device is shown in Figure 6-5. The field strength meter is basically a RF voltmeter. It has the ability to tune to a specific frequency, or channel. The output of this instrument is a reading in microvolts of the received signal. Many of these instruments will also provide an audio output of the station sound signal.

Figure 6-5 Antenna signal strength and orientation is easily done using a field strength meter. (Courtesy of Leader Instrument Corp.)

Field strength meters are useful for antenna orientation and selection. Sufficient signal must be present at the receiver antenna terminals if a quality picture is to be displayed on the picture tube. The relative strength of the electromagnetic field radiated from the transmitting antenna is of great importance to antenna and cable distribution systems technicians. Received signal strength values are necessary knowledge for the design of antenna distribution systems in business and apartment buildings.

High-voltage probe. This device is actually a high-voltage meter. It is made of a high-dielectric plastic. It has a meter that is calibrated in kilovolts built into the handle. This probe is used to measure the color TV high voltage. Most receivers have a high-voltage adjust circuit. It is necessary to mea-

Figure 6-6 This type of high-voltage tester is used to measure CRT second-anode voltage levels.

sure and monitor the picture-tube high voltage when making this adjustment. One model of a high-voltage test probe is shown in Figure 6-6.

SIGNAL-PRODUCING DEVICES

A second group of test equipment is generally classified as signal-generating or signal-producing devices. The purpose of any signal generator is to develop a stable reference signal. This signal is used for test purposes in the receiver. There are a variety of test signals used in television. These include IF alignment, RF alignment, video, color, and audio generators. Generators capable of producing each of these signals are available as separate units. Most technicians prefer to purchase multipurpose test generators. The most common types of these fall into three categories. They are sweep generators, a so-called "analyst," and color generators.

The television analyst. This generator has several functional outputs. The unit pictured in Figure 6-7 has the ability to produce an RF signal modulated with video information. The output frequencies from this section are estab-

Figure 6-7 A multifunction test generator, called an Analyst, is used to substitute signals in the receiver. (Courtesy of Dynascan Corporation.)

lished to equal several TV channel frequencies. These frequencies are in both the VHF and UHF range. It will also produce an IF signal. The analyst contains a flying-spot scanner tube. Transparencies are placed in front of the scanner. The picture on the transparency becomes the video signal for the analyst. A standard test pattern, as shown in Figure 6-8, is provided with the analyst. This or any other type of visual may be used as a test pattern.

Analyst units of this type are useful because they have the capability of producing all the signals found in a receiver. Several outputs are obtained from the analyst. These include video, sync, 4.5-MHz and 1-kHz audio, and color signals. Each of these is useful when a substitute test signal is required. Information of specific methods of using this, or other similar testers, is provided in Chapter 7.

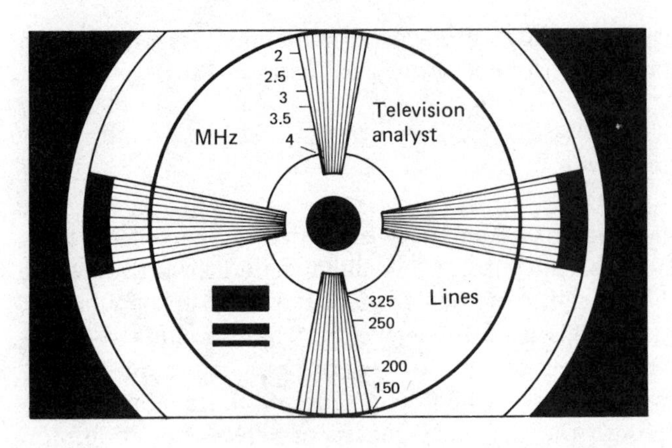

Figure 6-8 This test pattern is used to check frequency response and adjust scanning controls in the receiver. (Courtesy of Dynascan Corporation.)

The B&K unit shown in the illustration has two other useful features. One of these is a bias voltage source. This is a variable dc voltage. It may be used to substitute for AGC or other similar control voltages in the receiver. A second major feature of this unit is its sweep substitution section. This analyst has the capability of providing all sweep signals used in a receiver. It will also check windings on and yokes for shorted turns. Signals are available to drive the HOT without using the receiver's horizontal output transistor stage.

Color generator. The complexity of the color signal makes the purchase of a color generator an absolute must. Color generators are available with a variety of signals. The range of functions will also vary. In many instances the field service technician uses a color generator for receiver setup. This requires a modulated signal at the IF frequency of the receiver. Usually, channels 3 and 4 are also available as modulated RF signals. This enables the technician to check for tuner problems in the receiver. One model of color generator is shown in Figure 6-9. The switches on the left side of the generator select either the RF channel or the IF signal. The knob above these

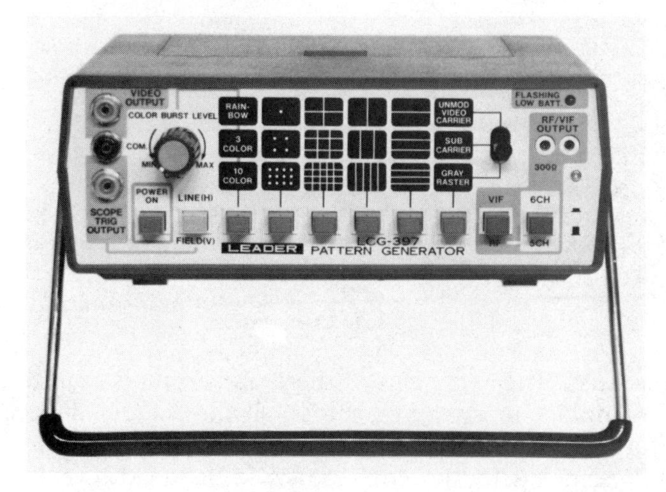

Figure 6-9 A color bar and line generator is essential for proper color receiver setup. (Courtesy of Leader Instrument Corp.)

pushbutton switches is an output-level control. On the right side of the generator is a function switch. Several test patterns are available from this generator. These patterns are shown in Figure 6-10. One test pattern produces a white-level blank screen. This is useful when adjusting receiver purity levels. Other functions include horizontal and vertical lines, a crosshatch pattern, horizontal and vertical dots, and color bars. All of these are used for setup and adjustment of the receiver. This type of generator may also be used in the test bands for the same purposes.

Sweep-alignment generators. The television signal requires a bandwidth of 6 MHz. Tuned circuits in the receiver IF are not broad-banded enough to amplify signals equally in this spectrum of frequencies. A system of stagger

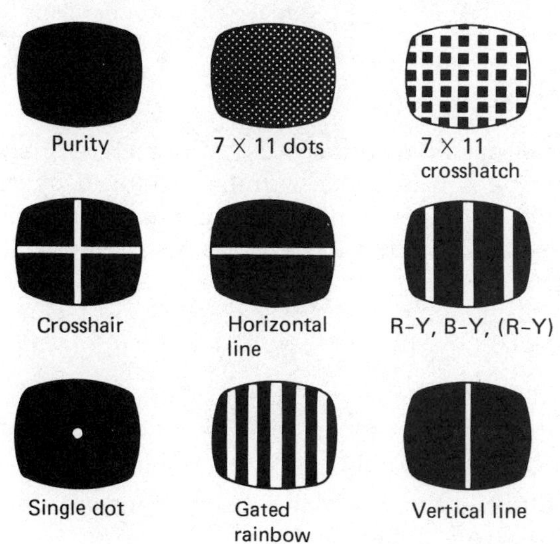

Figure 6-10 The color generator develops signals such as these. (Courtesy of Dynascan Corporation.)

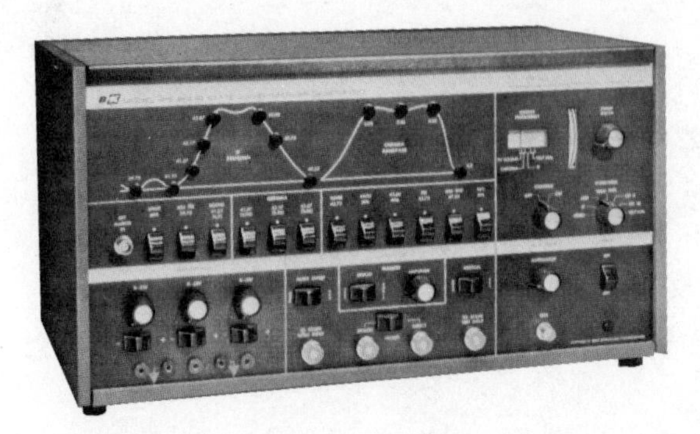

Figure 6-11 A sweep generator is used for alignment of the IF section of the receiver. (Courtesy of Dynascan Corporation.)

tuning IF transformers is used. The alignment of these transformers requires a generator capable of scanning, or sweeping across, all the frequencies. A sweep generator is used for this purpose. It is illustrated in Figure 6-11. Sweep generators produce an IF signal with the correct 6-MHz bandwidth. Additional generators operate in this unit. Each of these is capable of producing a "marker" frequency. The markers are used to align specific transformers in the receiver. They may be switched on or off as required in the alignment process. A picture of the traditional IF response curve with its marker frequencies is shown in Figure 6-12.

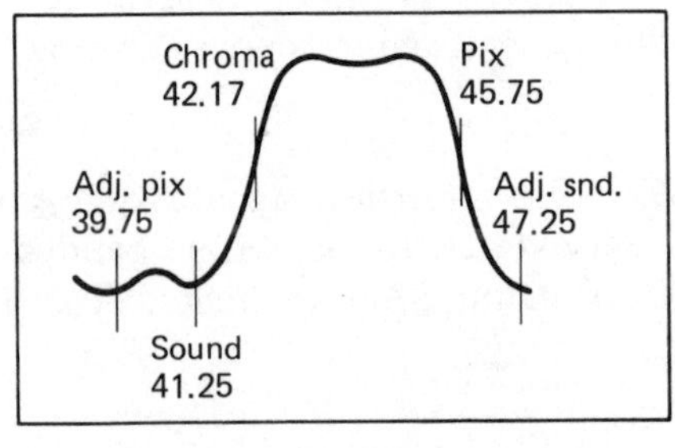

Figure 6-12 The output of the sweep generator develops a response curve such as the one illustrated. (Courtesy of Dynascan Corporation.)

The three generators discussed in this section will produce almost all of the signals required by the technician. Other units, produced by manufacturers other than those whose products are illustrated, are also available. The technician should evaluate various units and field test them if possible before making a final selection. One word of caution: As a beginner in television servicing, do not attempt to do an alignment on a receiver until you have acquired some experience in aligning other circuits.

TESTERS

A group of instruments found in most service organizations are classified as testers. These units are used to evaluate the condition and/or operation of specific components. Each is usually named so as to identify the type of test it is used to make.

Transistor testers. Many people feel that each transistor must be tested. The tests include leakage, shorts, and gain. Several types of transistor testers are available. The range of these includes laboratory-type testers on one end of the range to a pocket leakage-gain tester at the other end. Both have their place in service work. Most technicians, however, prefer to use the smaller type of unit illustrated in Figure 6-13. This unit will measure for leakage, shorts, and gain. In addition, it will identify the leads on the transistor.

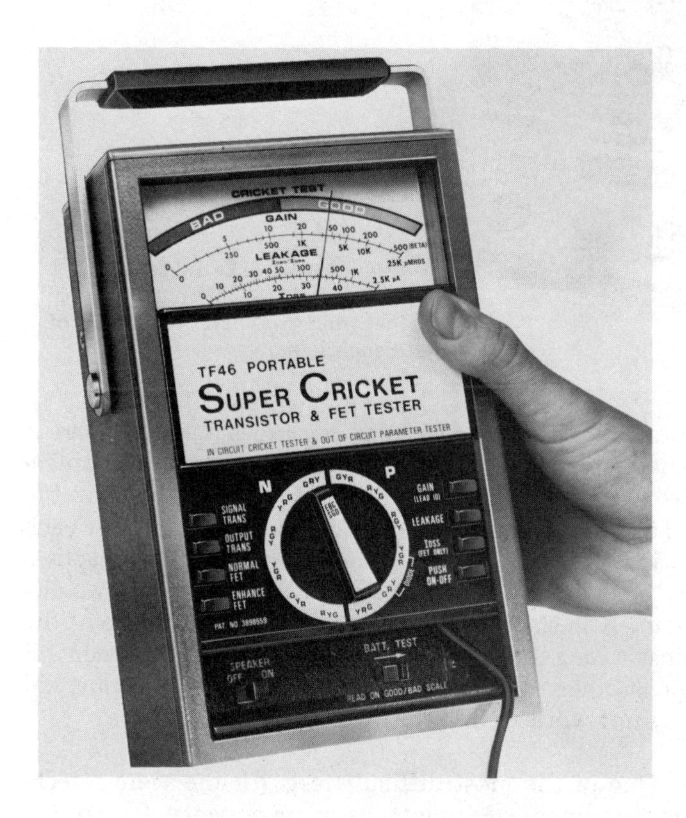

Figure 6-13 One type of "in-circuit" instrument is used to identify leads as well as test transistors. (Courtesy of Sencore, Inc.)

Tube testers. Many service organizations purchased tube testers. There were several excellent reasons for doing this. First, it was more convenient to test tubes than to circuit analyze. Second, there is an excellent profit margin in the sale of tubes. Tube testing could be done in front of the customer. In some cases tubes whose operation was marginal were replaced even though their operation was not related to the service problem. The ethics of this type of sale will not be discussed. The problem with a good many types of tube testers is that they fail to duplicate receiver operating conditions during the test. The result is almost a go/no go type of test. Either the tube is completely bad or it is completely good. There is little in between area for evaluation.

Tube testers are becoming things that belong in the past. There are still a good many tube-type receivers on the market. Tubes are still being manufactured and sold. If one wishes to purchase a tube tester, a model similar to that shown in Figure 6-14 is typical of the better ones available today.

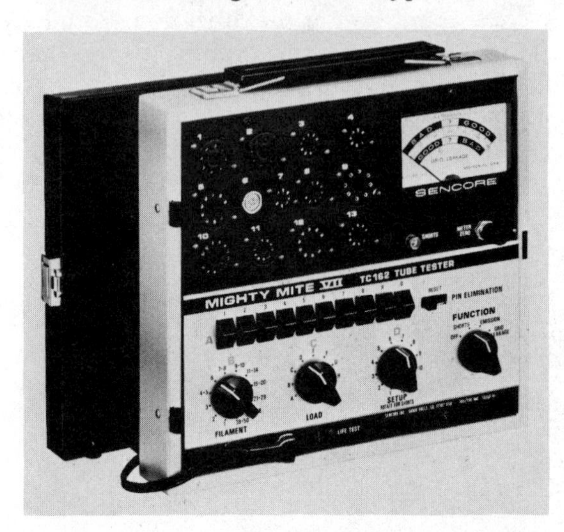

Figure 6-14 The tube tester, although still in limited use, is rapidly becoming obsolete. (Courtesy of Sencore, Inc.)

CRT testers. Picture tubes are really cathode-ray tubes (CRTs). There is a need for a picture-tube tester. The technician needs to evaluate the operation of the three electron guns in the color tube. A CRT tester, such as the one show in Figure 6-15, will aid in this evaluation. A comparison of each electron gun's emission is important. Some of the CRT testers have reactivation capabilities. It may be possible to remove a short from between two elements on an electron gun. It is also possible to reactivate the cathode of the gun. The odds are against a long-term repair using this type of equipment. It does help to show the customer that an effort is being made to use his old CRT, even though it may not work well.

Component testers. One of the most difficult tests for the average technician to make is the measurement of inductance or capacitance. One of the best types of units that has been available for many years is the *R-C-L* bridge. This unit, pictured in Figure 6-16, will measure any of these quantities. It will measure and check capacitors for leakage as well as the quantity of capacitance.

A second type of tester for measuring capacitance is illustrated in Figure 6-17. This is a portable unit with a digital readout. Since capacitor evaluation is a major factor in electronic servicing, this unit has filled a need.

Figure 6-15 One style of tube tester that is still used frequently is the CRT tester-restorer. (Courtesy of Sencore, Inc.)

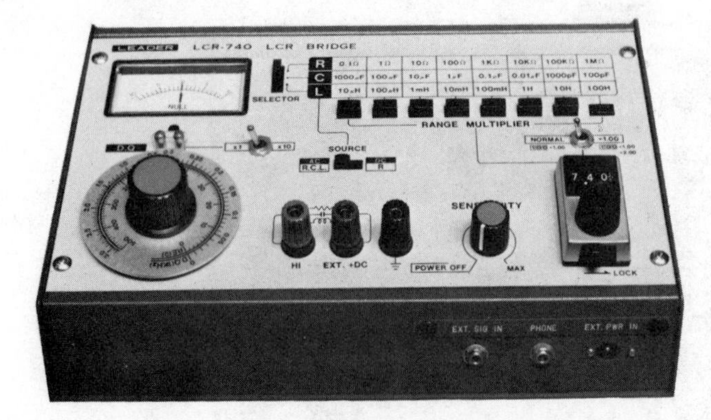

Figure 6-16 This bridge-type tester will evaluate resistance, capacitance, and inductance. (Courtesy of Leader Instruments.)

Figure 6-17 One newer style of tester is this digital readout capacitance meter. (Courtesy of Dynascan Corporation.)

PROBES

Each test instrument requires some sort of wire, or probe, to connect between the test instrument and the circuit to be tested. In many instances a pair of wire leads is all that is required. In other cases a more sophisticated test probe is required. Each technician needs to know when and why to select the proper probe. Failure to do so will produce invalid results or may damage test equipment. This section of the chapter deals with the variety of test probes that are available.

Isolation and direct probes. Most multimeters use this type of probe. The probe unit contains an isolating resistor and an on–off switch. The resistor value is on the order of 100 Kilohms (kΩ). It is used to isolate the tester from the circuit. This reduces any loading effect of the meter to a very low value.

The "direct" position bypasses the resistor. It is the same as using two pieces of wire. Often the common lead to the meter forms a shield around the "hot" lead. This eliminates the effects of any stray magnetic fields influencing the reading.

The probe used with an oscilloscope is also a direct-isolation type. When in the isolation position the voltages measured are attenuated. The result is a 90% drop in voltage across the probe. These probes are often called 10:1 probes. They permit 10% of the measured value to enter the scope. This extends the range of the scope to 10 times its normal range.

Demodulator probes. Almost all signal-measuring equipment has frequency limitations. The presence of a signal indicates that the carrier has delivered it to the receiver. The important factor is the modulation waveform. A demodulator probe is a detector. It extracts the modulation information from the carrier. The result is a signal that is within the measuring capability of almost all scopes and meters.

High-voltage probes. An accessory that is used with a multimeter is the high-voltage probe. This probe is available in several ranges. These include 0 to 10 kV, 0 to 25 kV, and 0 to 40 kV. They are, in essence, an additional voltage dropping resistor. The multimeter is set at some specific voltage range. If the probe drops the measured voltage to $\frac{1}{15}$ of its value, the reading on the multimeter is multiplied 15 times to determine the measured voltage.

There are other specialized probes that are available. Selection of these is based upon individual needs. Individual needs will also dictate the purchase of other types of test equipment. Some of the more common items found in service organizations are discussed in the following section.

DEGAUSSING COILS

One of the problems that affects the color picture tube is outside magnetic fields. The electron beam is positioned on the face of the tube by an electromagnetic field in the deflection yoke. Any other magnetic field will also influence beam position. During shipping the CRT will be influenced by stray magnetic fields. Part of the installation procedure is to neutralize any of these magnetic fields that develop on the tube.

A coil of wire, called a *degaussing coil*, is used to neutralize these unwanted magnetic fields. The unit of magnetic strength is called the *gauss*. The coil of wire shown in Figure 6-18 is a degaussing coil. Alternating-current (ac) power is used to develop a constantly changing field. This will demagnetize the face of the picture tube.

Figure 6-18 This degaussing coil is used to neutralize magnetic fields that develop around the front and sides of the color CRT. (Courtesy of Triad-Utrad, Inc.)

ISOLATION TRANSFORMER

Safety is of great importance to the technician. His or her life depends upon safe operating techniques. One device that should always be used is the isolation transformer. This unit, pictured in Figure 6-19, is used to isolate electrically the receiver under test from the power lines. Many receivers have one side of the power line cord connected to the receiver chassis. Test equipment may have the negative test lead connected to the chassis of the tester. It is possible to have a 120-V difference between the two chassis if power cords are reversed. This could be dangerous if one were to put one hand on each instrument. The use of an isolation transformer eliminates this shock hazard.

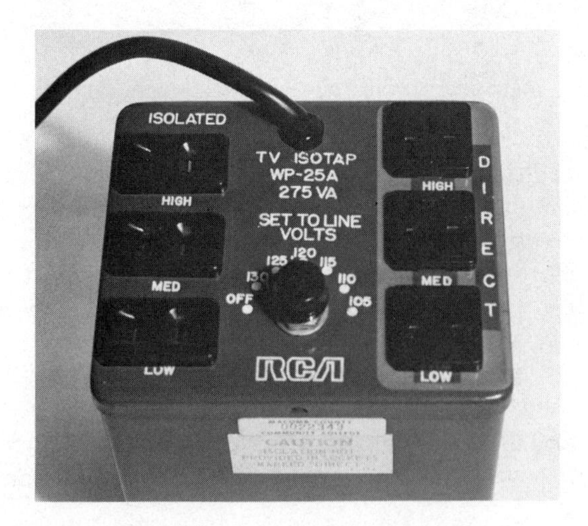

Figure 6-19 An isolation transformer is essential for the servicing of color receivers.

TEST JIGS

One of the problems facing the service technician is that he often will become a furniture mover. A receiver often has to be taken into the shop for repair. Moving the cabinet and its contents would seem to be the way to do this. There are problems when this is done. One is that it may require two people to move the receiver. A second is that the cabinet may get scratched or damaged during the moving process. A third problem is that the receiver requires a complete setup when it is returned to the customer.

Receiver manufacturers have resolved these problems. This was done by designing the receiver so that the picture tube and its related deflection components could be left at the customer's house. Only the chassis and tuner are removed for service. When this is done, another picture tube and yoke are required. Test jigs, such as the one illustrated in Figure 6-20, are used for substitutes. The jig contains a picture tube and a deflection yoke. Some jigs also have speakers and high-voltage meters. A set of cables is used to interconnect the chassis and test jig. Adapters are available to match the variety of connectors used by different receiver manufacturers.

Figure 6-20 This style of test jig eliminates the need to remove the cabinet and CRT from the customer's home. (Courtesy of Phillips-Sylvania, Inc.)

TUNER SUBBERS

A common problem found in older switch- or turret-type tuners was dirty contacts. The exact cause of this problem was not always apparent. There was also a reluctance on the part of the service technician to repair a tuner.

A positive method for localizing troubles to the receiver tuner was needed. In most receivers the tuner is a separate unit. It is connected to the chassis with cables. A shielded cable carries the IF output signal to the first IF amplifier in the chassis.

A solution to the tuner testing problem is to substitute the receiver tuner with one that is reliable. Tuner substitution boxes, or tuner subbers, are designed for this purpose. One such subber is shown in Figure 6-21. It is a self-contained unit. Its output cable connects to the receiver IF input. Using a tuner subber will localize the problem to either the chassis or the receiver tuner.

Figure 6-21 A tuner substitution unit is used to diagnose receiver front-end or IF problems.

The repair technician uses test equipment to localize a problem. Signal generators are useful to create a constant-value test signal. Testers, such as picture-tube or transistor testers, are best used to verify results found by signal tracing or voltage measuring. Attempting to use testers to isolate a problem is a slow and often time-consuming effort. Chapter 7 describes applications of test equipment.

QUESTIONS

List test equipment in each of the following categories:

6-1. Measuring devices
6-2. Signal generators
6-3. Testers
6-4. Probes

6-5. What is the advantage of the oscilloscope over the meter?
6-6. When is a tuner subber used?
6-7. What outputs are available from a TV analyst?
6-8. Why is an isolation transformer used?
6-9. Why is a test jig utilized?
6-10. Why are test generators used in repair procedures?

Chapter 7

Utilization of Test Equipment and Information

Owning a service bench that is filled with the latest types of test and measuring equipment is very impressive. Having the knowledge and skill to select the proper piece of equipment is also very impressive. Being able to use the proper test equipment correctly is also impressive. However, the most important factor is knowing how to interpret the results obtained from the test. A good technician uses a lot of deductive reasoning as the troubleshooting plan progresses.

Chapter 5 discussed a general philosophy for troubleshooting. The technician has to make many judgments when localizing a defect. A good understanding of basic electronic theory is required for this procedure. Application of Ohm's law and Kirchhoff's voltage and current laws are used constantly. The memorization of these laws is not enough. The technician is constantly applying them in the repair process. Some examples of these two laws have been presented in earlier chapters. A review of them will help focus on their importance.

BASIC TROUBLESHOOTING

There are different ways of approaching a troubleshooting problem. One must always keep in mind that the set did work at one time. One must also go on the basis that the set has not been redesigned by someone attempting to repair it. The most important factor in any sequence of repair is to think. The problem area should be identified very early in the troubleshooting process. Chapter 6 described this process. The technician is not supposed to do this work without any help. One of the best aids is the schematic diagram. This diagram will help identify blocks. It provides specific information about the circuit wiring in each block of the receiver.

Once the area of trouble is located the technician has to know how to locate the specific component that has failed. When a modular receiver is being repaired, the malfunctioning module must be identified. The process of identifying a malfunctioning module or circuit requires the use of test equipment. Selection and application of proper test equipment speeds up the repair process. Two basic methods of localizing a problem area are signal injection and signal tracing.

SIGNAL TRACING

Signal tracing requires the use of two pieces of test equipment. One of these is a signal generator. The other is an oscilloscope. The signal generator is used to develop a constant signal. The specific type of signal used will depend on the circuit that requires tracing. One method uses a composite video signal generator. The output of the generator is connected to the receiver antenna terminals. This is shown in Figure 7-1. The receiver accepts this signal as if it were being received at the antenna.

Logic must apply when troubleshooting. Some people feel that one

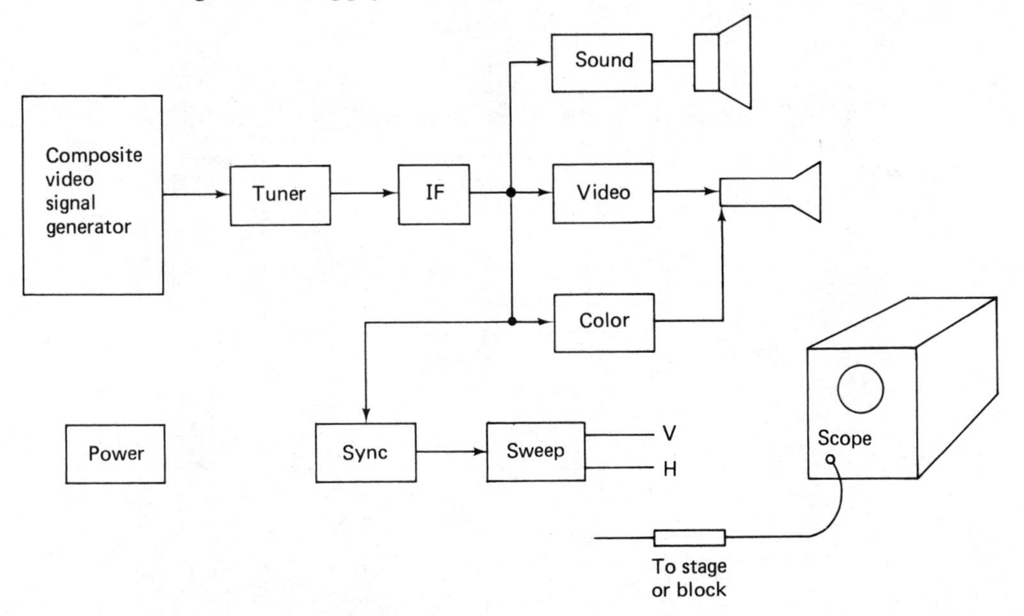

Figure 7-1 An oscilloscope is used to trace a valid signal through the receiver. Schematic information showing typical waveforms is compared to the observed waveforms.

must trace the signal from the antenna and move, stage by stage, through the receiver to the output. This is a waste of your time. It is better to connect the test signal and observe the operation of the receiver. Eliminate all working sections first. For example, if the raster and sound were observed without any picture information, the sweep and audio blocks can be eliminated. The area of trouble would be localized to the video and picture-tube circuits, as shown in Figure 7-2.

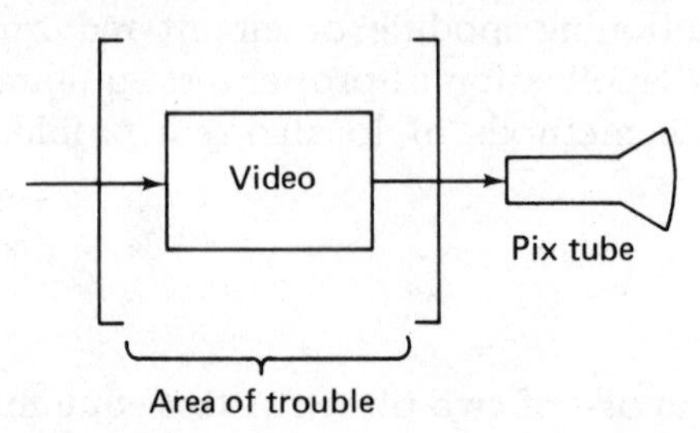

Figure 7-2 Trouble areas are localized by eliminating those sections that are working properly.

Troubleshooting to this point has eliminated sound, sync, and sweep circuits. The next step is to trace the signal through the suspected area. The very first step in this process is to *think.* Stop, put down any tools or test probes, and consider the best way to quickly localize the defective component. A procedure that is quite common uses the partial diagram of Figure 7-3 for reference.

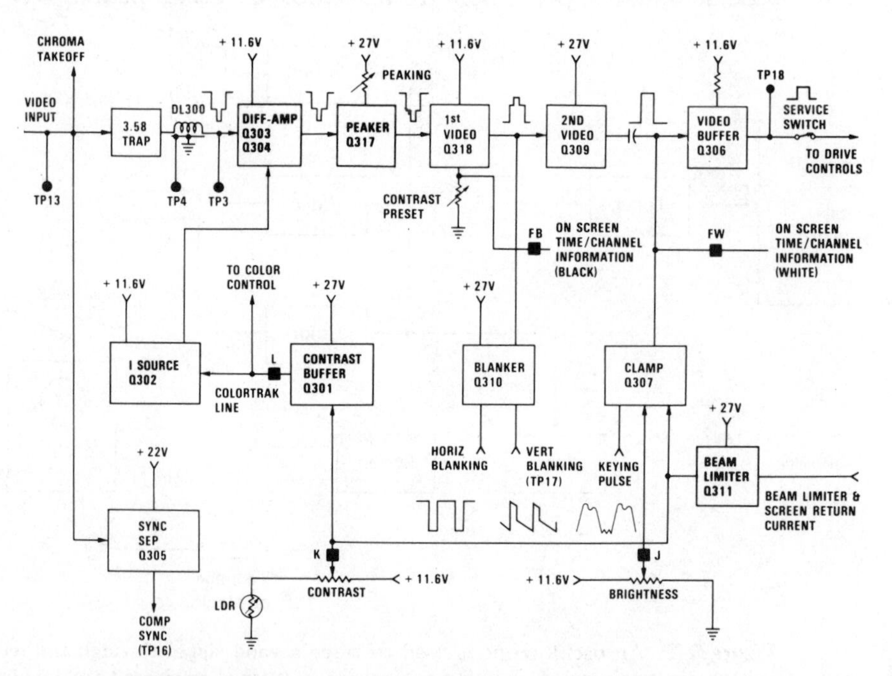

Figure 7-3 Service literature often shows polarities and test points in the receiver. (Courtesy of RCA Consumer Electronics.)

The first step in the thinking and planning process is to identify the signal path circuit. This is a linear flow path circuit. The rule for troubleshooting a linear path circuit is to make the first test at or near the midpoint. In this circuit the midpoint is transistor Q_{318}. The oscilloscope probe is placed on either the base or collector connection of this transistor. The presence of proper signal at this point indicates that all stages between the test point and the input are working properly. Making this one midpoint test eliminated the three or four checks that would have been made if you proceeded stage by stage from the input.

The area of trouble is reduced to include the first, second, and video buffer stages. The next step is a repeat of the first approach. The balance of the trouble area is split in half. This would be at Q_{309}. A check with the scope at the output of this transistor shows no signal. The area of trouble is reduced to either Q_{318}, Q_{309}, or Q_{319}. Another test, this time at a point between the two transistors, will localize the trouble to one circuit. Instead of making five checks you made three. The result is a faster, more efficient, troubleshooting procedure. This procedure is the signal-tracing system.

SIGNAL INJECTION

A second method for localizing a problem is the signal injection system. In this system a correct signal is injected at the proper point in the receiver. For example, the receiver sync is not functioning. In an attempt to localize the problem a sync signal is obtained from a signal source. This signal is correct in amplitude and phase. It is injected at the input to the sync processing blocks of the receiver. Restoration of the receiver sync indicates that the problem is before the sync separator. It is not receiving sufficient signal for proper operation. The technician then moves the area of suspicion back toward the input of the receiver.

An illustration of this system is shown in Figure 7-4. A television analyst type of signal generator is used. It creates all required signals for the receiver. The proper signal is selected and injected in the receiver at the input to a specific block. This quickly determines if the block is functioning properly.

This approach can be used with any block if the proper signal source is available. The diagram shown in Figure 7-3 is used to illustrate this procedure. A video generator is used to inject a good signal at various test points in the video strip of the receiver. The first step is to identify the type of signal flow path. Here, again, it is a linear flow path system. No signal is observed at the output of Q_{306} (or test point 18). A video signal is injected at or near the midpoint of the system. This is the base of Q_{318}. If an output appears, all is well between the base of Q_{318} and the output. The trouble area is between the input and the base of Q_{318}. The remainder of the system

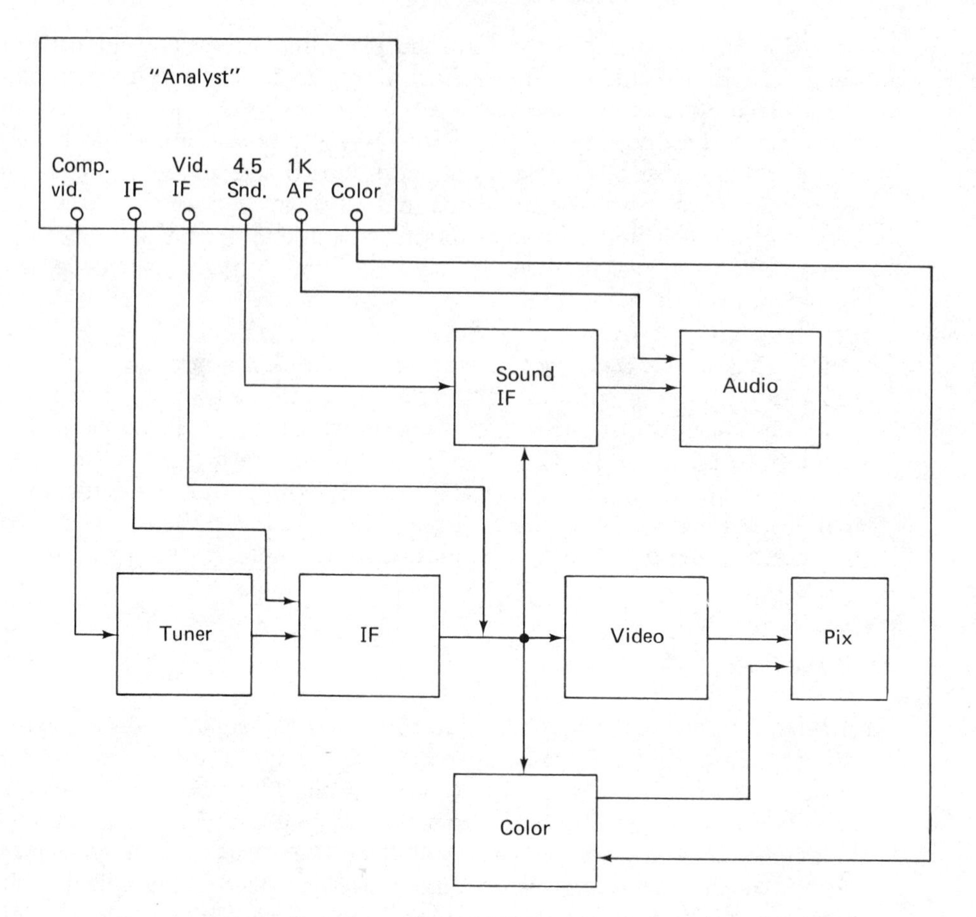

Figure 7-4 An analyst generator is used to inject correct test signals at various places in the receiver.

is split in half. A signal is injected at the output of Q_{303}. This system is used until one stage is isolated and identified as being defective.

The approaches used for signal tracing and signal injecting are similar in procedure. The basic similarity is the point at which the troubleshooting starts. The difference is the manner in which it continues. Tracing moves from the input toward the output circuits. Injecting moves from the output back toward the input of the block. Both methods are correct and proper to use. The advantage of using the oscilloscope is that waveform shape and amplitude are checked against the manufacturer's service data. This cannot be done when signal injection procedures are followed. Procedures for either of these techniques are discussed thoroughly in the manuals provided by test equipment manufacturers. These manuals contain a wealth of valuable information. They should be required reading for all users of test equipment.

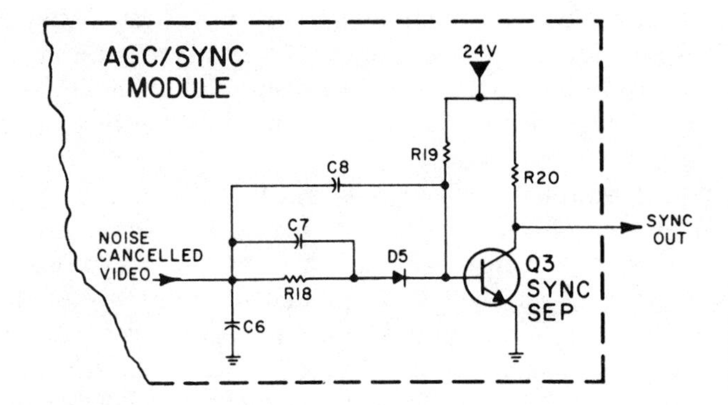

Figure 7-5 Analysis and troubleshooting of this circuit is described in the text. (Courtesy of Magnavox Consumer Electronics.)

LOCALIZING THE DEFECTIVE COMPONENT

Procedures described earlier in this chapter may be applied to a specific stage. There are two processes that happen at the same time in a working electronic circuit. One of these is signal processing. The second is dc operation and current flow. Analysis of a defective stage requires analysis of both of these processes. The circuit in Figure 7-5 is a good example of the kind of analysis done by a technician. Much of this thinking is done at a subconscious level as skill in troubleshooting is gained. The circuit containing R_{20} and $Q_{3\ E\text{-}C}$ is a series circuit. A source of 24 V will cause an electron current flow. Total current is unknown. As the current flows, voltage drops develop across the two resistances (R_{20} and Q_3). The probable voltage at the collector is one-half of the source voltage, or about 12 V. This last statement is based on experience. The circuit is a basic common-emitter amplifier. A small signal at the base

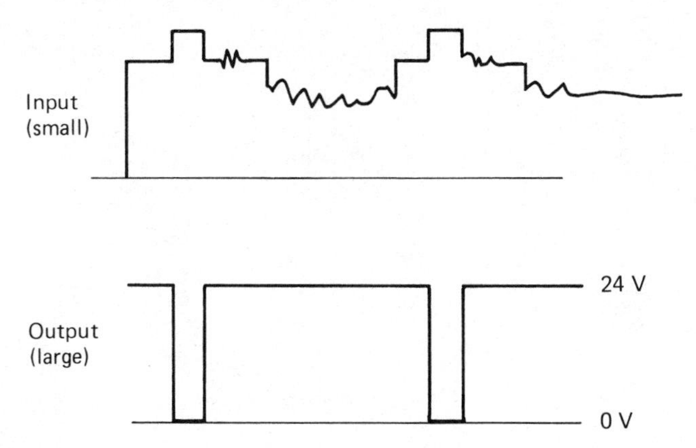

Figure 7-6 Sync separation as it occurs in the receiver. Part (input) is the composite video signal. Part (output) shows the sync pulse after it is removed from the signal.

will produce a larger signal at the collector. The collector, or output, signal is inverted in this type of circuit. The transistor, being used as a sync separator, should clip all video information. Its output waveform should appear as sync pulses without any video information on the balance of the waveform. The input and output waves are shown in Figure 7-6.

The input circuit bias is established by $Q_{3\ B\text{-}E}$ and R_{19}. Since the emitter of Q_3 is at common, or 0 V, the base voltage, with no signal applied, is 0.7 V. This is based on knowledge of how a transistor works. It is also based on experience. Almost all transistors used in TV receivers today are made from silicon. A silicon transistor has a junction forward voltage drop of about 0.7 V. Signal applied to this junction is not able to reduce the voltage in this circuit to less than 0.7 V. It will make the voltage rise above the 0.7-V figure. In this way the transistor responds only to a positive-going signal. The level of composite video applied to the base circuit will not permit conduction in the transistor until the signal rises to the level of the sync pulse.

All of the foregoing information is based on being able to identify a basic amplifier circuit. There are only three basic amplifier circuits. These are common emitter, common base, and common collector. Each circuit has its own set of characteristics. These relate to signal processing as well as voltage and current gain. In addition, one must know dc current flow characteristics

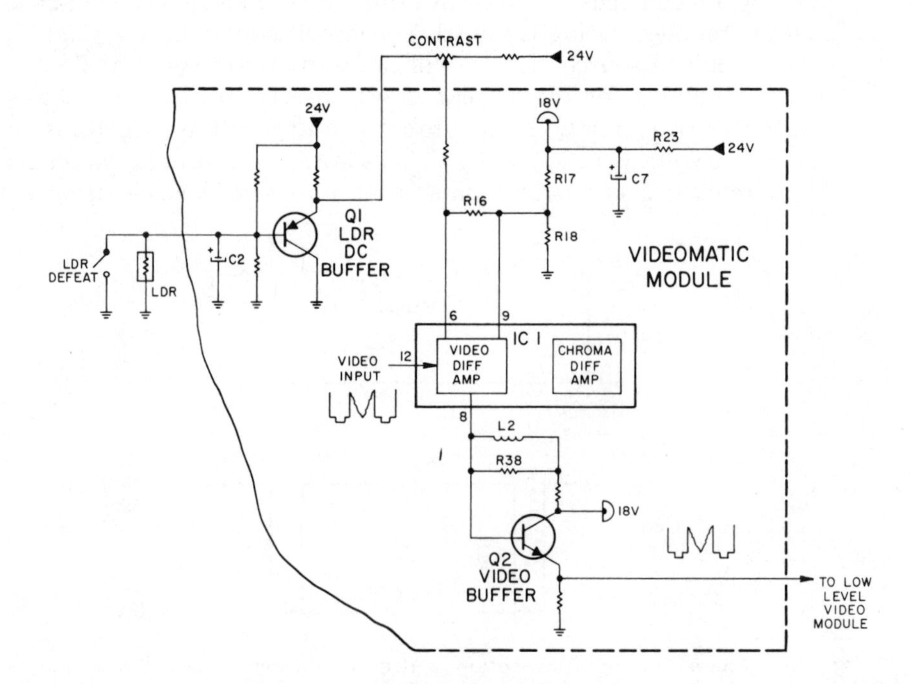

Figure 7-7 This video amplifier requires diagnostic work in addition to that described previously. (Courtesy of Magnavox Consumer Electronics.)

for the series and parallel circuit. These characteristics are based upon rules established by Kirchhoff.

A second circuit to be analyzed is shown in Figure 7-7. This circuit is more complex. It contains an IC in addition to the transistor Q_2. Information about the IC is not available, but it has a name, "video difference amplifier." One can assume that it will process a signal and the amplitude of the signal. Many IC circuits use the term "gain" amplifier rather than either voltage or current amplifier. This is really a better term. The definition of an amplifier is that it is a device that uses a small amount of power to control a larger amount of power. In many instances the gain in the circuit is in the form of a current gain. This cannot be measured with ordinary test equipment. We are used to observing voltage gain in an amplifier on the oscilloscope. A current gain is not an observable factor. The technician may falsely assume that a stage is malfunctioning if a voltage gain is not observed. Knowledge of circuit operation is important to analyze properly the information obtained from the receiver.

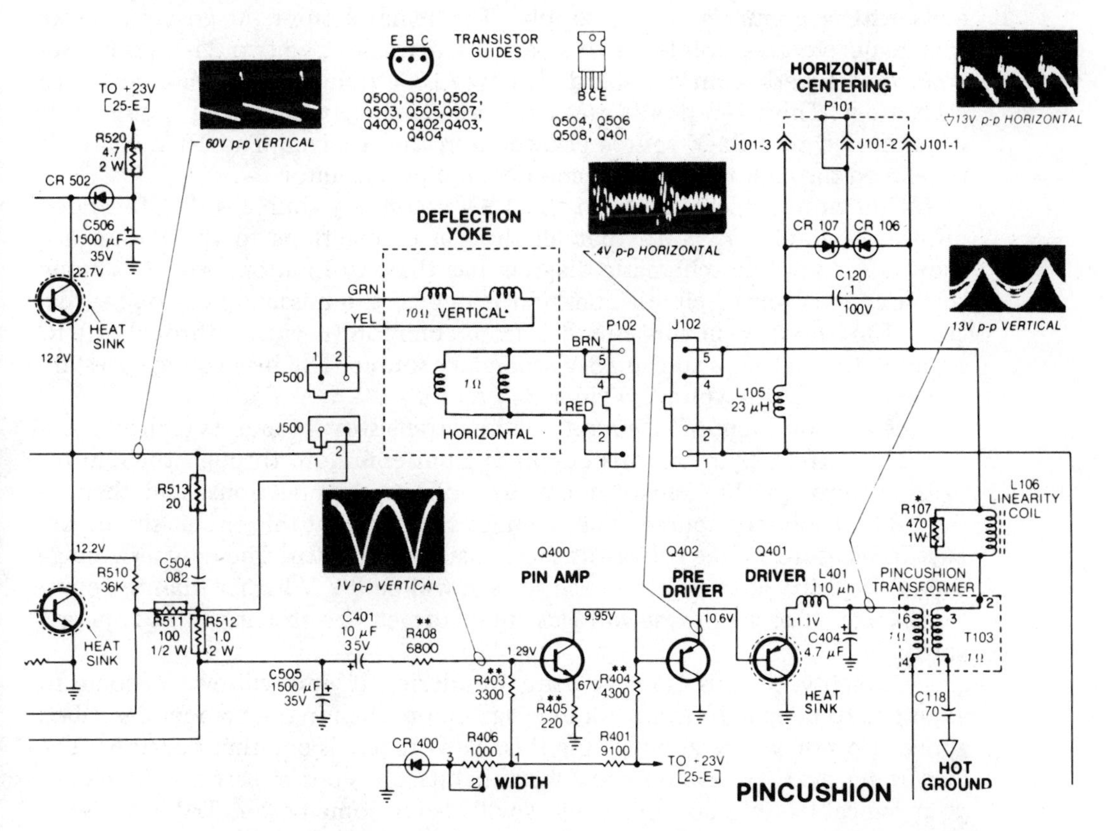

Figure 7-8 Many schematic diagrams supplied by manufacturers contain voltage and waveform information. (Courtesy of Magnavox Consumer Electronics.)

This circuit may be analyzed in two different ways. Both are correct and required. The easiest way to start is to consider signal processing. Signal information is included with the schematic diagram. This is in the form of waveshape information. Transistor Q_2 is a common-collector or emitter-follower circuit. Its characteristics include a voltage gain of 1 and no signal inversion from input to output. Checking with an oscilloscope should verify these facts. No information is available about the circuitry in the IC. The signal out at pin 8 should have the same shape as the input signal at pin 12. This deduction is based on the characteristics of the common-emitter amplifier Q_3. One could conclude that the input and output waveforms have the same shape and polarity. Since values of the amplitude of these signals are not provided, we cannot measure and compare this factor.

The second method of evaluating this circuit is to consider the dc current path. This should be done after signal analysis. After all, if signals are correct, there is no need for further analysis of this circuit. The earlier test results may eliminate these steps. This circuit is really two parallel wired circuits with a common source supply. The manufacturer, Magnavox, shows four major voltage points on this partial schematic. The two 24-V points are connected together on the board. The two 18-V points are also connected to each other. The 24-V points are considered as the common supply source. It is connected to a 24-V source elsewhere in the receiver. The 18-V source is developed through resistor R_{23} and decoupling capacitor C_7.

One circuit operating from the 18-V secondary source is IC_1, the video difference amplifier. Note that all the pin connections to the IC are not shown. A complete schematic diagram has this information. We can assume that a connection to circuit common is one of the missing parts of the diagram. Current flow in this circuit is from common (negative) through the IC at pin 9, through R_{17} to the 18-V secondary source. The bias voltage is established by use of the voltage divider R_{17} and R_{18}.

The second amplifier circuit contains transistor Q_2 and its emitter resistor. The current path for this circuit is from common, through the emitter resistor, through the transistor emitter–collector connections, and then to the 18-V secondary source. The technician looking at this circuit should expect to measure source voltage at the transistor collector. The emitter voltage should be fairly low and the base voltage will be 0.7 V higher than the emitter voltage. If any of these voltages are incorrect, the circuit will not operate correctly.

Probably by this time you are wondering if you will ever become so expert as to be able to think about a service problem in the manner described above. Do not worry about doing this. The process is one that develops. The idea is to start to develop good work habits. As your experience increases, your logical troubleshooting analysis will also become better. Try to review a troubleshooting procedure after the problem is corrected. See if you can develop a better way of solving the problem for the next time. It might help if you discussed the problem and your approach to the solution with other technicians. After all, why do you have to rediscover that which others are using?

TECHNICAL INFORMATION

Information required for the rapid repair of a receiver may be obtained from several sources. There are two major sources for this information. One of these is the receiver manufacturer. Almost all of the major manufacturers have supply outlets. These are usually located in or near major cities. A phone call to one of these outlets will establish the availability and cost of this information.

A second major source is information published by the Howard Sams organization. This company publishes a series of packets of technical information. They are called "Photofacts."* Each packet contains schematic, alignment, and replacement parts information for a group of consumer electronic devices. This material is available from many local independent electronic parts distributors. Many libraries also have this information available.

The information included in the service literature available from either source includes waveforms, operating voltage values, and component information. A section of one of the schematic diagrams is shown in Figure 7-8. A full schematic is too large to reproduce in this book and still be readable.

Do not attempt to repair any TV receiver unless you have the service data available. The extra time spent in obtaining this valuable information pays off in time saved when making the repair. Each manufacturer has its own way of using the components in the circuit. Parts layout varies greatly from one set to another. This is even true for sets made by one manufacturer. Analysis time is not wasted time. It is actually time well spent. This time will ultimately earn money for you as you develop good work habits.

One word of caution about schematic diagrams. There are times when the information given is incorrect. This is not a frequent occurrence, but it does happen. If, in the process of repairing a set, you run into a situation where there is an apparent error, use your knowledge of electronic theories to analyze the circuit and develop more accurate values. As a courtesy to others, write to the publisher of the service information so that corrections can be made.

QUESTIONS

7-1. Describe the signal-tracing process.

7-2. Describe the signal injection process.

*Trade name for H. W. Sams publication.

7-3. What equipment is used to locate a specific defective component?
7-4. How do Kirchhoff's laws affect circuit analysis?
7-5. Describe, in brief terms, the signal analysis process of troubleshooting.
7-6. Describe the current path analysis process of troubleshooting.
7-7. Where can one obtain technical service literature?
7-8. What test equipment is used for signal tracing?
7-9. What test equipment is used for signal injection?
7-10. What information is obtained from service literature?

Chapter 8

Repair Parts

Once the defective component or module is identified, the next problem facing the technician is obtaining a replacement. If one is employed by an organization that has its own supply department, this is an easy accomplishment. All one has to do is to request the replacement item and install it. Most of the people that service TV receivers are not so fortunate. For reasons that are beyond the understanding of this author, most television technicians prefer to operate in small one- or two-person organizations. The operating capital and parts backup stock is usually very limited. These people must resort to getting components as required from outside sources.

All parts are ordered from a source. Basic items, those used almost on a daily basis, should be stocked in all service facilities. The important thing that must be done by the service technician is to know where and what to purchase. Equally important is to know when to change a component.

There are two schools of thought on the replacement of parts. One of these is to "shotgun" the defective stage. Using this approach, the technician removes all components that *might* be bad. These are replaced and, hopefully, the set is repaired. This approach is *not* one that is recommended by this author. There are problems that develop during the shotgun approach. One of these is maintaining the larger inventory required when using this method. A second is the question of who is going to pay for the installation of these unnecessary components. The third is the problems that are present when a foil circuit is overheated and the foil is destroyed. Another problem is the probability of installing the new parts incorrectly and the resulting possible damage to the good components in the circuit.

A much better method of repair is to use the skills and knowledge developed during your training to isolate defective components. Material presented in Chapters 6 and 7 shows how to isolate a problem component in the receiver in general terms. Material in this chapter discusses these procedures in more specific terms. Also covered are the very important factors of parts selection and procurement.

The partial schematic diagram shown in Figure 8-1 is used as a reference for troubleshooting a transistor amplifier circuit. All voltage values given are from the manufacturer service data. The problem is isolated to transistor Q_{101}. The collector voltage on this transistor measures 18 V instead of the 12.5 V given on the schematic. The inexperienced technician's first thought at this point is to replace the transistor. *This is not correct.* There are several other components in this circuit. Any one of them could produce the condition described above. Additional checks must be made before any component is removed and replaced.

When source voltage is measured at the collector of this transistor the indication is that:

1. No current is flowing in the circuit.
2. The reason for a no-current condition is an open somewhere in the circuit.
3. The open has to be between common and the last point where source voltage is found.

This leaves only two areas to check before the transistor is judged to be bad. The base and the emitter circuits have to be evaluated. Base voltage is measured. It should be near the value given in the schematic. If it is too high, the transistor is saturated. The problem would then be a shorted blocking, or coupling, capacitor C_{107}. The capacitor is checked with an ohmmeter. A second component that is suspect is the emitter resistor R_{104}. If this resistor is open, there is no current path to the transistor. It is also checked with an ohmmeter. If all components checked appear to be near their correct value, the transistor is evaluated. Either transistor tester or an ohmmeter is used for this test. If all other tests seem to be proper, the problem is usually the transistor.

Another condition that occurs in this type of circuit is low collector voltage and a lack of complete amplification. This is an indication that the transistor is biased at too high a level. The probable causes for this condition are:

1. Emitter bypass C_{114} is open. This causes signal degeneration because some signal develops across the emitter resistor.

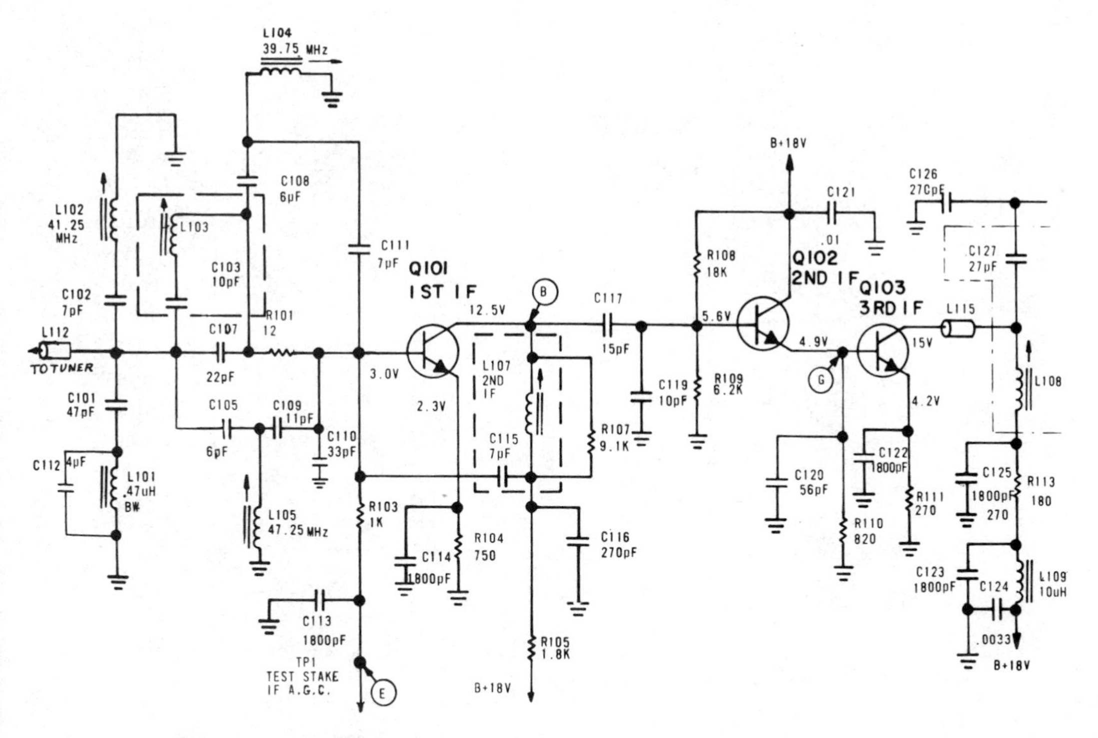

Figure 8-1 Transistor amplifier circuit. This circuit is used in the text as a reference for troubleshooting.

2. Transistor leakage. Measure the base-emitter voltage at these elements. The voltage drop for a silicon transistor is 0.7 V. Values higher than this indicate a leaky transistor.

Integrated-circuit troubleshooting uses the same thinking, but applied to a different device. The schematic shown in Figure 8-2 is used to illustrate this approach. This circuit uses three sections of one IC for audio processing. It is checked as follows:

1. Check for an output from the IC. It is possible to have a bad speaker. Some receivers also have a headphone jack. This, too, could be bad.
2. Check for correct input signal at pin 15 of this IC.
3. If no audio is present, check operating voltages on the IC.
4. Check voltages at volume control. Most of the newer sets use a variable dc-to-control stage gain. This replaces the more traditional method of varying the signal level through a potentiometer.

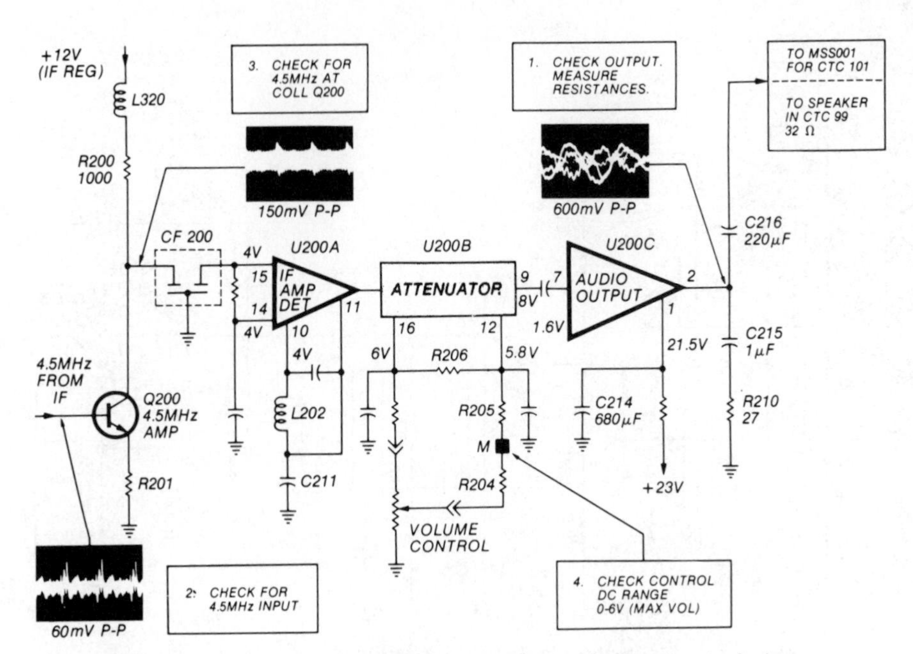

Figure 8-2 IC circuit used as an example of troubleshooting techniques. (Courtesy of RCA Consumer Electronics.)

If control and operating voltages are proper, the problem is localized to the IC. It is replaced.

INTERMITTENTS

One of the hardest parts of servicing is to locate a defect that is not always present. These are called *intermittents.* Often, the problem can be localized to a specific area. Monitoring the set and hoping to catch the problem may not be the answer. Often, the technician will expedite matters. This type of problem is usually caused by a temperature change. A rise in operating temperature will cause some component to change value. This changes the operating conditions of the receiver. After this type of failure the component will cool and the set will function correctly again.

Heating and cooling will help isolate this problem. A soldering iron tip held near a suspected component will usually heat it enough to cause it to fail, as it does in the set. The other extreme is to cool the suspected component rapidly. A spray coolant is available for this purpose. A stream of gas is directed onto a specific part. The gas cools the part rapidly. If the set has failed, due to an intermittent, the cooling gas will restore the component to its working condition. The gas is sold under trade names such as Freeze Mist and Circuit Cool.

PARTS SOURCES

Once the defective component is identified, a replacement has to be obtained. There are several sources for components and modules. Each has to be evaluated by the service technician. Evaluation includes access, cost, and exactness of the replacement item.

Factory outlets. Factory outlets for specific manufacturers are found in several parts of our country. Some of these are locally owned and operated. Others are owned by the receiver manufacturers. These outlets carry replacement components in addition to the line of manufactured goods. In most cases the exact replacement components are available from these outlets. They are identified by the manufacturer's part number.

The advantage of purchasing from these outlets is the exactness of the replacement component or module. The disadvantages are that many of these outlets sell only to franchised dealers, and the location of one or two outlets in one state is not always convenient for those who do not live or work near the outlet.

Independent outlets. There are a number of independent parts outlets throughout the nation. These companies sell a large variety of replacement components. They do not stock these by a set manufacturer's part number. They are stocked either by component value or by the stock number of the manufacturer of the part.

Many of the replacement items are identical to those used by the set manufacturer. In fact, some are made by one source and sold to set manufacturers and independent parts distributors. Solid-state devices, such as transistor and ICs, are sold as a replacement line. These are universal parts rather than identical parts in many instances. Their installation may require extra time to obtain results equal to the original component. This is discussed further in the next section.

Remanufacturers are also a part of the independent outlet picture. The first major component to be remanufactured was the picture tube. As tuner problems increased, these, too, were remanufactured. The latest item to be included is the module. Most remanufacturers have outlets in major cities. Some of them also have a mail or parcel delivery service available.

The availability of parcel delivery has helped to increase the number of order-by-mail outlets. Some of these companies sell original parts. This is particularly true for semiconductors. There are times when receiver manufacturers sell their surplus stock. These parts are often available through mail-order or independent supply outlets. Each service organization develops its favorite supply sources. The purpose of this section is to make newcomers to the field of television or electronic servicing aware of what is available.

PARTS SELECTION

Several decisions are made during the process of purchasing any item. Price, availability, and correctness are only some of the factors involved in this process.

Price. Most distributors have quantity prices on replacement items. A general rule is: The more one buys of an item, the less expensive it becomes. An example of this is the price of a 0.01-microfarad (μF) 600-V capacitor. The single unit price was about 36 cents in 1980. In quantities of 100 the price is 21.6 cents for each capacitor. This is a very popular size of capacitor. Many are replaced during the month or year. It would be to the advantage of the technician to purchase more than one of these at a time. The additional money earned as they are sold helps increase the profit margin of the business. This is also true for rectifier diodes. Individual units sold by one manufacturer have a retail price of $1.40. The dealer cost is about 84 cents per unit. These diodes are often sold in packages of 10 for about $1.00. The unit price is now 10 cents instead of 84 cents. Profit on the sale of two diodes will pay for all 10 and there is money left over. Profit in the balance of the 10 diodes is $13.00. This is a legitimate profit for the technician. Buying at the proper price is one of the most important factors in a successful business.

One other factor to be considered with price is warranty. Almost all manufacturers have a warranty on their product. A low price without any warranty could be very costly to the technician if any of the parts failed or were bad when purchased. There are firms that sell untested parts on an "as is" basis. Most of us do not have the capability to test each part adequately before installing it in the circuit.

Replacement parts. Most people in the repair field would prefer to replace one part for another one that is exact. Part numbers are used to identify the component. There are times when the selection of a specific component is very critical. The service technician needs to know a lot about the replacement-parts market. One factor relates to semiconductor part numbers. Two sets of part-numbering systems are in use. One is based on Japanese manufacturers' identification numbers. These numbers use a 1S . . . system for diodes and a 2S . . . system for transistors. The transistor system uses one number, followed by two letters, two to four additional numbers, and possibly a last letter. The identifier 2SA924C is an example of this system. The second letter in the sequence has a specific meaning. It supplies type and application data. This information is shown in Figure 8-3. The last letter in the identifier is also significant. This provides information relating to the beta range and other operating parameters. These factors are important when replacing these transistors with ones that are identified using this system. These transistors are available from set manufacturers and from some buy-by-mail suppliers.

Identifer	Type	Application
2S A	PNP	Small signal
2S B	PNP	Power
2S C	NPN	Small signal
2S D	NPN	Power

Figure 8-3 System used to identify transistors that have Japanese-style part numbers.

The second set of transistor part numbers uses the system established by American Joint Electrical Device Engineering Council, or JEDEC. This system uses 2N . . . numbers for transistors. An example of this system is the number 2N3055 for a transistor. Identification of the electrical characteristics and application is not possible using this system. There are some special application numbers that are also recognized in this system. In addition, each manufacturer uses some "house numbers" for items that are not registered with the JEDEC system. Most of the JEDEC-assigned number transistors may be purchased from electronic distributors carrying a specific product line.

There are hundreds of semiconductors identified by the JEDEC system. Inventory of all of these by each distributor would require a large investment. This is not necessary for most replacement semiconductors. Most of the major manufacturers have developed a universal replacement line of semiconductors. The major operating parameters of all semiconductors were identified. They were then programmed into a computer. Certain key values for a general replacement type of semiconductor were also fed into a computer. The program used identified one transistor type that would be equal to, or better than, a select group of JEDEC-numbered semiconductors. These universal replacement transistors are given a family name to identify them. Phillips-GTE (formerly GTE-Sylvania) uses an ECG . . . series, RCA uses an SK . . . series, and GE uses a GE . . . series of semiconductors. These universal replacement semiconductors will work well in most instances. There are some circuits in which they do not want to work correctly. This should be kept in mind when selecting and purchasing these universal replacement semiconductors.

Convenience has its price. One should compare the cost of universal replacements and JEDEC-numbered (or Japanese-numbered) semiconductors. One should also compare the price of quantities of commonly used transistors and single lot prices. These factors are a part of the evaluation of supply sources discussed earlier in the chapter.

TOLERANCES

The best way to start this section is to make the statement "things are not always as they seem to be." This statement is very true for electronic parts.

All electronic components are built to a set of specifications. The specifications offer a range of values that fall within the allowable tolerances. Resistors, for example, often have a ±10% tolerance. This means that a 10-kΩ resistor may have any resistance value between 9 and 11 kΩ and still be considered as a good 10-kΩ resistor. There are occasions when the tolerance of a resistor is ±1%. This critical tolerance value does not allow such a wide variation in acceptable values as does its ±10% counterpart.

The technician selecting a replacement part, or evaluating the part in the receiver, needs to understand the importance of the tolerance rating. The tolerance rating provides a range of acceptable values. It is possible, and correct, to replace a component with a 5% tolerance with one that has a 10% tolerance *if* the 10% unit measures within the range of 5% values. For example, a 1-kΩ resistor rated at ±5% can range in value from a low of 950 Ω to a high of 1050 Ω. Any resistor whose resistance measures between 950 and 1050 Ω may be used to replace the original unit regardless of its tolerance.

The least known, but widest variation in value tolerance is used for capacitors. Many technicians do not really understand the capacitor and its action. Replacement therefore seems to demand the exact value unit. Almost all electrolytic capacitors have an extremely large tolerance. One of the capacitor manufacturers has rated electrolytic capacitors within the 150 working volt range as -50% to +100% tolerance. This means that a 100-microfarad (μF) capacitor will range from a low of 50 μF to a high of 200 μF. From the technician's point of view, almost any value between those extremes can be used successfully.

Many of the low-voltage-rated electrolytics used in solid-state receivers have a tolerance range of from -10% to +50% of the marked value. This is better than the higher-voltage units. It still allows a lot of variation when selecting replacement capacitors. Nonelectrolytic capacitors also have a tolerance rating. The exact amount of tolerance will depend on the manufacturer's specifications.

What all this means to the technician is that exact value units are not required when replacing a component. The term "close" would be better than "exact" when selecting a replacement component. Many electrolytic capacitors are made as multisection units. There may be as many as four capacitors in one container. Replacement parts, unless obtained from the set manufacturer, may not be the exact values as the original. Most replacement-part manufacturers will offer a list of substitute values for the original unit. These will work very well in almost all situations. There is always an exception to the previous statement. In some rare occasions only the exact-value unit must be used. This is particularly true in oscillator and wave-shaping circuits. These components must be selected with care. Replacement values have to be very close to the original value if the circuit is to operate correctly. Often there will be a resistor or inductor, whose value is variable, in the circuit. These variable components are used to adjust for capacitor variance and maintain the correct circuit operation.

Another factor to consider when replacing components is the voltage rating. The two major components that have a voltage rating are capacitors and resistors. Capacitor voltage ratings are very obvious. These ratings are usually printed on the body of the capacitor. It is possible to substitute a unit with a higher voltage rating if there is room for the capacitor in the receiver. Usually, the higher-rated units are physically larger. The most important factor is the amount of capacitance. This must meet circuit values. The voltage rating shows the dielectric insulating value.

Resistors also have a voltage rating. This value is not commonly known. The information shown in Figure 8-4 gives this rating. It applies to low-wattage resistors only. The ¼-watt (W) (0.125) resistors have a maximum voltage rating of 150 V regardless of resistance value. Two watt resistors may be used in circuits with voltages as high as 750 V. Exceeding these voltage values will cause a component failure. Keep in mind that these are dc voltage values. Circuits that have peak-to-peak values as well as dc voltages may require special resistors. Manufacturers' service literature will provide this information.

Power rating	0.125 W	0.25 W	0.5 W	1.0 W	2.0 W
Voltage rating	150 V	250 V	350 V	500 V	750 V

Figure 8-4 Voltage ratings used with resistors. Operation at voltages that exceed these values may cause resistor failure.

One area that requires some special consideration is the group of "flameproof" components. Certain circuits in some receivers have a tendency to start a fire when they fail. These components are marked on the technical service literature. Replacement of these parts with others must be done with care. There are special flameproof resistors available for these circuits. The technician has the obligation to replace these parts with ones that will not cause a fire when they fail. Some capacitors are also included in this group. Be certain to watch for information about critical components on the service literature. The partial schematic diagram shown in Figure 8-5 illustrates how these are marked.

In summary, the replacement of a defective component should follow the values established by the receiver manufacturer. Replacement components whose values vary from those stated by the receiver manufacturer are acceptable only if the replacement component is within the tolerances established by the manufacturer.

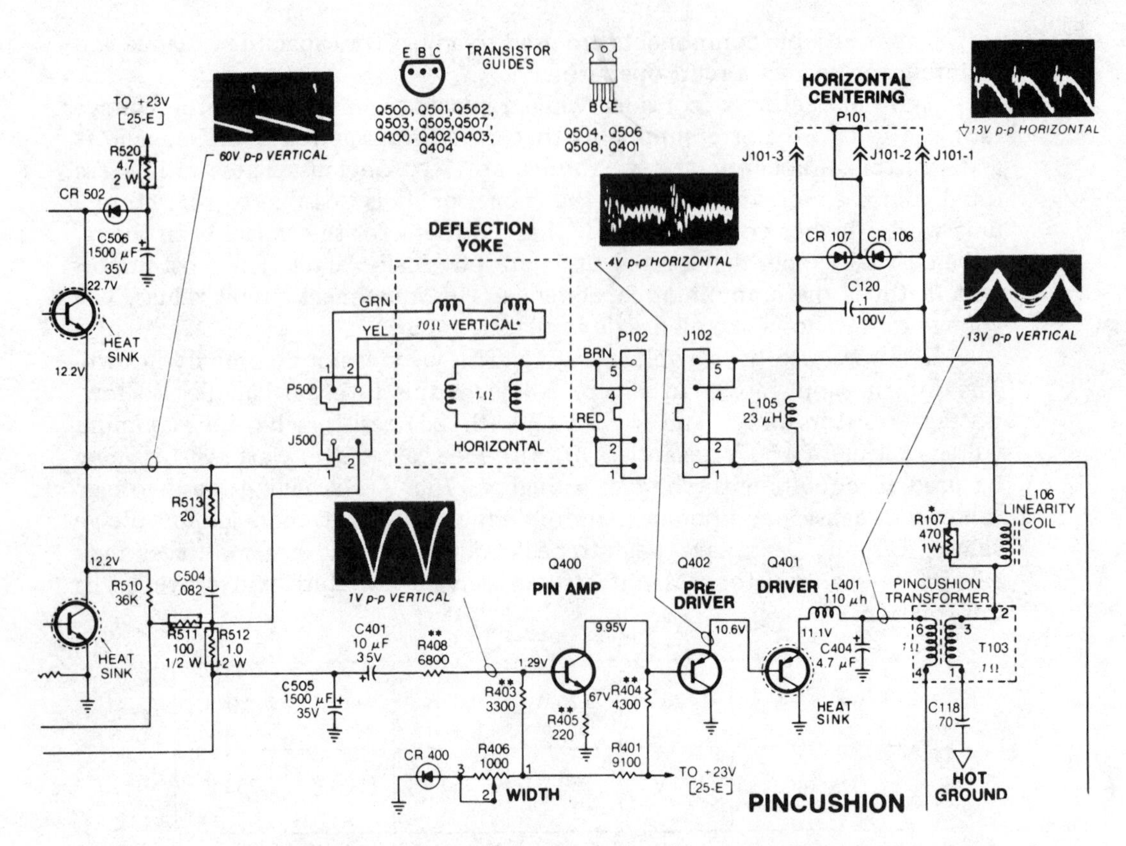

Figure 8-5 Critical components are now identified in the schematic diagram by using an asterisk and shading the area around the component. (Courtesy of RCA Consumer Electronics.)

The wise technician tries to determine why a component failed as a part of the repair process. It is foolish to replace a damaged resistor without first determining the cause of the resistor failure. Resistors, for example, seldom burn up unless another component in the circuit has failed. A burned resistor usually indicates that an excess of current tried to flow in the circuit. The cause of this kind of problem is the reduction of total circuit resistance. This is usually due to failure of another component in the circuit. Replacement of the defective resistor without removing other defective components is foolish. The result of such action is that the replacement component will immediately fail in the same manner as its predecessor. Nothing is gained when this is done. Actually, time and money are lost.

The effective technician determines why the failure occurred as a part of the troubleshooting process. The schematic diagram is an excellent and necessary tool in this procedure. The kind of thought processes required uses Figure 8-6 for illustration. A visual inspection of the circuit shows that the

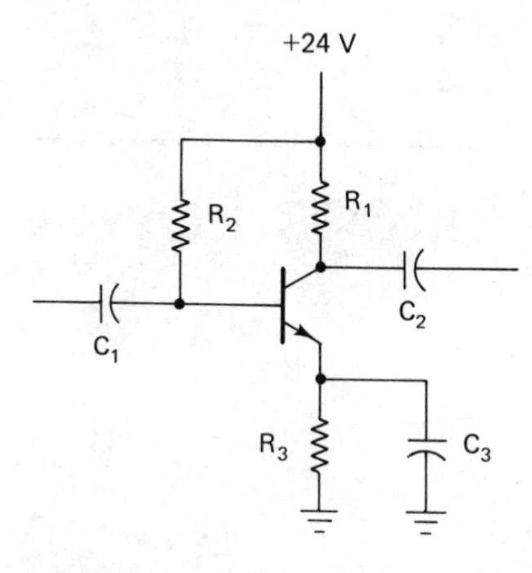

Figure 8-6 Transistor circuit used in the text as an example for troubleshooting.

collector load resistor R_1 is charred. This indicates that an excess of current flowed in the circuit. Two components could cause this condition. They are the transistor and the emitter bypass capacitor, C_3. An ohmmeter check of the components is done while both are still in the circuit. The technician is attempting to identify the defective component before removing any parts from the receiver. The results of the ohmmeter test show zero resistance between emitter and circuit common. The process identified a short-circuited capacitor, C_3. Replacement of this capacitor and the damaged resistor repaired this circuit.

It is possible that the transistor was damaged during the destruction of its collector load resistor. An ohmmeter check between emitter and collector will indicate if the transistor is open or shorted. In a circuit that has damage of this nature, the second test is an excellent suggestion. Circuit failure, of course, would be seen immediately when the set was operated for a post-repair check.

QUESTIONS

8-1. Discuss the disadvantages of "shotgun" servicing.
8-2. What problems are present when one measures source voltage at the collector of a NPN transistor?
8-3. Describe basic rules for troubleshooting IC circuits.
8-4. Briefly describe troubleshooting techniques for intermittent circuit problems.
8-5. Where can one obtain replacement parts?
8-6. What tolerances are acceptable for components such as capacitors and resistors?
8-7. Of what importance is a voltage rating of a component?
8-8. Why are flameproof components used?
8-9. What is the advantage of a universal replacement transistor?
8-10. What is the disadvantage of a universal replacement transistor?

Chapter 9

Low-Voltage Power Sources

The purpose of the power source, or power supply, is to provide the proper operating voltages and currents for each section of the receiver. There are many different methods of obtaining the proper operating power. Each receiver seems to use a different method. All power sources produce the same output. This output is, in almost every case, a dc operating voltage. The exception to this is the ac power used to heat the cathode of the picture tube. The dc operating power values will be different when comparing a variety of receivers. Each manufacturer designs the power sources for a specific application.

There are several different ways of describing power sources. One of these relates to the level of output voltage. High-voltage power sources are used to develop the operating voltage for the second anode of the picture tube. The exact value of high voltage depends upon the requirements of the picture tube. In general, the level of voltage increases with the size of the tube. A small picture tube may require a high voltage on the order of 600 V. A large 25-inch tube usually requires about 24 kV for proper operation. The power source used to supply these operating voltages is called the high-voltage power source. The material in this chapter covers low-voltage power sources.

Low-voltage power sources provide the operating power for other circuits in the receiver. The terms "high voltage" and "low voltage" are relative. One receiver may require a low voltage of 12 V and a high voltage of 600 V. Another receiver will require low voltages on the order of 400 to 600 V and a high voltage of 20 kV. As you can see, the terms are used to relate to applications of operating voltages in the receiver rather than to specific voltage levels.

PRIMARY AND SECONDARY SOURCES

Another method of describing power sources is related to the source of input power. Primary sources are connected directly to the ac power sources or to a battery. Secondary sources are developed from operating circuits in the receiver. Both provide the necessary operating power. A source is often described using terms from each of the previously described sources. Still another method of describing power sources uses the name of the type of rectifier system as a part of its title. These systems are discussed later in the chapter.

Primary sources. A block diagram for a typical primary power source is shown in Figure 9-1. Input to the system may be from one of two types of sources. One of these is the ac main. This is the power supplied from the local power company generator. The second source is a battery pack. Some receivers are able to use either source. Others are designed to operate from only one of the two sources.

The block after the input is the circuit protection block. This block's main purpose could be stated as a design for failure. A very specific current-limiting device is the heart of this block. It may be a fuse. Another common device used in this block is the circuit breaker. It, too, is designed to fail at a specific current value. A third kind of circuit protection device is called a wire link. The wire diameter is selected based on normal current flow. An excess of current causes the wire to melt. Each of these three devices is wired in series between the input source and the rest of the receiver. Failure of any of them will produce an open circuit. No current flows in the receiver under these conditions.

The output of the circuit protection block may take one of these paths. This will depend upon the specific type of source and circuit. A battery source will usually be wired directly to a filter network. An ac source may be wired to the rectifiers in a "transformerless" set. The third path sends power to a transformer. The transformer secondary is designed to provide the required number of outputs at the required voltage–current levels. The first two paths are illustrated in the diagram by dashed lines.

The output from the transformer is usually sent to a rectifier system. In this block the ac voltage is changed into pulsating direct current. In some receivers one transformer may have multiple ac outputs and a corresponding number of rectifier systems. Filament power, being ac, is usually one of the secondary windings of the transformer.

The pulsating dc is not suitable for use in the receiver. Further process-

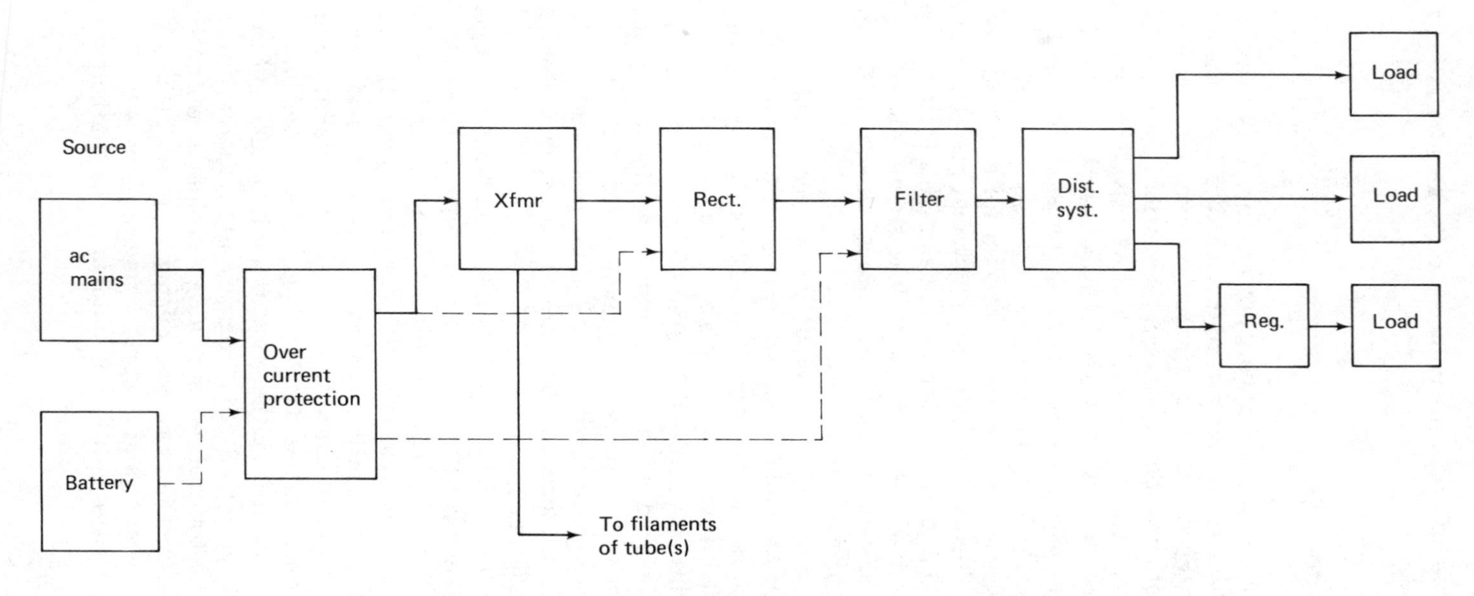

Figure 9-1 A primary power source will often have a block diagram similar to this one.

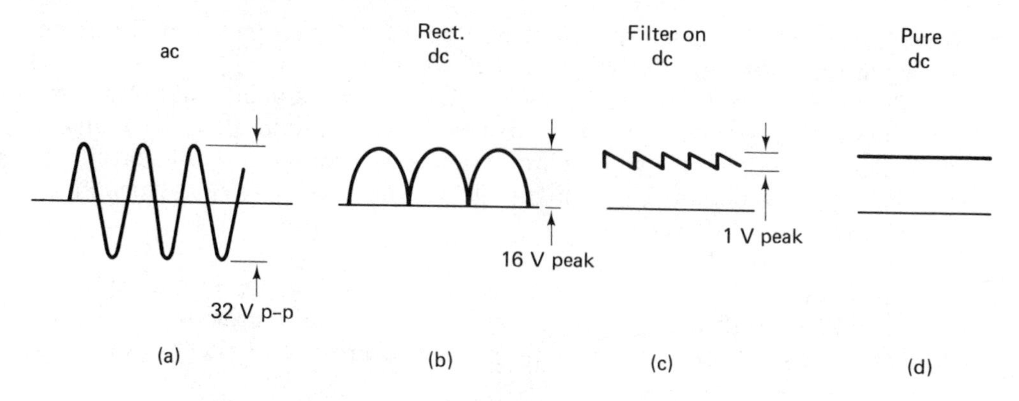

Figure 9-2 The steps illustrate how voltage variations, called "ripple," are eliminated in the power supply section of the receiver.

ing is required. This processing is called *filtering.* Variations in the level of dc due to rectifier action is called *ripple.* Ripple voltage variations should be less than 10% of the total dc voltage levels for proper circuit action. The filter block reacts to voltage and current variations to produce a smooth dc output. This process is illustrated in Figure 9-2. Ripple voltage is measured with an oscilloscope. The ideal dc waveform is shown on the right of this drawing. The raw ac wave is shown on the right in part (a) of the figure. Rectified, but unfiltered dc is shown in part (b), and part (c) shows the filtered dc with an allowable amount of ripple voltage. The waveform shown in part (c) is typical of those observed when testing a working power source. The rectified, but unfiltered dc cannot be observed. This is because the filter section is wired into the circuit.

Power is distributed to each section of the receiver from a distribution section. Each of the loads has its specific power requirements. A typical system is shown in Figure 9-3. Each line from the power source to the load has

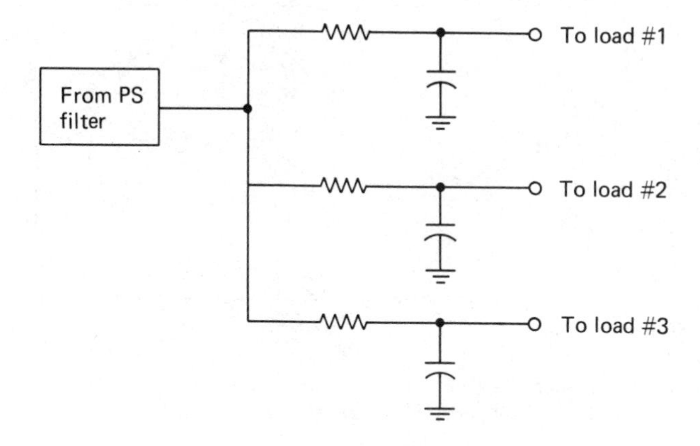

Figure 9-3 A typical power supply distribution system often looks like this schematic drawing.

its own decoupling capacitor. This is used to keep signals in one circuit from getting into another circuit through the power source.

Some electronic circuits in TV receivers have very critical voltage requirements. A regulation circuit is necessary to maintain a specific voltage level. This block is connected between the power source distribution block and the load it is designed to control. One or more of these regulator blocks may be found in a receiver.

Secondary sources. These power sources often use a high-frequency ac as the input to a rectifying and filtering system. This permits the use of smaller and lighter-weight transformers. Filter capacitors are also smaller because less

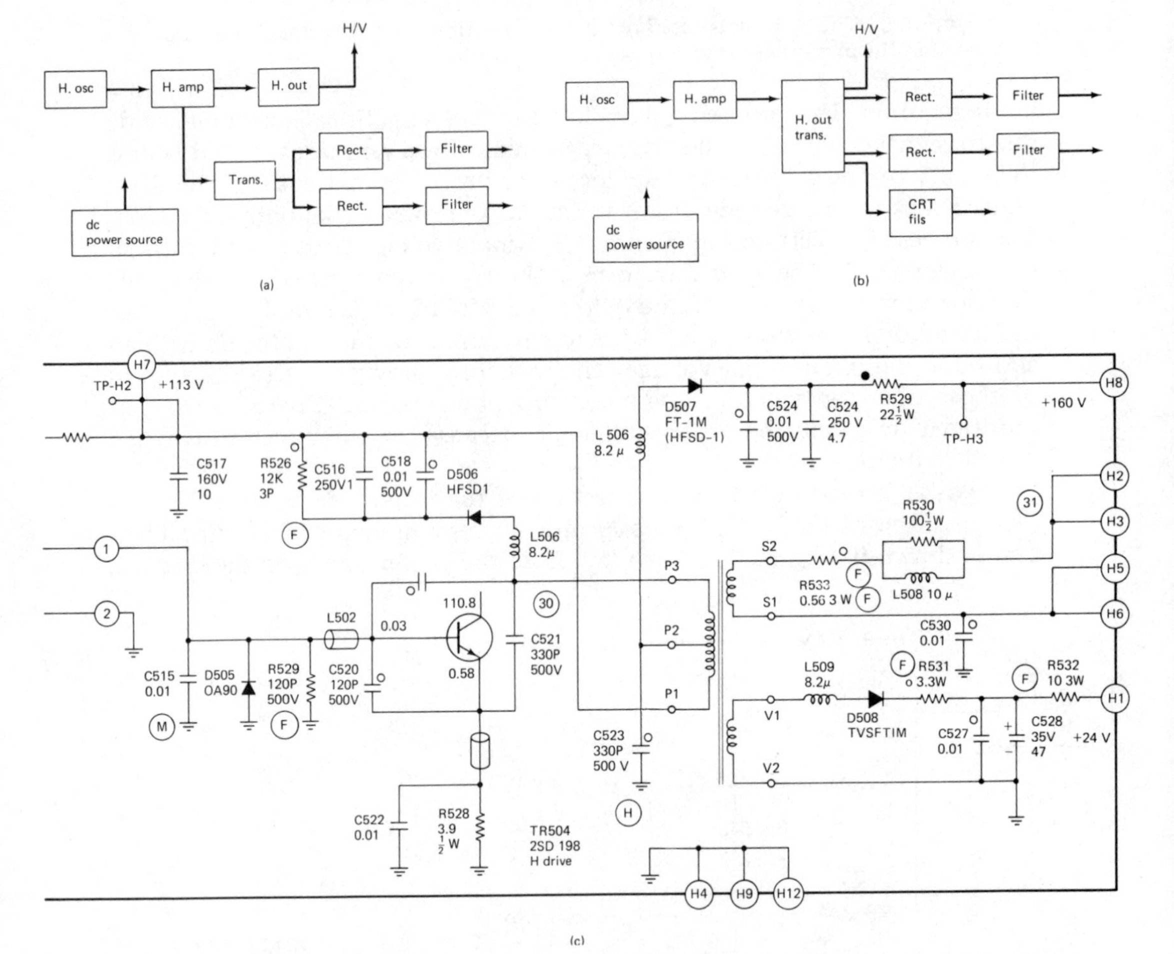

Figure 9-4 Scan-derived power sources use the horizontal pulse and may develop from (a) the amplifier or (b) the HOT. A schematic for one such system is shown in (c).

capacitance is required at the higher frequency. The high-frequency ac source that is readily available in a TV receiver is the 15,734-Hz signal found in the horizontal sections.

Figure 9-4 shows one of the common systems used to develop the scan-derived power. Power in the horizontal amplifier block is used for this. There may be one or more small transformers in this block. The actual schematic diagram for this section is shown in part (c) of the figure. Transformer T_{503} is used as the power transformer. It has one primary and three secondary windings. Terminals P_1 and P_3 are used with diode D_{506} to develop a 113-V source. Capacitors C_{516}, C_{517}, and C_{518} form the filter network. A second output is connected to terminals P_3 and P_2. This winding develops a 160-V supply. Diode D_{507} and capacitors C_{529} and C_{525} are also part of the circuit. The third dc source uses terminals V_1 and V_2 along with D_{508}, C_{527}, and C_{528} in order to form this circuit.

A second system uses windings on the flyback or horizontal output transformer. Extra windings, as shown in Figure 9-5, are added to this transformer to establish the proper operating power sources. In this receiver the picture-tube filament voltage is also developed in this transformer. A partial schematic diagram for this circuit is shown in part (b) of the figure. Rectifier diodes CR_{401}, CR_{403}, CR_{404}, and CR_{405} are used to produce pulsating dc

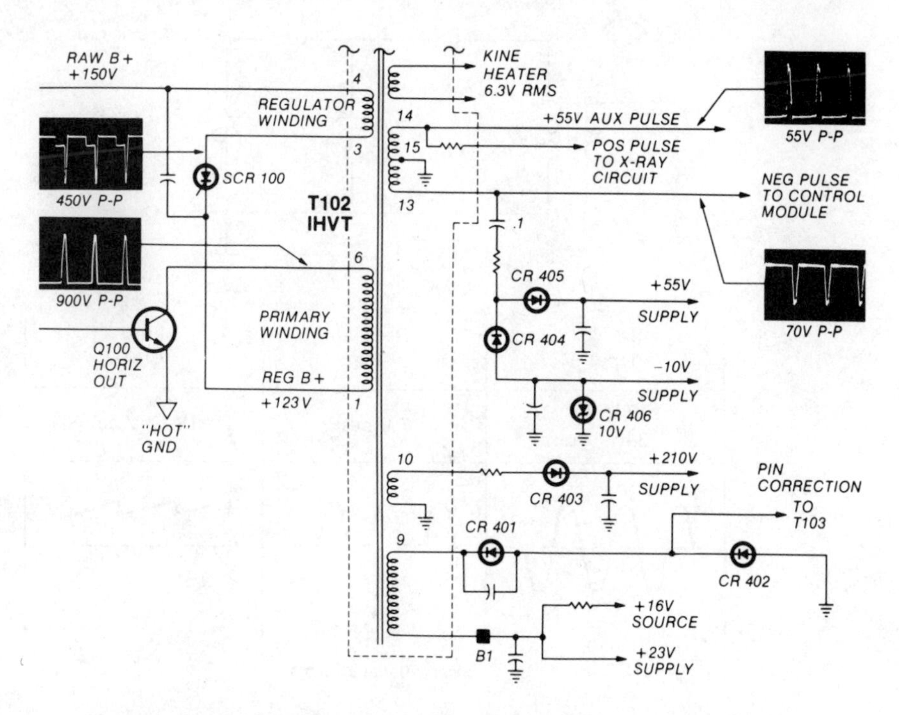

Figure 9-5 Schematic diagram of a HOT scan-derived power supply. (Courtesy of RCA Consumer Electronics.)

from the output windings of the transformer. C_{405}, C_{409}, and C_{416} are used as filter capacitors. These values are much less than the 500- to 1500-μF capacitors required for 60-Hz power sources.

RECTIFIER SYSTEMS

Each of the rectifier systems described in this section is found in a television receiver. Some receivers may use only one system. Other receivers utilize two or more of these systems. Both primary- and secondary-type power sources use them. The difference will be found in the characteristics of the rectifier diode and the values of filter capacitors.

Half-wave rectifiers. The name for each of these systems accurately describes the action. This system rectifier is half of the ac wave. Its output is one-half of the input wave. A typical half-wave rectifier is shown in Figure 9-6. The output voltage is on one-half of the total time for one input wave cycle. One significant factor is the change in voltage level that occurs. In reality there is no change. The difference is in the system used to describe each value. Almost all 60-Hz ac is described in root-mean-square (rms) values.

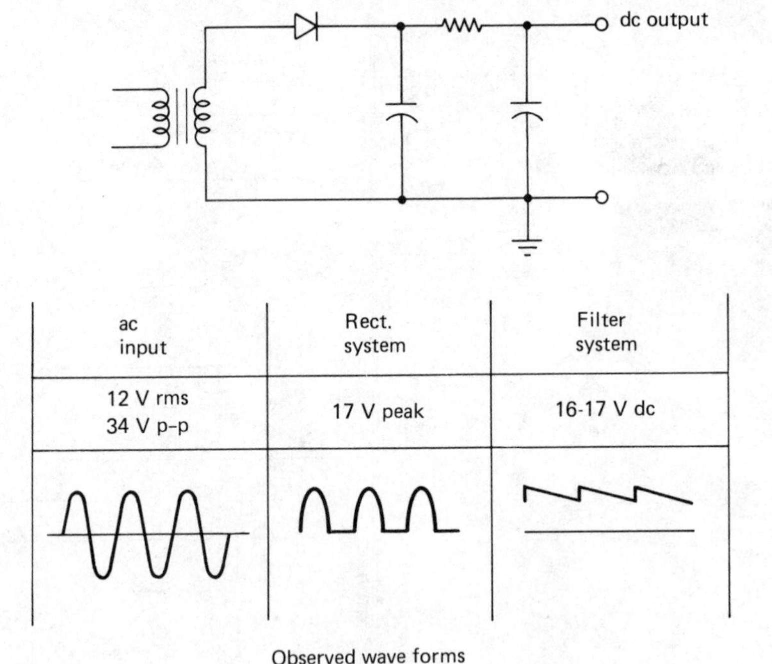

Figure 9-6 Half-wave rectifier system with voltage and waveform information.

This, in the example, is shown as 12 V. Rectifier systems react to half of the wave. Usually, the ac wave is described in peak-to-peak terms. Converting rms to peak to peak (12 x 2.828) gives a peak-to-peak voltage of almost 34 V. One half of this voltage is processed by the rectifier system. This develops the 17-V dc level. Filter action keeps the voltage at this point.

Two voltage values that are important must be mentioned at this time. Both are related to the peak voltage. Rectifiers have a peak reverse voltage (PRV) rating. This is sometimes called a peak inverse voltage (PIV). The terms are interchangeable. What this means is that when the rectifier is not conducting, almost all the applied voltage develops across its terminals. The PRV rating must be high enough to keep the rectifier from breaking down. A minimum value for a rectifier diode PRV rating is at least $1\frac{1}{2}$ times the peak voltage in the circuit.

The second voltage rating is that found on a capacitor. The capacitor also responds to peak values. Voltages above this rating will destroy the capacitor. Care must be taken to be certain that the voltage rating on any capacitor is higher than any voltages encountered in the circuit.

Half-wave rectifier action is similar to the action of a switch. Diodes when foward biased have a very low resistance. When reverse biased the diode resistance is very high. Basic electrical laws show that a low-resistance component will develop a small voltage drop when current flows. The diode and the load form a series circuit. When the diode is forward biased during one-half of the cycle, almost all the applied voltage develops across the load resistance. The result is a flow of current through the load. Work is performed when this occurs. During the other half of the cycle, the diode is reverse biased. Its resistance is many times greater than the load resistance. Almost all of the applied voltage develops across the diode. No current flows in the circuit due to the very high diode resistance. Therefore no work is done in the load. The frequency of the ripple voltage in this circuit is 60 Hz.

The circuit designer's concern is to develop a constant voltage across the load. This is accomplished by use of a filter network. Energy is stored in the filter during one part of the cycle. When the reverse-biased portion of the cycle is on, the energy stored in the filter is released to the load. This action tends to maintain a constant dc output voltage at the load. Work is therefore performed during the entire input wave cycle.

Full-wave center tapped. There are two distinct characteristics that identify this power source. One of these is the use of a center-tapped transformer. The second is the use of two diodes instead of one. A schematic diagram for this supply is shown in Figure 9-7. Each diode conducts during one half of the input ac wave cycle. The cathodes of both diodes are connected together in this type of circuit. The output voltage for this circuit is the same as that of the circuit used for the half-wave rectifier system. The advantage of this system is that smaller filter capacitors are needed. The ripple frequency of

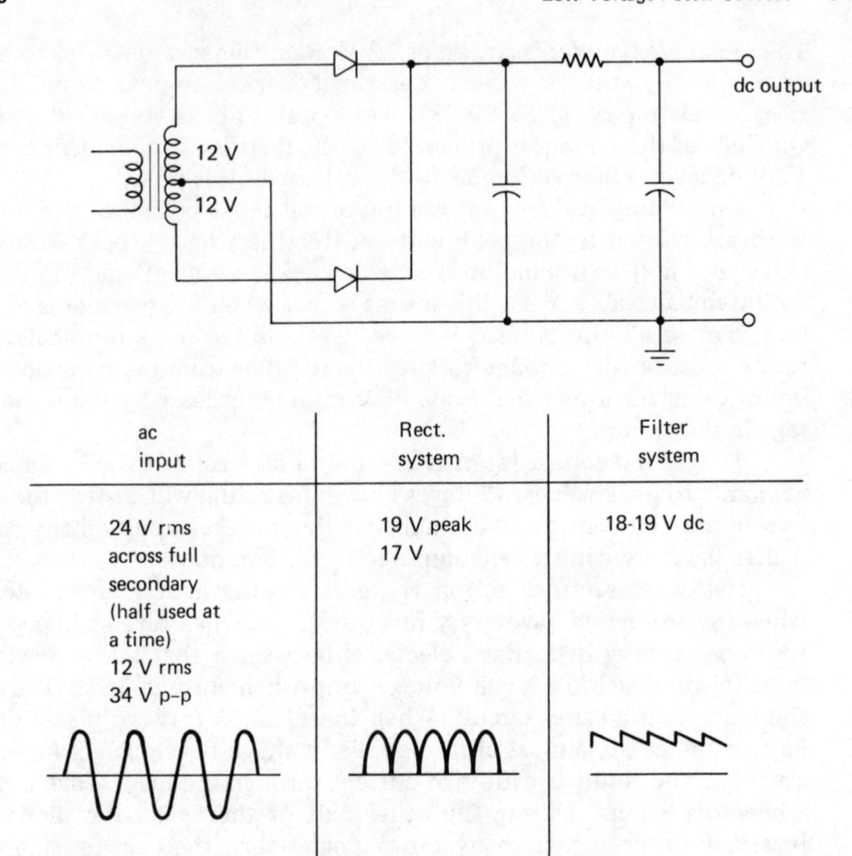

Figure 9-7 Full-wave center-tapped rectifier system.

the output of this system is 120 Hz. A higher average output voltage is also obtained when a full-wave circuit is used.

Full-wave bridge. A circuit that takes advantage of the full secondary winding voltage is the full-wave bridge rectifier system. Four diodes, as shown in Figure 9-8, are required in this circuit. Only two of the diodes conduct during each half-cycle of operation. Diodes D_1 and D_3 operate for one half and diodes D_2 and D_4 during the other half of the cycle. This type of circuit uses the full secondary voltage of the transformer. Its output voltage is therefore double that obtained from a center-tapped rectifier system. This circuit has the advantage of the higher output voltage. It also requires the smaller-value filter component since its output frequency is 120 Hz.

Voltage doublers. This type of circuit takes advantage of the charging of a capacitor. It is identified by the series capacitor C_1. Current flow during one

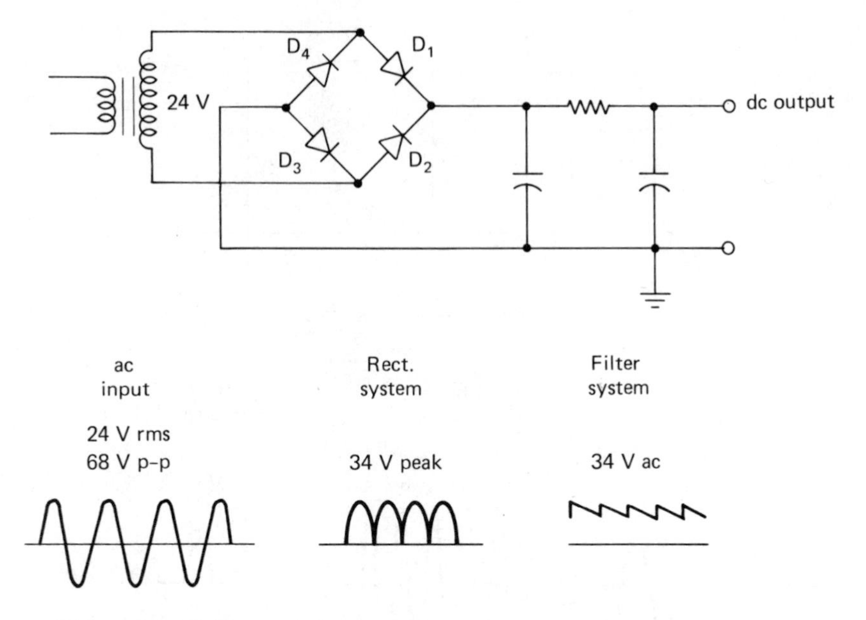

Figure 9-8 Full-wave bridge rectifier system. It is identified by the use of four diodes.

half of the cycle is through D_2 and C_1. This action charges C_1 to the full applied voltage. During the second half of the cycle, current flow is through the load and D_1. The voltage developed in the transformer and the stored charge developed in C_1 add to produce a source voltage double that of the transformer secondary. This circuit is shown in Figure 9-9. This circuit develops a 60-Hz ripple frequency. Large-value capacitors are required in the filter section. It also has poor output voltage regulation.

The full-wave voltage doubler is shown in Figure 9-10. Its identifier is the two series capacitors C_1 and C_2. During one half of the cycle, capacitor C_1 is charged through D_1. During the second half of the cycle, capacitor C_2 is charged through D_2. The two voltages add to provide an output voltage double that developed by each capacitor. This is called a full-wave circuit because capacitor charging current flows during each half of the cycle. The output frequency of this system is 120 Hz.

Sources of power. The systems described in this chapter are used with a variety of rectifier input voltages. Some may be operated directly from the ac power line. This is true for all except the full-wave center-tapped system. All may operate using a horizontal pulse in the input source. Each system requires diodes that will operate correctly at the frequency of the input power. The size of transformers (if used) and filter capacitors will be reduced as the frequency of the system increases.

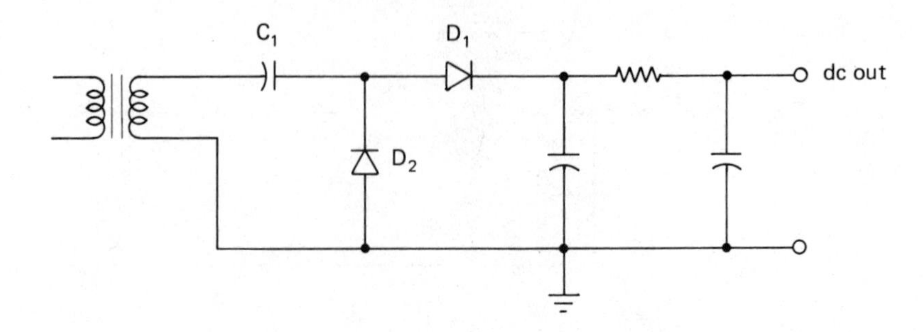

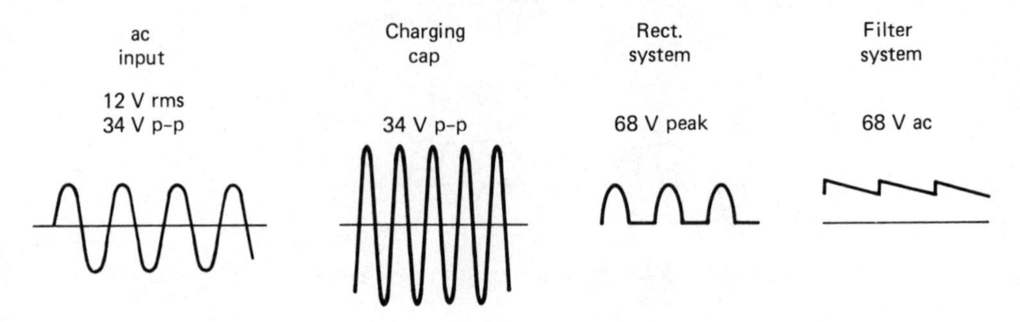

Figure 9-9 The half-wave voltage-doubler system uses a series-connected input capacitor.

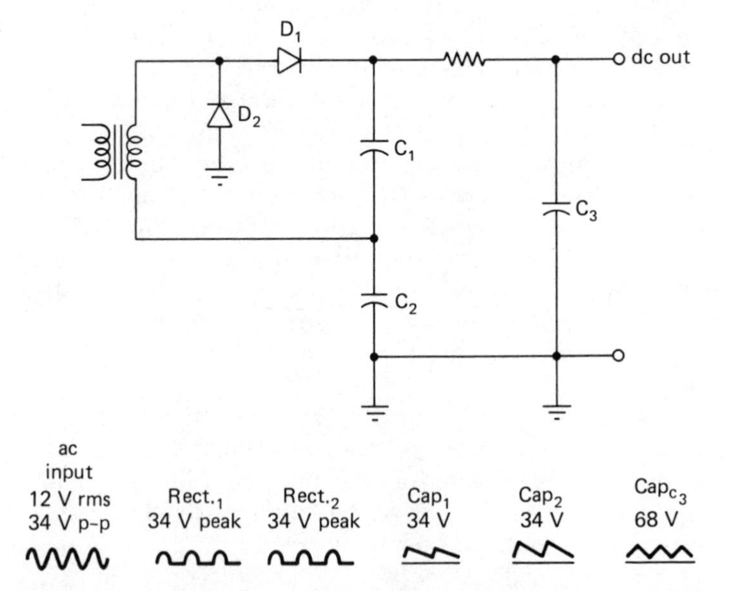

Figure 9-10 The full-wave voltage-doubler rectifier system is identified by the series-connected capacitors C_1 and C_2.

REGULATION

Operational voltages for solid-state circuits are more critical than those required for tube-type circuits. A voltage-regulator circuit in the receiver is used to provide these stable voltages. One regulation system block diagram is shown in Figure 9-11. This system, used in Magnavox receivers, develops a startup voltage for the horizontal oscillator. Once started, the pulses from the horizontal oscillator are used to develop the required operational voltages for the rest of the receiver. There are several subunits in this block.

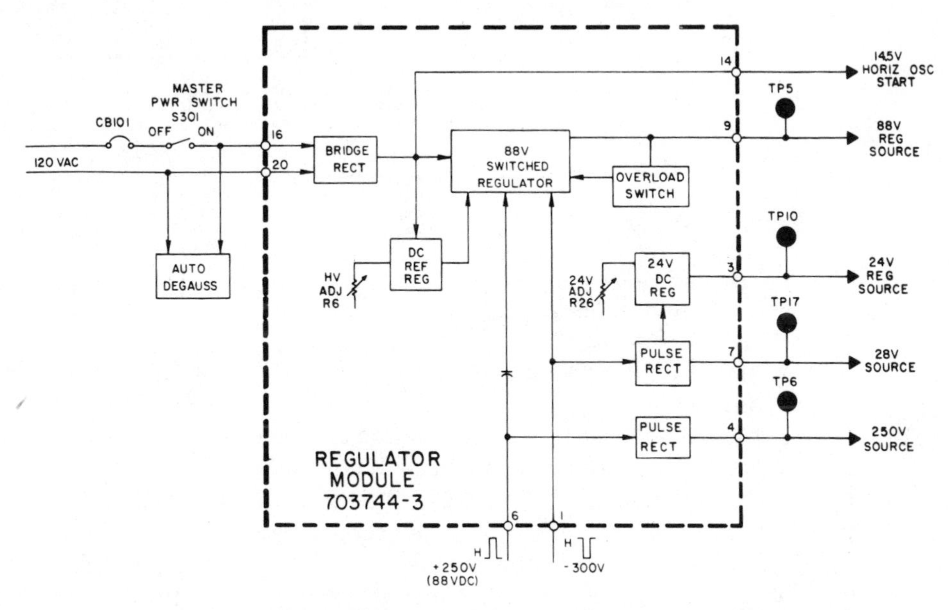

Figure 9-11 This solid-state regulated power supply uses a startup source and pulses from the HOT to develop the power levels required. (Courtesy of Magnavox Consumer Electronics.)

Reference regulator. The schematic for this section is shown in Figure 9-12. A bridge rectifier is used to develop the 145-V startup voltage for the receiver horizontal oscillator. This provides the necessary horizontal pulses for regulation. Transistor Q_1 is used as a voltage regulator. Its base voltage is developed by diodes Z_1 and D_{14}. Once the horizontal oscillator starts, a 200-V source is developed. This 200-V source is applied to the cathode of D_{14}, biasing it off. The 200-V source is then used as voltage source for Z_1. Transistor Q_1 is used to establish a reference voltage of 88 V. This transistor is a voltage regulator. Pulses from the horizontal output are supplied to the emitter circuit of Q_1. One of these is a positive pulse of 250 V. The other is a negative pulse with a value of 300 V. These two pulses are added to the dc

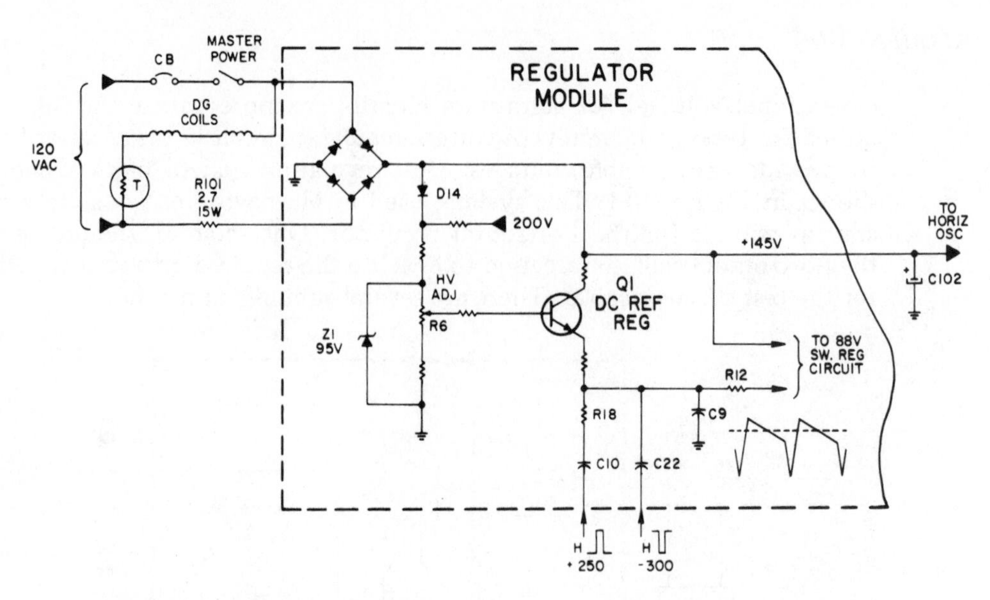

Figure 9-12 Regulator circuit used in the solid-state supply. (Courtesy of Magnavox Consumer Electronics.)

output of the regulator transistor. The result is the modified sawtooth waveform illustrated.

These pulses and the 145-V source are connected to a switching regulator circuit. This circuit is shown in Figure 9-13. This circuit is controlled by switching transistors Q_4 and Q_5. These are biased so that when the first one is on, the other is off. When the second one is on, the first one is off. These transistors control transistor Q_6, which is the regulator. Transistors Q_4 and Q_6 are always in the same operating mode. The positive and negative pulses switch Q_4 and Q_6 on and off at the horizontal rate. This maintains the required 88 V.

The action of this circuit is illustrated in Figure 9-14. The dashed line at A is used to represent the 88-V regulated value. Normal conduction time for Q_6 is shown at T_1. If the output voltage should drop to level B, the on time for Q_6 increases. This charges the output capacitor C_{204} and restores the operating voltage to the correct level. This action occurs at the horizontal rate. It is very fast. The result is a well-regulated output voltage.

The final portion of this unit is shown in Figure 9-15. Three output voltages are obtained. One is a 250-V source. It is developed by adding the 250-V pulses from the flyback transformer to the 88-V dc supply. These pulses are rectified and filtered to produce the required 250 V. Negative 300-V pulses are rectified and filtered to create a 28-V source. Transistor Q_{10} is used in a regulating circuit to develop a 24-V source. Transistor Q_9 is used to adjust the bias in transistor Q_{10}. This permits some variation in the regulated value.

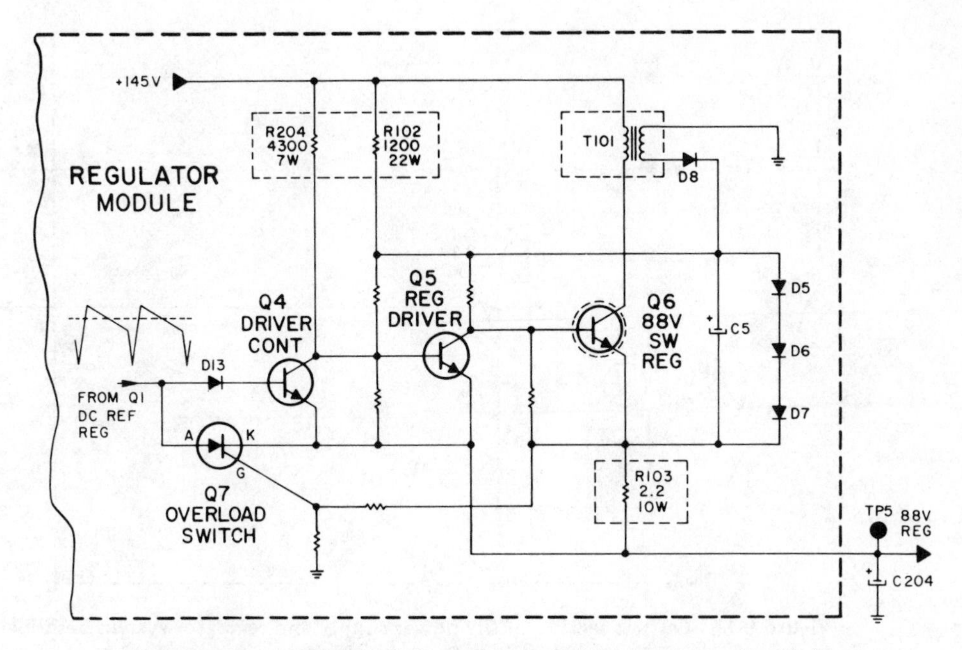

Figure 9-13 Schematic diagram for a switching regulator circuit. (Courtesy of Magnavox Consumer Electronics.)

Regulation occurs in this circuit when ripple voltage from Q_{10} emitter is inverted by amplifier action in Q_9 and returned to the base of this transistor. The out-of-phase ripple cancels almost all the ripple that is present in this circuit.

One additional circuit found on this board is an overload protection and shutdown circuit. The silicon-controlled rectifier (SCR) Q_7 and resistor R_{103} are used for this purpose. These are illustrated in Figure 9-14. Under normal operating conditions Q_7 is shut off and has no effect on the circuit. If load current in the horizontal output circuit occurs, the voltage drop that develops across R_{103} causes the SCR to turn on. It acts as a short circuit between base and emitter of Q_4. This transistor, in turn, cuts off Q_6. This disables the 88-V source and shuts down the receiver.

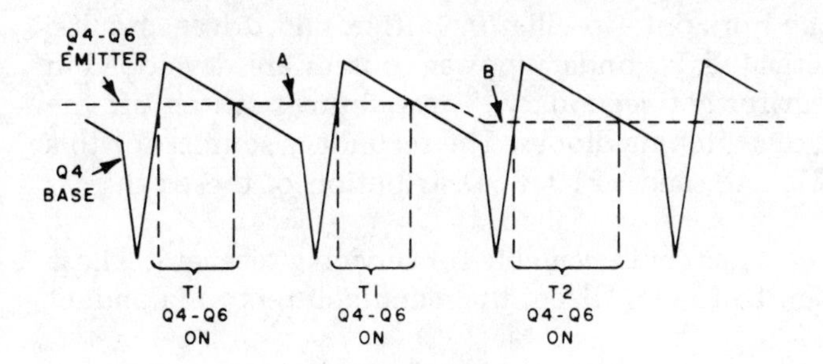

Figure 9-14 Voltage development in a switching regulator source. (Courtesy of Magnavox Consumer Electronics.)

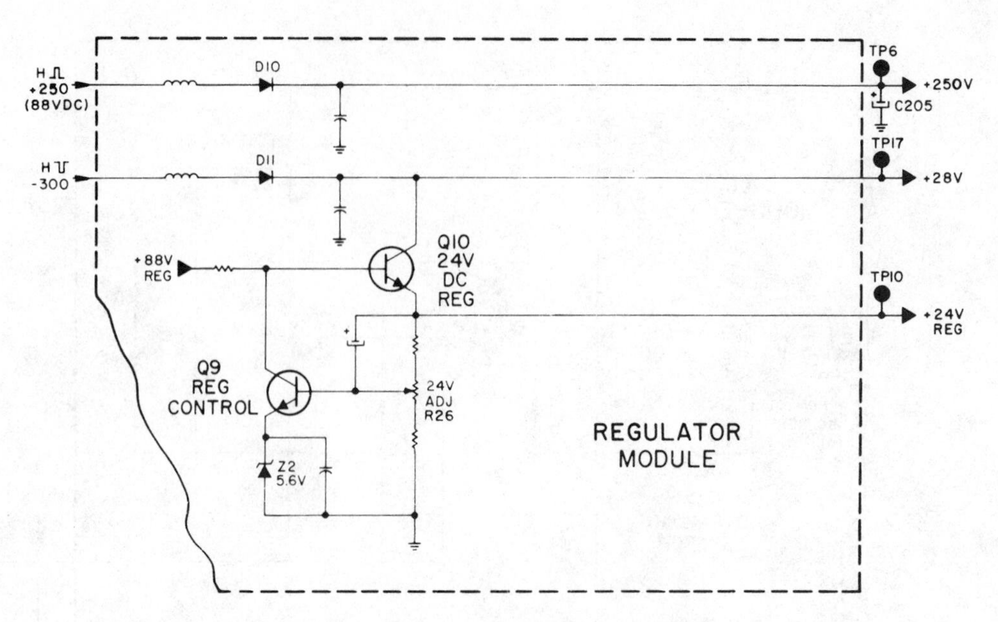

Figure 9-15 Output section of the power supply and regulator system described in the text. (Courtesy of Magnavox Consumer Electronics.)

TROUBLESHOOTING PROCEDURES

A major problem in the repair of a combination power source, regulator, and shutdown system is the location of the problem. There are several possible causes for a set that has no power. The first step is to review the schematic to determine the possible sequence of events that occur in order to provide power. A diagram of a power supply interface system is shown in Figure 9-16. Arrowheads are used to show the direction of action that occurs in this system. The input to this system is from a +150-V source. This voltage is used to operate the horizontal output circuit. It is a regulated source. The regulated value is 114 V. The horizontal output circuit will not operate until a signal is developed and fed into it.

A start circuit is used to develop the required 22- and 27-V sources. These are used to start the horizontal oscillator, buffer, and driver circuits. Once these circuits are activated, secondary power sources are developed in the horizontal output transformer secondary. These sources switch off the starting circuits by means of switching diodes. The secondary sources for this receiver are +11, +22, +27, -40, and +210 V. Distribution of these values is shown in the diagram.

There are some very excellent aids available for servicing receivers. These are available from the manufacturers. Often, the manufacturers will conduct a service training seminar at a local distributor. Take advantage of these seminars. The information presented at them will help in the diagnosis and repair of receivers. One bit of information that comes from a seminar is the initial troubleshooting procedure for this receiver (RCA chassis CTC 93). When the receiver is first turned on, the technician should listen for audio operation. This includes speaker noise of any kind. If any sound is heard, it is assumed that the starting +150-V horizontal regulator, and horizontal output circuits are working properly. If the sound system continues to function, the +22-V regulator circuit is also working. This is based on the analysis that the horizontal circuits have to be functioning to develop the operating power for each of the circuits above.

Almost all power source circuits use a combination of linear path and separating path systems. When troubleshooting these circuits, the technician must make an analysis based on the methods of troubleshooting discussed in Chapters 5 and 7. A good starting point is to identify the point of separation of the system. This is often at the output of the primary rectifier filter system. A voltage and/or waveform measurement at this point will either isolate the problem to the primary source or eliminate this section of the receiver.

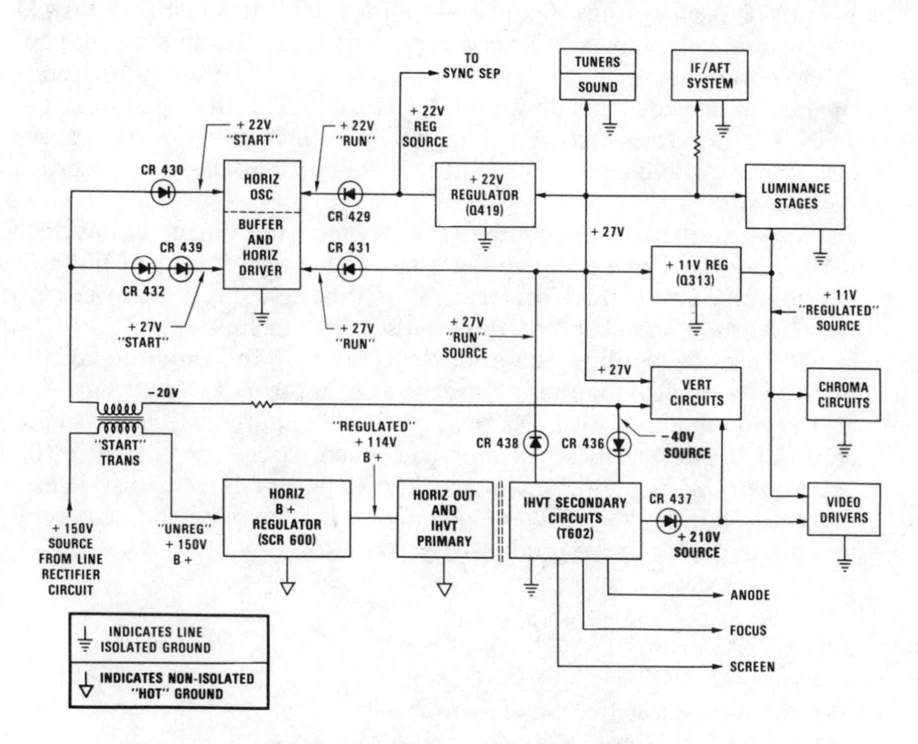

Figure 9-16 Troubleshooting aids developed by RCA for use in their CTC93 chassis. (Courtesy of RCA Consumer Electronics.)

If the area of the problem is identified as the primary power section, several checks can be made. Use the techniques for linear circuit analysis. The first check would be to be certain that power is available at the input to the circuit. This point is where the ac line cord connects to the receiver chassis. If power is present at this point, the next step is to split the circuit near the middle and make another check. This point is usually at the rectifier. Continue to test and eliminate working sections until the problem is isolated.

Circuit breakers used in TV receivers have been known to fail for no apparent reason. A test with an ac ammeter will determine if this problem exists. Connect one lead from the meter to each side of the circuit breaker. The breaker must be open to make this test. Turn on the receiver and measure the current flow. If it does not exceed the value shown on the service literature, the problem is solved by replacing the defective circuit breaker.

If the output voltage is correct at the point of separation, the problem is to be treated using the techniques for this system. A check should be made to determine if the startup voltages are present. The next test is to measure the output voltages from the secondary power sources. Another step that normally follows this is to test the shutdown protection circuit. Each receiver manufacturer has a specific procedure for this. This must be followed if valid checks are to be made. It is very important to do this in order not to damage other components in the receiver. Keep in mind that transistors will not operate under adverse conditions for more than a small fraction of a second. They are not forgiving of the technician's mistakes. Power transistors are also relatively expensive. The shutdown circuits are designed to protect these components.

The kind of tests required for servicing the power supply include the measurement of operating voltages and the measurement of pulses from the horizontal output transformer. An oscilloscope must be used to observe the amplitude and shape of these pulses. The technician is required to know how to use the oscilloscope in order to measure the amplitude of the pulses. Correct frequency for these pulses is also important. How to measure this is also required knowledge. Neither of these requires extensive training. It is essential, however, to know how and when to use the correct piece of test equipment. Also important is the knowledge of how to use it in order to make the correct test. These techniques will increase the efficiency of the technician and improve productivity.

QUESTIONS

9-1. Why are scan-derived power sources used?

9-2. Describe the action of a diode in a rectifier system.

9-3. What is the purpose of a rectifier filter system?

9-4. Voltage multipliers depend upon the charging action of what device?

9-5. What is the frequency of the scan-derived power system?

9-6. Briefly describe the action of a zener diode regulator.

9-7. Briefly describe the action of a transistor regulator.

9-8. Where should one start to troubleshoot a power supply?

9-9. Describe the action of a circuit breaker.

9-10. Why should the oscilloscope be used when troubleshooting a power supply?

Chapter 10

Tuners

MECHANICAL TUNERS

A major function of the tuner in a television receiver is to select and amplify the picture and sound signals received from a broadcast station. The second function, which is as important, is to convert the modulated carriers into a correct frequency IF signal. The original styles of tuners used for this purpose were bandswitching types. One style used a series of inductors that were wired to segments of a rotary switch. There were three or four sets of inductors and sections to the switch. The RF amplifier and oscillator both required a resonant circuit for each channel. The mixer output, being one frequency, only required one tuned circuit. Tuning was accomplished by mechanically selecting the proper inductance. The schematic diagram for this type of tuner is shown in Figure 10-1. All components used in this tuner are permanently wired into it.

A second type of switching tuner used a series of *strips*. One strip contained all the required inductances for a specific channel. The strips rotated on a drum. Each strip had a set of contacts that connected with a contact strip on the body of the tuner. All that was required for tuning was to rotate the drum type of holder and make connections to the strip for the appropriate channel. A schematic diagram for this style of tuner is shown in Figure 10-2. The dashed oblong boxes at the top of the schematic represent the set of inductors contained on a specific strip.

The introduction of additional channels created a tuning problem. It was not possible to easily add the tuning system for 70 UHF channels to either of

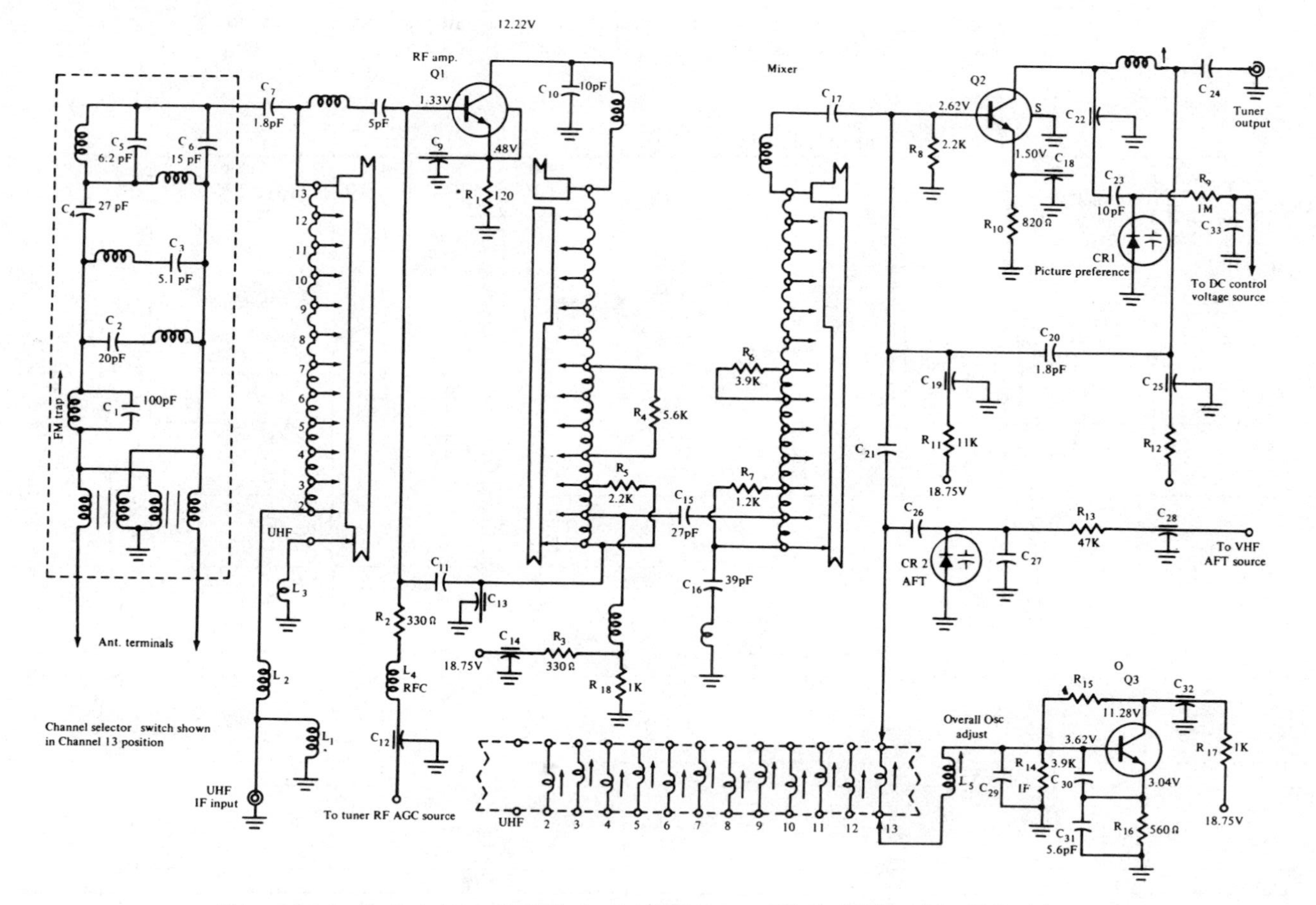

Figure 10-1 Switch-type fixed-inductor TV tuner. (Clyde N. Herrick, *Television Theory and Servicing: Black/White and Color, 2nd ed.*, 1976. Reprinted by permission of Reston Publishing Co., a Prentice-Hall Co., 11480 Sunset Hills Road, Reston, VA 22090.)

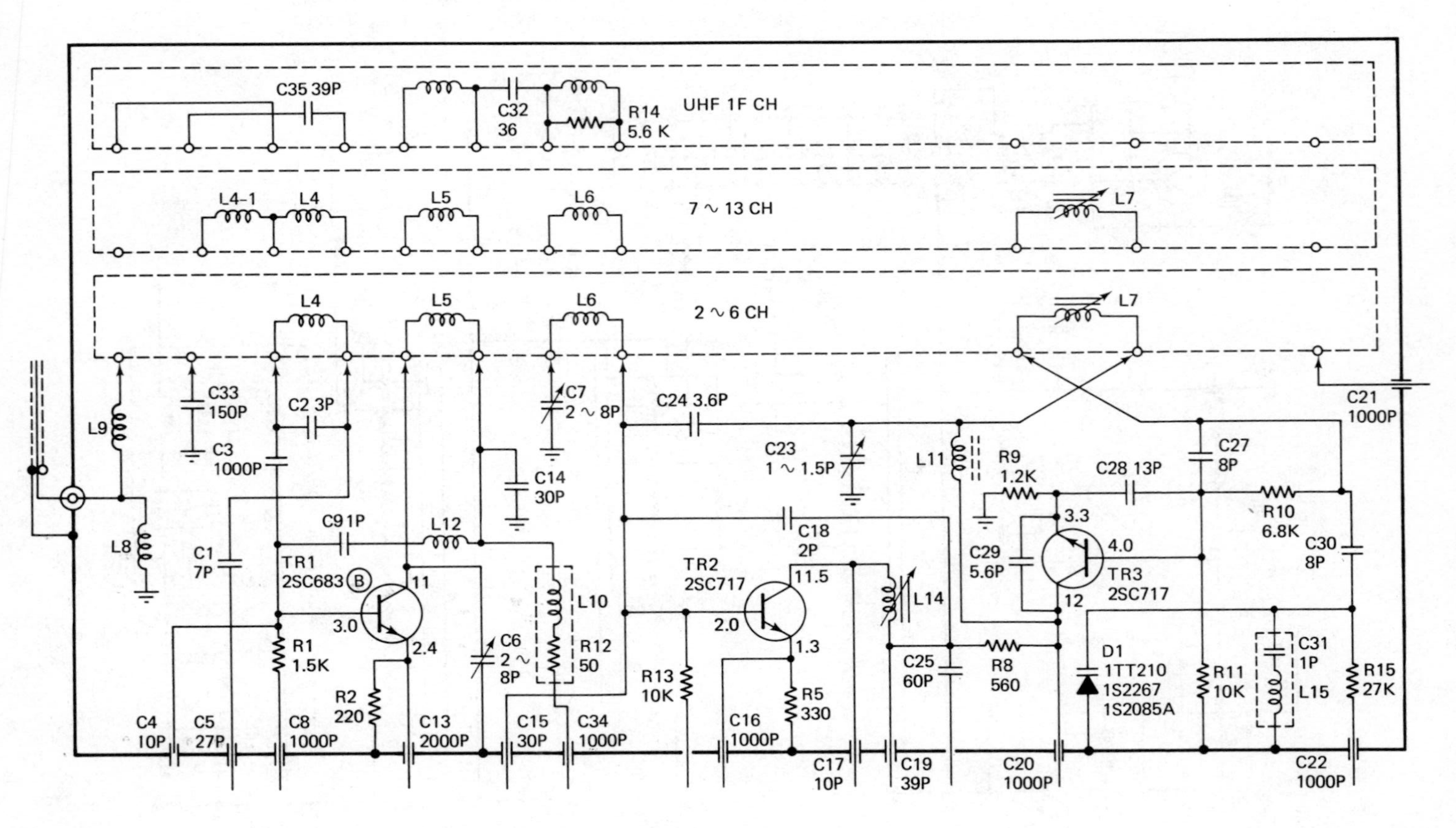

Figure 10-2 Schematic diagram of a turret-type switching tuner. The sections outlined by dashed lines represent specific channel strips.

these tuners. Another system was developed. This used a "piggyback" style of tuner. This second tuner, shown in Figure 10-3, tuned only the UHF channels. Its output is at the IF frequency of the receiver. When it is turned on, the oscillator in the VHF tuner is turned off. Inductors in the VHF tuner resonate at the IF frequency when tuning UHF signals. The RF amplifier and mixer in the VHF tuner act as IF amplifiers for UHF signals. These tuners required a mechanical tuning system. U.S. government regulations forced all TV receiver manufacturers to include UHF reception as a part of every receiver. They also required that each channel selector be indexed, or detented, so that tuning a station was easy to do.

A requirement for any television tuner is related to the bandwidth of the received signal. Each TV station is assigned a channel on which to operate. This channel has a bandwidth of 6 MHz. The receiver has to be able to receive with equal amplitude all signals transmitted within the 6-MHz segment. Unless this is done, a portion of the transmitted signal will not be processed properly. Each tuner has a fine-tuning circuit to aid in the reception of the total signal.

One major problem that appears in any circuit that is used to mechanically switch RF voltages appears as the switch contact assembly becomes dirty. Sometimes the problem is not actually dirt, but oxides of the metal that is used to make the switch's contact assembly. A good many of these switches are constructed on silver-plated metals. Silver has a tendency to oxidize. These oxides form a resistance layer on the metal. This, in turn, reduces the amount of signal voltage that is transferred from one circuit to another in the tuner. The basic theory behind this is that a voltage drop develops across any resistance as a current flows through the resistance. This is what happens to the low-level RF signal that is being processed in the tuner.

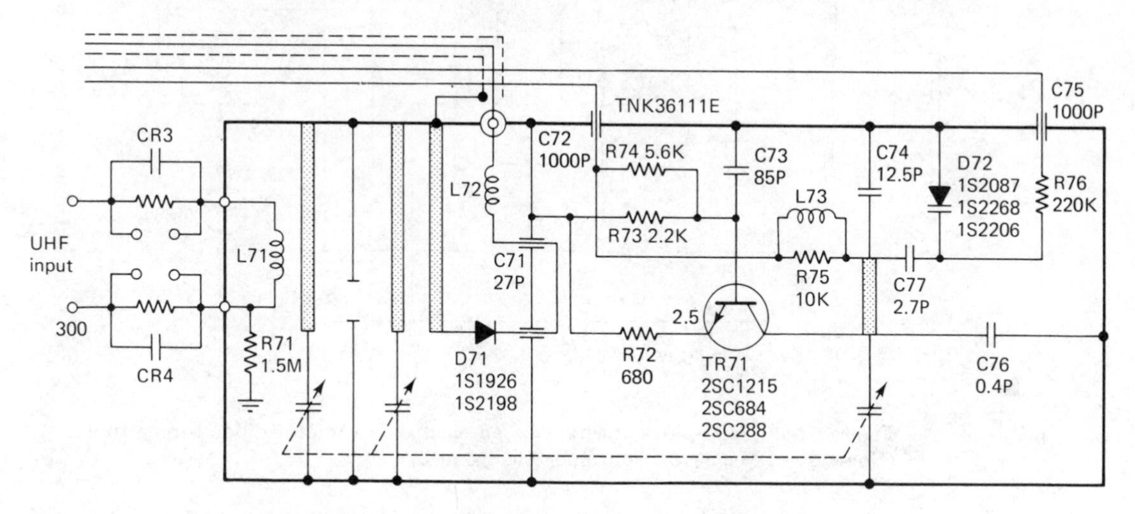

Figure 10-3 Schematic diagram of a UHF tuner.

Solid-state technology has developed a method of eliminating this problem. Special diodes, called *varactor diodes*, are used as variable capacitors. These diodes use a dc voltage in order to change their capacitance. The use of dc on the channel selector resolves the problem of oxidation on the switching parts. This is because there is less resistance to current flow as the frequency decreases. The frequency of a dc voltage is zero hertz. A partial schematic diagram for an electronic tuner that uses varactor diodes is shown in Figure 10-4. Each of the tuned circuits has a varactor diode instead of a variable capacitor. Fixed-value inductors are permanently wired into the circuits instead of switching in an inductor set for each channel. A dc control voltage is connected to a switching mechanism. Selection of a specific channel will establish a fixed level of dc for the varactor tuning diodes. Often, a series of variable resistances are used to fine-tune the circuit. A partial schematic for this is illustrated in Figure 10-5. The variable resistances are used to establish the correct resonance point for the circuit. This type of control is also used on newer receivers in place of the more traditional audio volume control or video contrast control. A dc voltage is used to control the gain of an IC amplifier circuit. This eliminates the noise associated with tuning these controls.

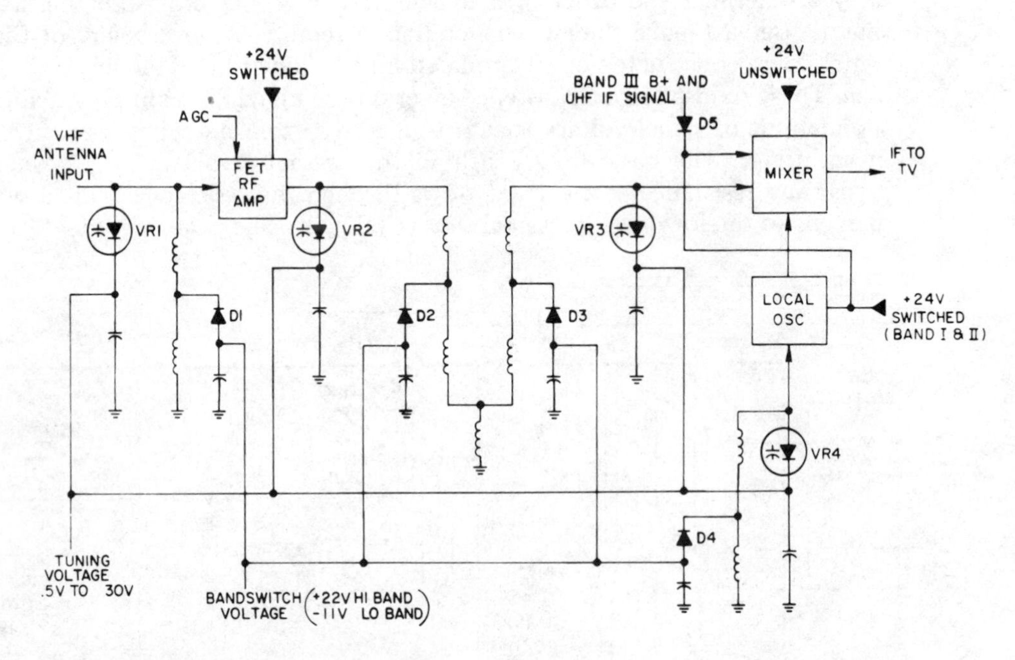

Figure 10-4 Semi-block diagram of an electronic varactor diode tuning tuner. (Courtesy of Magnavox Consumer Electronics.)

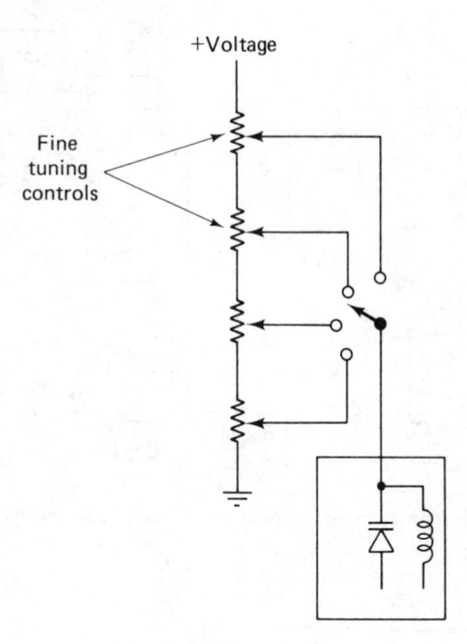

Figure 10-5 Fine tuning for the varactor diode tuner is done by adjusting variable resistors.

ELECTRONIC TUNERS

The development of the integrated circuit has done much to change the world in which we live. Miniaturization of electronic circuits has led to the introduction of the microprocessor. The ability to program information into a memory circuit leads to a wide variety of possibilities. Putting all these into a single package allows the receiver manufacturer to design an electronic control for the tuner.

A review of electronic tuning used in a television receiver immediately gives the impression of a very complex system. In reality the system is relatively simple. This is particularly true when one examines the individual units that are used to develop the complete system. A block diagram for a typical system is shown in Figure 10-6. This microprocessor unit has input and two outputs. The outputs from this type of system are in the form of dc voltages. The keyboard output develops a series of pulses. These pulses are used as encoding, or command, information for the microprocessor. The microprocessor changes these pulses into digital information. The digital information is then converted into a dc control voltage. This control voltage is used to select a channel.

There are three tuning units in this system. One controls all UHF channels. The second unit controls VHF channels 7 through 13 and the third unit selects channels 2 through 6. One of the outputs of the microprocessor is a

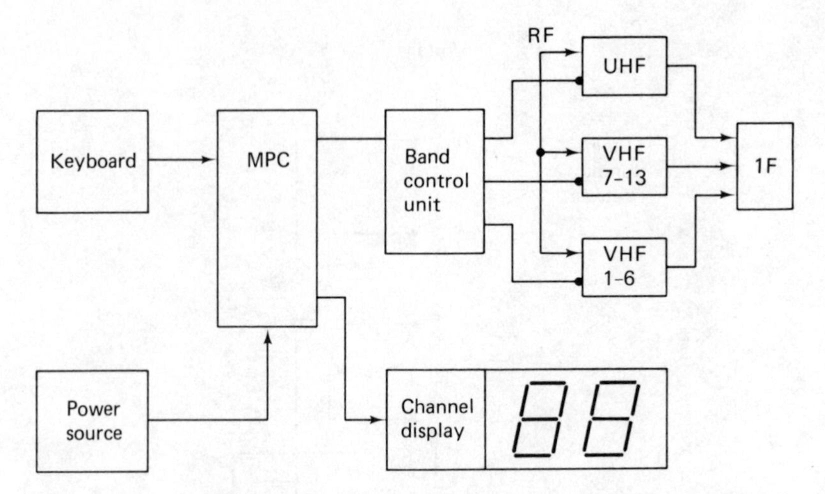

Figure 10-6 Block diagram of a microprocessor-controlled electronic tuner.

	CHANNEL	TUNING VOLTAGE		
		MIN	TYP	MAX
BAND I	2	4.1	4.5	4.8
	3	5.5	6.8	8.0
	4	8.2	9.3	10.5
	5	13.4	14.7	16.1
	6	21.0	22.0	23.0
BAND II	7	8.9	11.0	13.0
	8	9.6	12.1	14.4
	9	10.5	13.1	15.5
	10	12.0	14.4	16.8
	11	14.5	15.8	17.4
	12	17.0	18.0	19.3
	13	21.0	22.0	23.0
BAND III	14	1.5	2.4	3.2
	20	2.9	3.8	4.6
	25	4.0	5.0	6.1
	30	5.2	6.2	7.5
	35	6.5	7.7	9.0
	40	7.5	9.0	10.5
	45	8.6	10.4	12.3
	50	9.7	11.8	14.0
	55	10.7	13.0	15.3
	60	12.0	14.2	16.5
	65	13.3	15.5	17.7
	70	15.2	17.1	19.1
	75	17.4	19.0	20.8
	80	20.5	22.0	23.5
	83	23.5	25.5	27.5

Figure 10-7 Voltage chart used with a microprocessor-controlled tuner.

voltage that is used to select the proper tuning unit. Once this voltage is applied to the unit, the same voltage is used to tune the desired channel.

A typical tuning unit will require a maximum voltage of 30.0 V. The output from the microprocessor is some fraction of this value. A chart showing tuning voltage values is shown in Figure 10-7. Each channel is assigned a value of tuning voltage. This voltage is applied to the varactor tuning diodes in the tuner assembly. These diodes are used to resonate the tuned circuit to the desired frequency.

Another portion of the microprocessor is used to provide power to display the channel number. The system used for this will vary among receiver manufacturers. One manufacturer will have the channel display appear as a part of the video information. Others will incorporate a separate digital readout in the receiver. The information required for this display is also developed in the microprocessor control unit of the tuner. A separate section of the tuner unit converts the coding information for channel selection into a digital readout.

TROUBLESHOOTING THE TUNER

Diagnosis and repair of any type of television tuner is not difficult. There are, however, some physical problems that are encountered by the technician. Tuners are very compact. A picture of the internal wiring of a tuner is shown in Figure 10-8. One of the basic requirements for tuner repair is patience. Another is a very steady hand. A third is the use of tools that allow one to work in the confined area without damaging other components.

The first phase of any repair should be the diagnosis of the problem. The technician has to determine if the fault is really in the tuner. If this is true, the next phase of the diagnosis involves the elimination of those blocks in

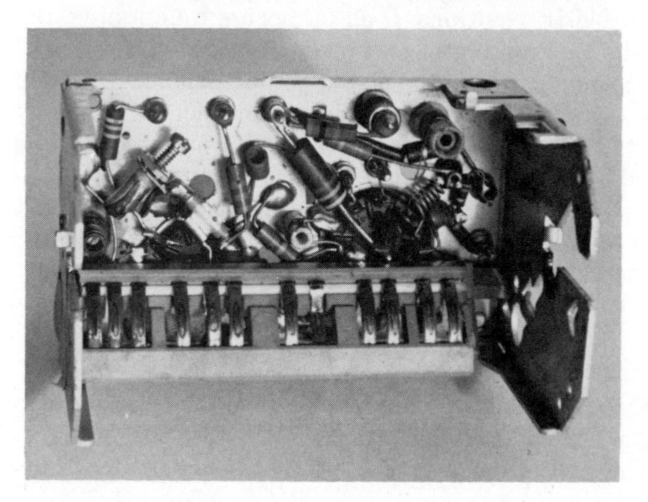

Figure 10-8 Underchassis view of a TV tuner.

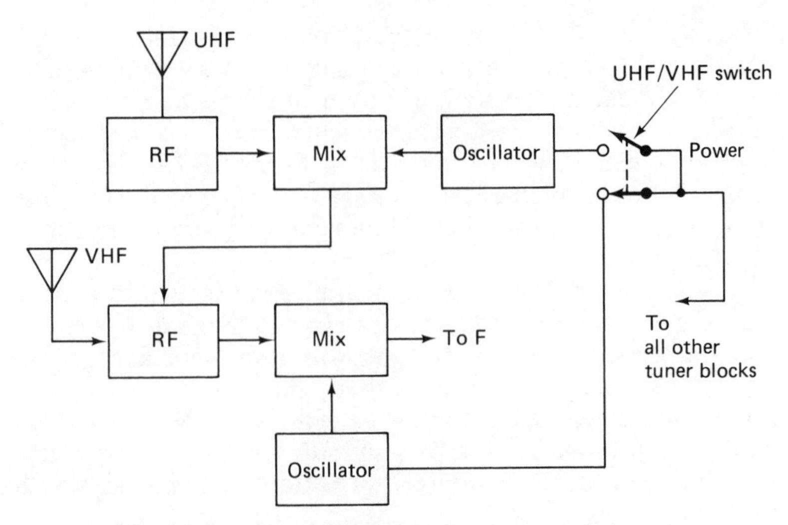

Figure 10-9 Block diagram for a nondigital tuning system.

the tuner that are working properly. Figure 10-9 is a block diagram of a nondigital tuner. Much can be learned from this diagram related to diagnoses of a defective block. The power source develops the required operating voltages for each block in the tuner. These blocks are always on except for the two oscillators. Only one of these is able to receive power at any one time. The UHF-VHF selector switch position determines this. This information is used to help eliminate working blocks in the tuner.

Rotating the channel selector with the receiver power turned on is the first step. If all VHF channels are working but there is no UHF reception, the VHF tuner may be eliminated from the brackets placed around the trouble area. If, on the other hand, UHF is working, but there is no VHF reception then the VHF oscillator is not operating. The VHF oscillator is the only block that is not employed in the UHF positions. If all others are functioning, the problem most likely is in the VHF oscillator. A third condition is poor operation on all channels. Turning and rocking the channel selector knob will help in this diagnosis. If a good-quality picture appears, the problem is probably related to dirty tuner switch contacts. If the received information stays at a low level for both UHF and VHF, the problem is in the VHF tuner RF amplifier or mixer blocks. One other problem that occurs is indicated by reception of only the low channels (2 through 6). This usually means that the VHF oscillator is not functioning at the higher frequencies.

NONDIGITAL TUNERS

Once the problem is localized to a specific area of the tuner, the approach to repair is very traditional. A schematic diagram for a solid-state VHF tuner is shown in Figure 10-10. The first thing to do is to study this schematic to de-

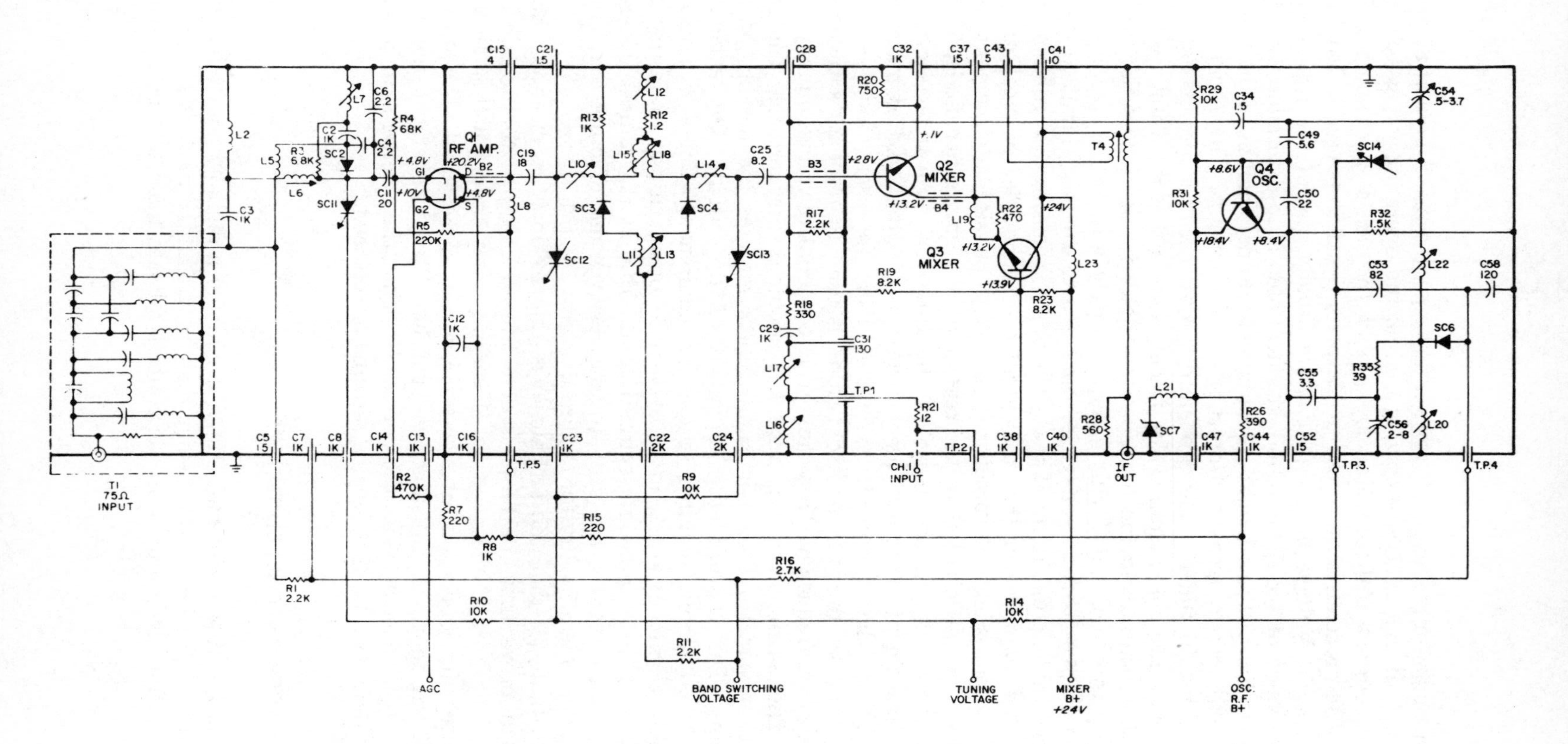

Figure 10-10 Schematic diagram of a solid-state VHF tuner.

velop a plan for the repair. Be certain that a good signal is present at the tuner input. Also check: *do not assume* that all the operating voltages are present and are correct. If these factors are correct, the next step is circuit analysis.

The procedures presented in Chapters 5 and 7 are used to troubleshoot any circuit. Voltage measurements are usually one of the first steps. If a wrong value of voltage is found, the next step is to determine why. Careful examination of the schematic diagram and proper analysis of the test results will help locate opens and/or shorted components. The oscilloscope has limited use in tuner repair procedures. Most oscilloscopes do not have the frequency response required for tracing signals through a tuner. One other test that may aid in the location of a defective section of the tuner is to "clamp the AGC line." This means to apply a dc voltage from an outside power source. This dc voltage is connected to the AGC input to the tuner. Its use will help determine if an AGC trouble is causing the problem.

One major repair that is beyond the capabilities of most technicians is the replacement of varactor diodes. The physical replacement of these diodes is no different than the replacement of any other diode. The problem involved is the alignment of the three or more tuner sections after the replacement is made. Varactor diodes are often sold as matched sets. This means that the electrical characteristic for each diode in the set is very close to the others. Each diode is used to resonate a tuned circuit at a specific frequency. They must react to a control voltage equally.

The alignment of the resonant circuit is very critical. The replacement of varactor diodes in some tuners requires the moving of the inductors. This disturbs the alignment of the circuit. The alignment of the tuned circuit requires specialized equipment. There is not enough of this alignment work in most service organizations to justify buying the expensive equipment required to do this work. Almost all electronically tuned tuners are sent to the receiver manufacturer or to a central independent facility for repair.

About all that the technician is able to do is to diagnose the problem to prove that the tuner is defective. A check of the dc control voltage will show if that voltage is varying as it should. Other dc voltages required for operation of the tuner are also measured. The only other suggested repair is to substitute another "test" tuner on the receiver. If the test tuner provides the correct IF signal, the receiver's tuner is proven to be defective. It must then be repaired or replaced.

MICROPROCESSOR-CONTROLLED TUNERS

A microprocessor electronic tuner is actually a digitally controlled varactor tuner. The portion of this device that is different is the digital signal-processing section. Its outputs are dc control voltages that are used to select the proper tuner unit and then to tune that unit to the correct channel. The process used to troubleshoot a microprocessor-controlled unit is not difficult. It is different from the type of troubleshooting encountered in other types of circuits. Each receiver manufacturer has specific instructions for diagnosis of tuners. These should be followed very carefully. In most cases the process involves checking voltages and waveforms.

Most of the repair procedure uses the tests to localize a fault to an IC. The repair process requires the measurement of operating voltages as an initial step. If the operating voltages are correct, the next steps are to check input and output waveforms. Service literature is required in order to know where to measure and what to expect to find when the measurement is made. In most cases if the output is wrong but input and control values are correct, the technician replaces the defective IC.

The replacement of a module or an IC is not difficult. The hardest part of the procedure is the determination that the unit is defective. The newness of the IC in receivers has introduced different methods of troubleshooting and repair. In reality the IC is little more than several diodes, resistors, and transistors in one circuit. Analysis and repair is no different from that used to repair receivers using discrete components.

QUESTIONS

10-1. Draw a block diagram of the VHF tuner.
10-2. What channels are tuned in the VHF tuner?
10-3. What channels are tuned in the UHF tuner?
10-4. Draw a block diagram of the UHF-VHF tuner combination.
10-5. What is the bandwidth of the TV tuner?
10-6. Why are varactor diodes used in tuners?
10-7. Briefly describe electronic tuner action.
10-8. What is the advantage of a varactor tuner over earlier switching types of tuners?
10-9. How does the microprocessor tuner differ from a varactor-type tuner?
10-10. Describe signal flow paths in the TV tuner.

Chapter 11

Antennas and Antenna Systems

One of the most important components in a television receiving system is the antenna. Too often, the antenna is overlooked and not considered when the receiver is analyzed for repair. The electromagnetic waves that are broadcast from the transmitting antenna have to be brought into the receiver. In addition, a certain quantity of signal has to reach the receiver. The amplifiers in the receiver are nondiscriminating. They will amplify any type of signal. If the incoming signal is weak, the amplifiers will process atmospheric noise as well as any signal information. The result is a picture that contains a lot of what is called "snow." This is shown in Figure 11-1. Snow is atmospheric noise that is converted into electronic signals. Viewing a snowy picture is like trying to look at something during a snowstorm. Much of the image is blocked by the falling snow. A reduced image is observed under these conditions. In the audio system a weak signal is typified by a rushing sound. This, too, is an indicator that not enough signal is available for processing by the receiver.

A good-quality television signal should have a signal level of about 1000 μV at the receiver antenna terminals. This value will vary for different receivers. Better receivers have more IF and RF amplification stages than do their less expensive counterparts. They require less signal. Signal level is measured by using a signal strength meter. A typical meter is shown in Figure 11-2. This device is a self-contained tuning system and IF system. The outputs of the field strength meter are a speaker and an indicating meter. The unit is tuned to the required channel. A meter reading indicates the relative strength of the received signal. Audio information is heard through the speaker. This aids in being certain that the proper signal is being monitored.

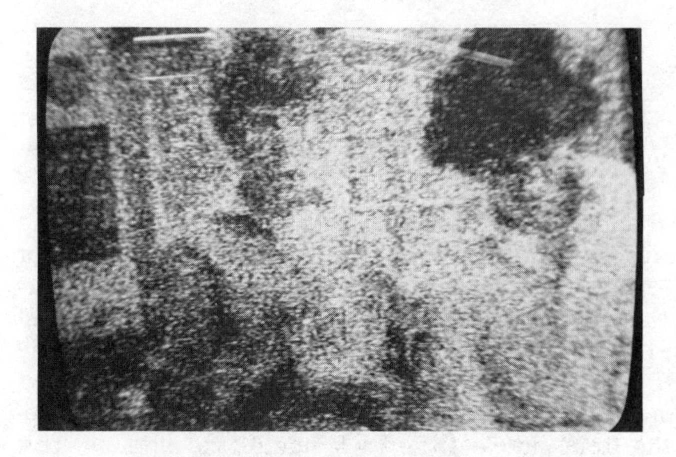

Figure 11-1 High levels of atmospheric noise and a low level of signal produces this "snow" in the picture.

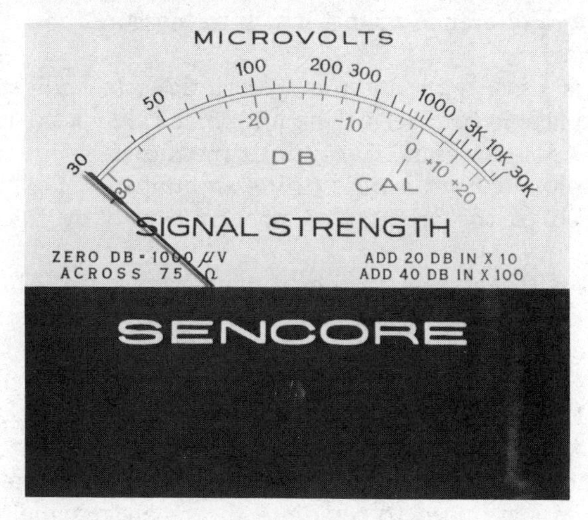

Figure 11-2 A signal strength meter is used to measure the amount of signal at the antenna. (Courtesy of Sencore, Inc.)

ANTENNAS

The electromagnetic energy broadcast by the transmitting antenna is induced onto the receiving antenna. The process of electromagnetic induction develops a voltage on the metal rods of the antenna. This voltage is commonly referred to as the "signal." Each antenna is tuned to the wavelength of the transmitted signal. Maximum signal is picked up by the antenna when its length is equal to one-half of the length of the electrical wave. This length is about 9 ft for channel 2 and close to 0.5 ft for channel 60. Antenna element length is reduced as the frequency of the signal increases because the wavelength of the signal also decreases as the frequency rises.

The reception of one channel's signals is accomplished by an antenna that is made to the specific electrical wavelength of the transmitted signal. The basic antenna is called a dipole. It is shown in Figure 11-3(a). This length

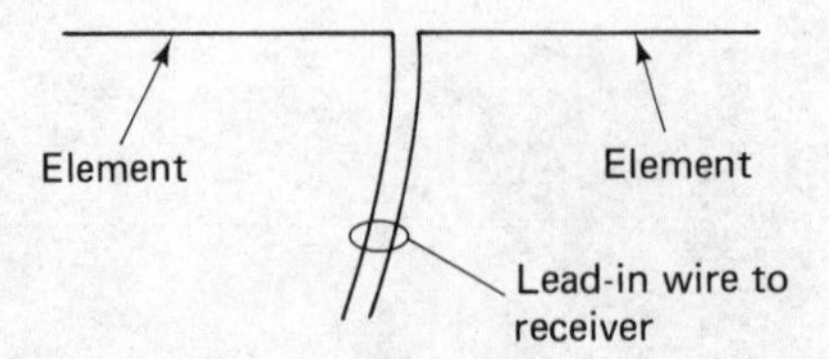

Figure 11-3 The dipole antenna has one pair of elements. These are connected to the receiver by use of a transmission-line lead in wire.

is calculated by use of the formula $L = 462/f$. Length L is figured in feet and f, the frequency, is figured using megahertz values. The reception pattern for this antenna is illustrated in part (b) of the figure. Maximum reception occurs when the elements of the antenna are at right angles to the transmitter antenna. The signal when one end of the dipole is pointed at the transmitting antenna is close to zero. The figure 8-shaped center part of the figure shows the receiving pattern for a dipole antenna. Dipole antennas have equal reception patterns on both the front side and the back side of the antenna. This type of antenna is used as a reference against which to measure the reception qualities of other antennas.

The term "gain" is used with antenna descriptions to indicate how much signal the antenna is able to receive. The reference value is the amount of signal received by a dipole antenna. The gain is measured in units of the decibel. This unit is developed from a ratio of the amount of signal received by the antenna compared to the amount of signal received by the dipole antenna.

Another term used with antennas is the front-to-back ratio. Almost all television antennas are directional. These antennas are designed to receive more signal when pointed at the transmitting station's antenna. The ratio of the amount of signal received in this position to the amount received when the antenna is pointed 180° away from the transmitter is the front-to-back ratio. Ideally, this should be a very high value.

The efficiency of an antenna is improved when additional elements are added to it. These elements are used to reradiate the received signal to one specific element of the antenna. This element is called the *driven element*. It is connected electrically to the receiver by use of a cable or transmission line. The length and spacing of the additional elements is determined by the frequency of the received signal. Elements that are placed behind the driven element are called *reflector elements*, as seen in Figure 11-4. These elements

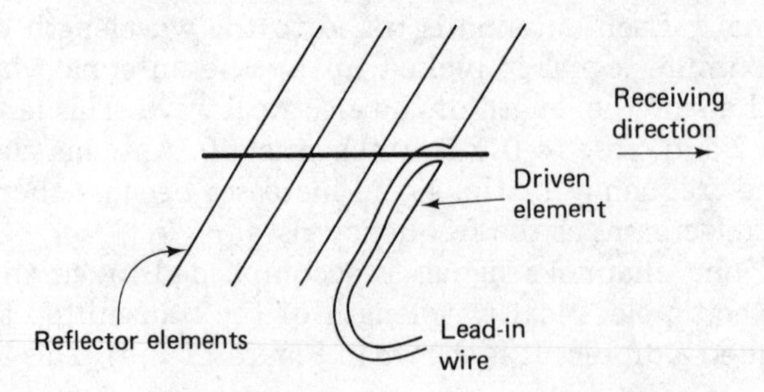

Figure 11-4 Additional gain is developed in the antenna by use of reflector elements.

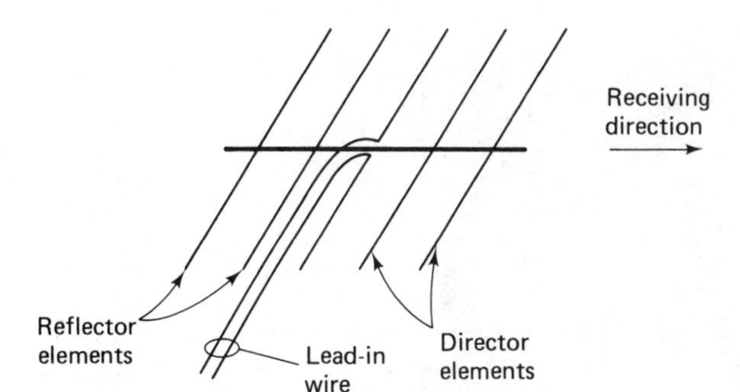

Figure 11-5 Most antennas use both director and reflector elements.

pick up the broadcast signal and reradiate it toward the driven element. Reflector elements are usually slightly longer than the driven element.

Another way of increasing the gain of the antenna is to add *director elements.* Director elements are placed on the front side of the antenna. These elements also pick up the signal from the broadcast station. The signal is reradiated or directed toward the driven element. Director elements are usually slightly shorter in length than the driven element. A typical antenna array is shown in Figure 11-5. It has director and reflector elements in addition to the driven element. This type of antenna is called a yagi antenna.

In locations that are able to receive only one channel, the yagi antenna is the best type to use. Its elements are designed to receive a maximum signal at one set of frequencies. In locations that have multichannel reception, another style of antenna is used. This multiband antenna is designed to receive all channels equally well. The antennas illustrated in Figure 11-6 are some of these all-channel types. In some areas the antenna is developed so that certain channels are given preference. These antennas will be tuned to receive all the channels, but they will specifically provide good reception for those channels in the geographic area.

The orientation of the antenna is important. The maximum amount of signal is induced onto the antenna when the receiving antenna is pointed at the transmitting antenna. This is not always possible. In some communities

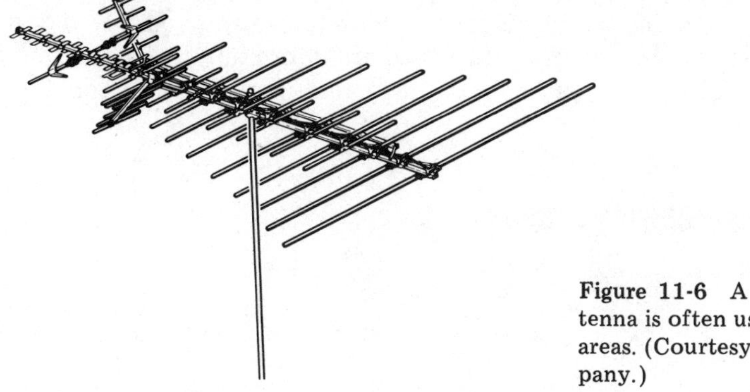

Figure 11-6 A multichannel antenna is often used in metropolitan areas. (Courtesy of Winegard Company.)

Figure 11-7 Reception of a reflected signal produces a "ghost" picture.

all transmitting antennas are clustered in one area. In other communities these antennas are separated. The result of the second set of conditions is that a compromise must be made as to antenna orientation. Another way of resolving this problem is to use an antenna rotator. Then the antenna can be turned to receive maximum signal from any direction.

A problem in areas that have tall structures is that the signal is often reflected off these structures. It then arrives at the receiving antenna a few microseconds after the direct signal arrives. This situation is shown in Figure 11-7. Both signals are on the same frequency. Both are received and processed by the receiver. The viewable result is a ghostly double image on the picture tube. These signals are called "ghosts" because of this effect. The only way to reduce the ghost image is by antenna orientation. A compromise position has to be found so that the second ghost signal is minimized.

FEED LINES

The signal induced on the receiving antenna has to be brought to the antenna terminals of the receiver. This is done with a transmission line. There are two basic types of transmission lines that are commonly used for this purpose. These are called twin lead and coaxial cable. Each is shown in Figure 11-8.

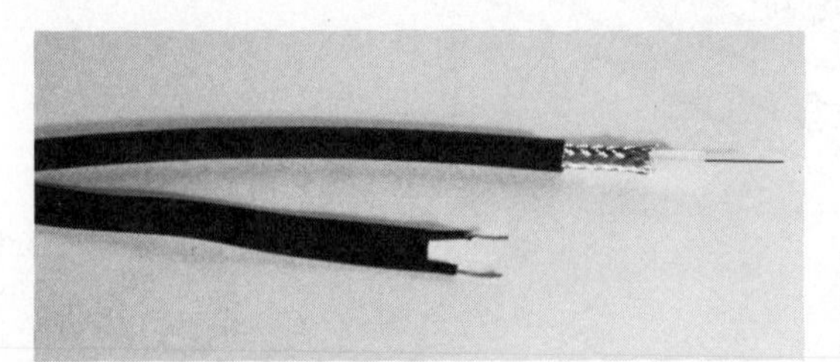

Figure 11-8 Two common types of transmission line are twin lead and coaxial cable.

Each of the transmission lines has some advantages and some disadvantages. Both are discussed in this section.

Twin lead. The nominal impedance of most receiving antennas is 300 Ω. Twin lead is available with 75-, 150-, and 300-Ω impedances. Almost all antenna installations use the 300-Ω type. Twin lead looks like a flat ribbon. It has two conductors. These conductors are spaced on ¼-inch centers. The covering for the wires and the material that holds it all in place is a polyethylene plastic. The system for twin lead is called a balanced transmission line. Both of the wires used in the twin lead are insulated. Neither is connected to circuit common. A center-tapped input transformer is used to couple the signal to the input of the tuner.

The advantages of twin lead are related to cost and convenience. It is the least expensive of the various transmission lines in use. It does not require special connectors. All one has to do is to bare the ends of the wires, form them around the terminal connector, and tighten the connectors. This leads to a simple installation.

The disadvantages of twin lead include weather and electrical noise. The plastic sheath around the conductors tends to deteriorate in the sunlight. This leads to the development of small cracks in the plastic. Dirt and moisture collect in these cracks. This reduces the efficiency of the transmission line to a high degree. The end result is that less signal is able to get to the input terminals of the receiver. The fact that both wires are "above common" electrically will make the pickup of electrical noise easy. This noise is then carried to the receiver along with the signal. Another factor to consider when using balanced-line twin lead is that it cannot be run along grounded surfaces, such as gutters. This tends to attenuate the signals and reduce input levels at the receiver.

The discussion about twin lead is not presented to discourage its use. Instead, one should be aware of the limitations of the wire. Use it carefully and it will do an excellent job. There is more twin lead being sold and installed than any other type of transmission line.

Coaxial cable. An unbalanced type of transmission line that is gaining wide acceptance is coaxial cable. Coaxial cable is two conductors. One of these is inside the other one. The inner conductor is centered inside the outer conductor by means of a plastic insulator. The outer conductor is covered with a plastic jacket. Usually, the inner conductor is a stranded copper wire. The outer conductor is a spiral-wrapped or woven braid. Coaxial cable used for television transmission line has a nominal impedance of 72 Ω.

The main advantage of coax is that it is able to shield outside electrical interference from being picked up by the transmission line. This is particularly true in urban areas, where there are many signals being broadcast that are undesirable for television reception. Another advantage is that coax is not bothered by weather.

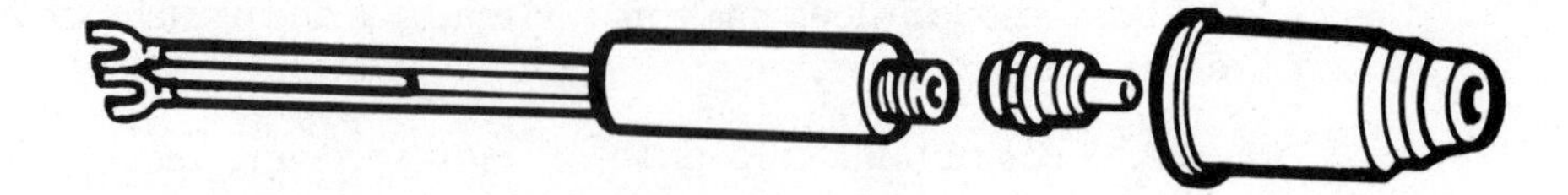

Figure 11-9 An impedance-matching transformer is used at the antenna and receiver with a coaxial-cable transmission line. (Courtesy of Winegard Company.)

The cost of coaxial cable is higher than that of twin lead. However, when one considers the *total* cost of a twin-lead type of installation, including standoffs and the extra time required to install them, the total cost is not too much higher. Most antennas, having a 300-Ω impedance, will require a matching transformer at the antenna terminals. This unit is small and mounts on the antenna. It is a matching type of transformer. It matches the 300-Ω antenna impedance to the 72-Ω coax impedance. One such transformer is shown in Figure 11-9. If the receiver is not designed to match the coax input, a second matching transformer is used.

Coax does not have to be kept away from radiating power lines as it is installed. It can be run next to grounded metal pipes or gutters without any attenuation of the signal. Standoffs are not required for installation. The coax may be stapled, taped, or even tied to any convenient support without fear of signal loss. One does have to be careful not to pinch the cable or to have too sharp a bend in it during installation. Either of these factors will introduce some losses into the transmission line.

Transmission-line selection. Coax is made in several sizes. Almost all of the cable used for television is of 72 Ω impedance. This cable is generally available in two physical sizes. The smaller of these is about ¼ inch in diameter. It is called RG59U. The second size is about ½ inch in diameter. It is called RG11U. Factors affecting the selection of either the proper coaxial cable or twin lead are frequency range, loss of signal in the line, and weather conditions. These are illustrated in Figure 11-10. This shows the line losses for 100 ft of transmission line. Losses for both wet and dry conditions are shown, as well as the effect of frequency on the signal. Two types of twin lead are used as reference. The flat type is the one most often used. Oval twin

Transmission line loss (db loss per 100 ft)

Type of line	100 MHz		500 MHz		1000 MHz	
	Wet	Dry	Wet	Dry	Wet	Dry
450-ohm open-wire	—	0.35	—	0.78	—	1.1
300-ohm flat	7.3	1.2	20.0	3.2	30.0	5.0
300-ohm tublar	2.5	1.1	6.8	3.0	10.0	4.6
RG-11U	—	1.8	—	5.0	—	7.6
RG-59U	—	3.8	—	9.4	—	14.2

Figure 11-10 Characteristics of different types of TV transmission lines.

lead has an oval-shaped web between the conductors. This reduces line losses due to weather and dirt effects.

The three frequency ranges are shown because these represent the range of frequencies used by television stations. Channel 2 uses frequencies of 54 to 60 MHz. Channel 13 has a frequency assignment of 210 to 216 MHz. The low end of the UHF band is 470 to 476 MHz and the upper end, at channel 83, has a frequency assignment of 884 to 890 MHz. Use of the proper transmission line will help to bring most of the signal from the antenna to the receiver. At times a trade-off must be done. The losses for twin lead at the higher frequencies are much less than for coax. However, if the twin lead gets wet often, then its losses are higher than coax. One must take this into consideration when making an installation.

DISTRIBUTION SYSTEMS

There are many times when one antenna is used to bring the signal to more than one receiver. In these installations a line-splitting device is used to separate the signal paths. This is required to maintain the impedance match between antenna and receivers. A mismatched system introduces losses of signal. This, of course, is undesirable.

One must consider the losses in any system when making an installation. A simple line splitter, or two-set coupler, could reduce the signal at each receiver to half of that measured at the antenna. A good-quality line splitter uses a transformer to keep both the impedance match and the proper signal level. When several receivers are to be installed, one should use an amplifier on the antenna line. There are several systems available for this purpose. A system using a line amplifier is shown in Figure 11-11. A master amplifier is installed between the antenna and the receiver system. Each of the components used in the system has a "loss." This is usually determined by the manufacturer. It is rated in decibels (dB). The loss for each device is added to determine a total loss for the system. The amplifier is designed with a gain. It is also rated in dB units. The amount of gain in the amplifier must be equal to, or higher than, the losses in the system. In other words, one balances the other. In this way the total required signal level is delivered to each receiver in the system.

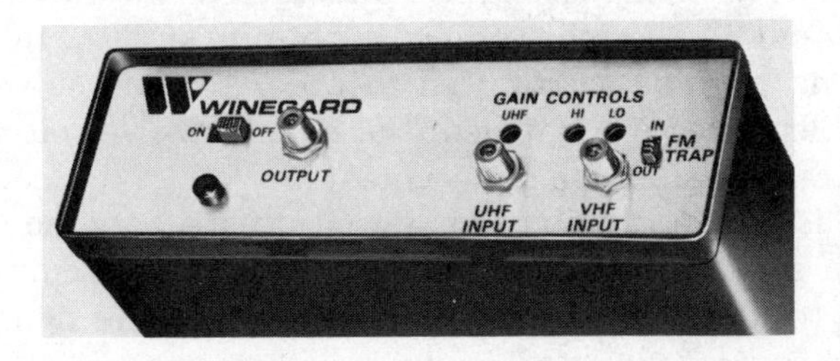

Figure 11-11 A line amplifier is used to distribute adequate signals to each receiver. (Courtesy of Winegard Company.)

Specific installation information as well as system design information is available from each of the manufacturers of these systems. When planning an installation that has several terminals, it is wise to discuss the requirements with the people who have this information available. The quality of reception after installation is dependent upon proper design.

TROUBLESHOOTING ANTENNA SYSTEMS

Diagnosis of the antenna system requires analysis of where the problem area is located. This is no different from the methods required for troubleshooting any other system. First, a word or two in general about problems in antenna systems. One must assume that the system was installed properly and did function as required. General problems related to antenna installations include badly oxidized antenna elements, broken lead-in connections, and deterioration of the transmission line.

There is a relationship between the condition of the antenna elements and the ability to induce a signal onto these elements. Corrosion and oxidation impede the induction of the signal onto the antenna. Most of the antenna elements in current use are made of aluminum alloys. This is done to reduce weight. Aluminum will oxidize easily. Antenna manufacturers often anodize the elements to seal them from oxidation. Manufacturers of large commercial communications antennas often provide a sealant coating to be installed after the antenna is assembled. This also reduces the effect of oxidation. Persons living in areas that have high levels of corrosive agents in the air have to replace antennas sooner than those that live in areas with low corrosive levels. The effect of corrosion on the elements of the antenna is a loss of signal at the receiver. The effect is snow in the picture and possibly a high noise level in the audio.

A second problem related to antennas is a broken transmission line. A complete break will also result in conditions similar to those just described. Often the break is not visible. This is because the plastic insulator on the lead-in does not break. In some instances only one wire will break. This is usually true when twin lead is used. Often, the break will be reconnected due to winds. A flashing in the picture is the viewable result of this condition.

The best way to test for either a bad antenna or transmission line is to use a field strength meter. The signal strength is measured at the antenna. It is also measured at the input to the receiver or distribution system. In some situations another antenna may be substituted in order to compare signal strengths. Once the problem area is identified, the repair involves either an antenna replacement or transmission-line replacement.

When the problem is in the distribution system, this is approached as with any signal-flow-path system. Locate the point where the system works properly. One may have to use a signal strength meter to measure signal levels at various tap-off points in the system. Analysis of the signal levels at different outlets will locate the problem area. Often, it can be found as a defective cable from the outlet in the wall to the receiver. In any event, proper troubleshooting analysis will help to find the problem.

QUESTIONS

11-1. How much signal voltage is required for a good-quality picture?

11-2. What is the electrical wavelength of a channel 2 signal?

11-3. What is the electrical wavelength of a channel 60 signal?

11-4. What is meant by the term "front-to-back ratio"?

11-5. What is the purpose of a director element?

11-6. What is the purpose of a reflector element?

11-7. What is a ghost?

11-8. Name the advantages of twin-lead transmission line.

11-9. Name the advantages of coaxial-cable transmission line.

11-10. What are two common antenna-related problems?

Chapter 12

The IF and Video Chain

One major breakthrough in the reception of electromagnetic signals occurred many years ago. This was the development of the heterodyne tuner. This tuner consists of the three blocks used in almost all tuners today. These blocks are the RF amplifier, the mixer, and the oscillator. All tuning is done in the oscillator and RF amplifier blocks. The output is a constant frequency signal. The output for a television receiver is 45.75 MHz. This is the video IF frequency of the receiver.

The advantages of this system are that only one knob is used for tuning and that much of the amplification of the signal is accomplished using fixed-tuned circuits in the IF amplifier. These make selection and amplification of the desired signal very easy. Before this system was used, each circuit had to be tuned in by the person operating a receiver. Fortunately for all of us, the heterodyne tuning system was introduced during the early days of radio reception. It was carried over to TV receivers. This system is used in almost every type of receiver currently in production.

IF AMPLIFICATION

IF amplifiers require several tuned stages of amplification. In practice, the stages are tuned to slightly different frequencies. This is done to pass the full 6-MHz bandwidth of the IF signal. A block diagram for a typical receiver is shown in Figure 12-1. The video IF information from the tuner is sent to the IF module. Several things occur in the module. These include IF amplifica-

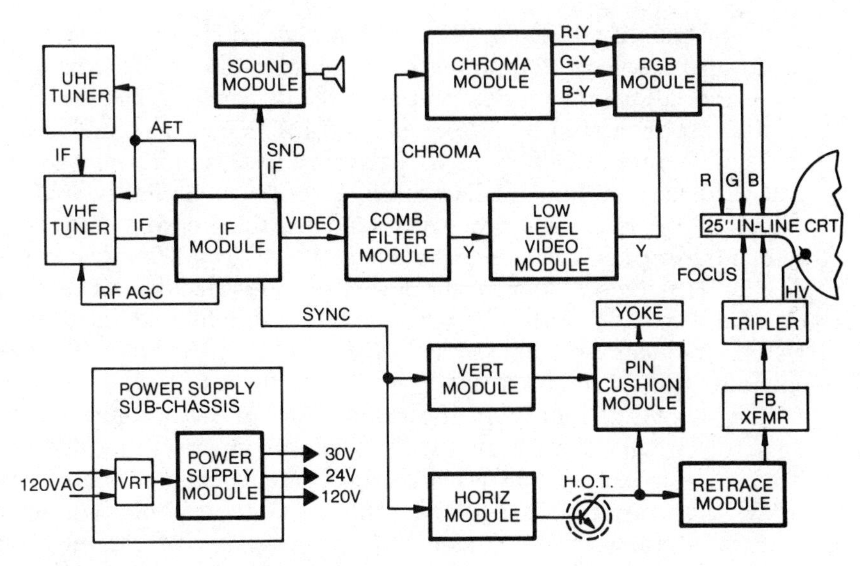

Figure 12-1 Block diagram of a modular TV receiver. (Courtesy of Magnavox Consumer Electronics.)

tion, video detection, AGC development, sync separation, and sound detection. An illustration of the blocks used in this module is shown in Figure 12-2. A resonant circuit is used between each stage of an IF amplifier. These circuits are stagger-tuned in order to pass the required 6-MHz signal. Almost half of the required signal amplification in the receiver is accomplished in the IF section.

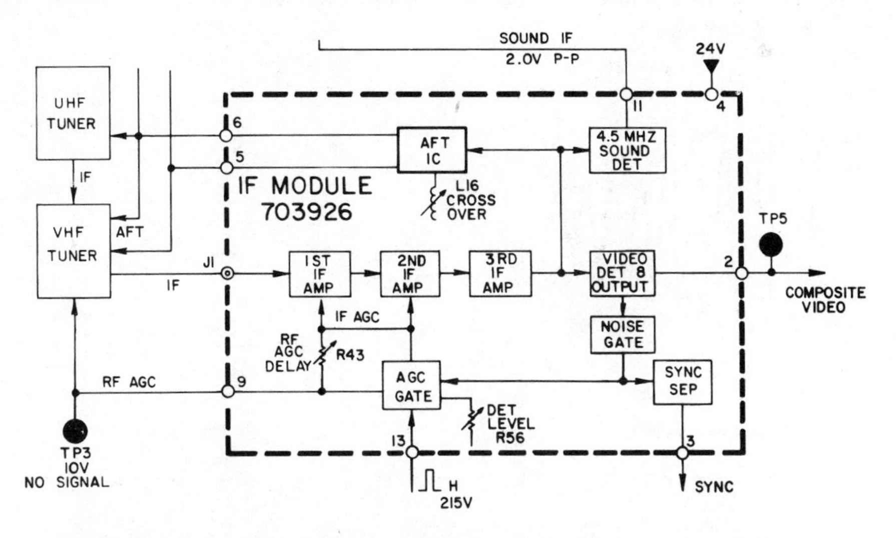

Figure 12-2 Block diagram of a solid-state IF amplifier system. (Courtesy of Magnavox Consumer Electronics.)

The amplified IF signal is sent to two different blocks. This signal is made up of several components. It includes a video IF carrier with video, sync, and color information and a sound IF carrier. The video and sound IF carriers are separated by a frequency of 4.5 MHz. This composite IF signal is sent to a sound detector block. The input to this block contains a mixer circuit. The circuit may utilize a mixer diode to heterodyne the video and sound IF carriers. This process produces a third carrier frequency. It is developed from the difference between the 45.75-MHz video carrier and the 41.25-MHz sound carrier in the IF section. This signal has a carrier frequency of 4.5 MHz. This carrier contains amplitude-modulated video information as well as frequency-modulated sound information. It will be processed by the sound section of the receiver.

A second output from the video IF amplifier is connected to the video detector circuit. The circuit used in early production receivers is very much like a power supply half-wave rectifier circuit. The action observed in this block is shown in Figure 12-3. The input to this block is an amplitude-modulated carrier. The upper and lower halves of this signal are mirror images of each other. Only one of them is required for the video signal-processing blocks. The detection process rectifies this signal. The output circuit of the detector includes a carrier frequency filter. This removes any of the remaining carrier so that all that remains is the modulation portion of the signal. It contains video, sync, and color information. Later production receivers do not use this diode video detector system. A circuit called a synchronous video detector is used. The reason for this is that the synchronous detector system reduces unwanted signals developed in other types of detectors. These unwanted signals produce interference patterns in the receiver video circuits.

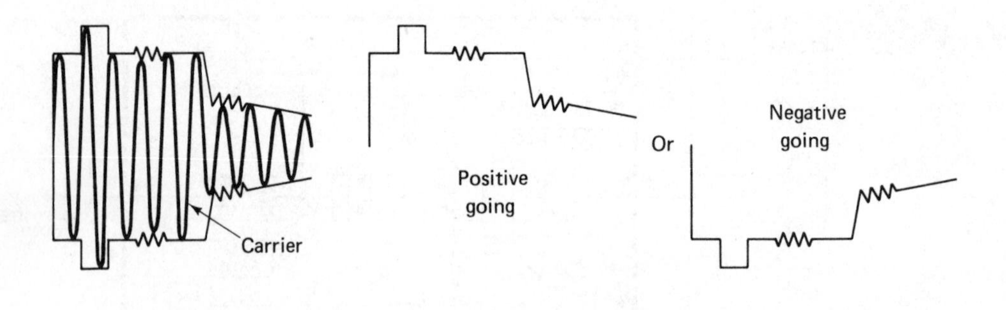

Figure 12-3 Composite video IF signal before and after demodulation occurs.

THE BASIC CIRCUIT

A very traditional system for an IF video chain is presented first. This system uses discrete components. The illustration shown in Figure 12-4 shows the first two stages of this system. There are six tuned circuits in this portion of the system. Those marked "trap" or "null" are used to tune out any undesired frequencies that are produced by mixer action in the tuner. The two stages are wired as common-emitter circuits. They will provide a fairly large voltage gain for the system.

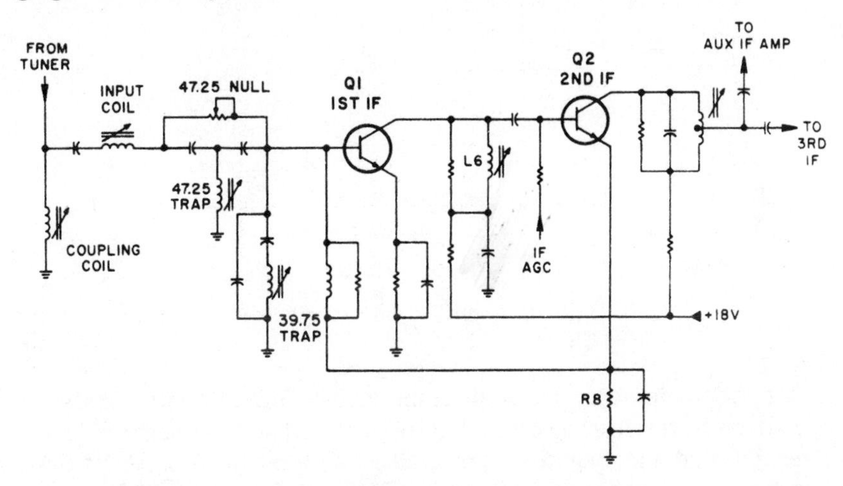

Figure 12-4 Schematic diagram of the first two stages of an IF amplifier. (Courtesy of Magnavox Consumer Electronics.)

The second stage in this system is directly controlled by IF AGC voltage. The base of transistor Q_2 is connected to the AGC line. The gain of the first stage is also controlled by AGC voltage. The voltage for this stage is developed as current flows through the Q_2 emitter resistor R_8. The control voltage for these two stages is developed in the AGC amplifier. It provides about 3.0 V to the base of Q_2. AGC voltage developed by the received signal in the video detector will reduce the gain of these stages.

The output of the second IF amplifier is coupled to the third IF amplifier. This stage is illustrated in Figure 12-5. This is the intermediate section of the luminance systems. It contains the third video IF amplifier, the video detector, and the video driver stages. There are two traps at the input to the third video IF amplifier. These tune out any of the 41.25-MHz sound IF carrier. This is necessary to keep sound information out of the picture information in the video amplifiers.

The third IF amplifier is also a common-emitter circuit. It is coupled to

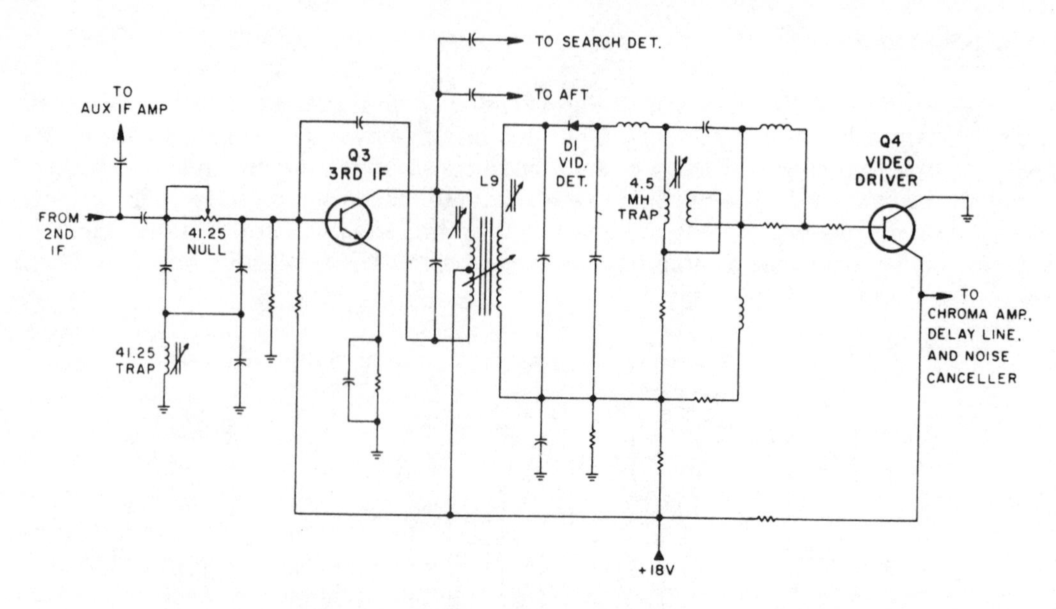

Figure 12-5 Third IF amplifier and video detector stages. (Courtesy of Magnavox Consumer Electronics.)

the video detector through transformer L_9. The video detector circuit is a half-wave rectifier system. The polarity of the detector diode D_1 develops a signal that has negative sync pulses. The purpose of the 4.5-MHz trap is to remove any remnants of the 4.5-MHz sound IF signal that may have developed during detection. A filter network is included in the detector circuit to remove any of the IF carrier signal that remains after detection. The composite video signal is then coupled to the video driver stage.

Video amplifiers. On a color TV receiver the video signal is sent to the video amplifier section. Here it is amplified to the full level required for a black picture. A delay line is installed in the video amplifier circuit after the takeoff point for the color signal. This delay line slows the arrival of the video signal at the picture tube by about 1μ sec. It establishes sufficient time so that the color information can be processed. Both video and color information have to arrive at the picture tube at exactly the same moment if a good-quality picture is to be displayed. The color-processing blocks require a few microseconds of time more than is required in the video section. This is the reason for the use of a delay line.

There are several methods for the processing of video information. The specific circuit is dependent on the manufacturer of the receiver. These next sections will explain some of these basic systems.

The video driver stage is a common-collector or emitter-follower amplifier. This produces a signal with unity voltage gain that is negative going.

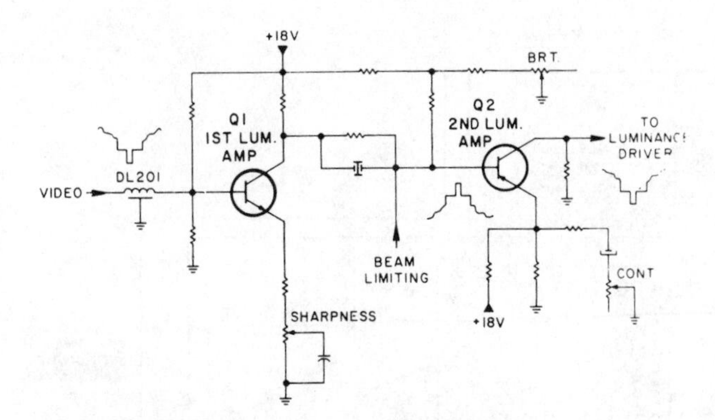

Figure 12-6 Video amplifier circuits in the receiver. (Courtesy of Magnavox Consumer Electronics.)

Emitter-follower amplifier stages do not invert signals. The output of this stage is coupled to the chroma and video circuits.

The first two stages of video amplification are shown in Figure 12-6. The video signal first passes through a delay line. Its purpose is to hold back the signal for a few microseconds. This allows the video and color information to arrive at the picture tube at the same moment of time. This is necessary due to the longer time required to process the color signal through several additional stages. The luminance or video signal is amplified in two common-emitter amplifiers. The output of this section is then processed by the luminance driver Q_3. This stage is an emitter follower. It is directly coupled to the output stage, Q_4. The signal is amplified and inverted in the output stage. It then goes to the cathodes of the picture-tube electron guns. One other feature of this section is blanking. Blanking pulses from the vertical and horizontal stages are applied from Q_5 to the base of the driver transistor. This biases the signal to black level during this time period. The viewable result is a black portion of the picture border.

IC CIRCUITS

The transition from discrete components to integrated circuits used essentially the same basic circuit as is used for sets with discrete components. The difference in the receivers is in the use of an IC to replace one or more stages of amplification. A schematic of an IC IF amplifier is shown in Figure 12-7. This receiver uses plug-in modules for major circuits. The system on this module uses one IC chip. There are three separate amplifier units in the chip. The signal is picked up from the tuner module at the IF input jack, J_2. It is then coupled through transformers T_1 and T_2 to the input of $U_{1\text{-}A}$. This input is shown as terminal 3 of the IC. The output from this amplifier is terminal 6. It is connected internally to the input to the second amplifier, $U_{1\text{-}B}$. A transformer, L_3, is used to tune this circuit to the IF frequency.

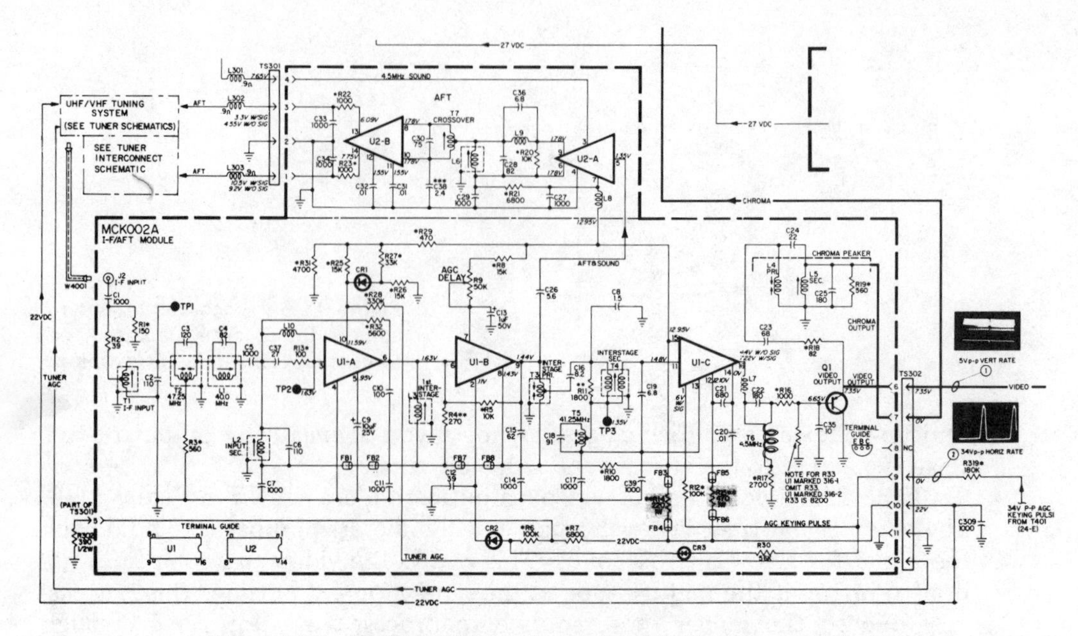

Figure 12-7 IC IF amplifier. (Courtesy of RCA Consumer Electronics.)

The output of the second IF amplifier is connected through T_3 and T_4 to the final stage, $U_{1\text{-}C}$. This stage also acts as the video detector. Its output, at terminal 16, is a composite video signal. The signal is processed to the next module and enters at pin 8 of the module connector. It then passes through the delay line DL and goes to the base of the first video amplifier, Q_3. Additional amplification is done by transistors Q_1 and Q_3. The output of Q_3 is then coupled off the board at terminal 15. It continues to the three color-output stages on another module. This signal may be mixed with color information and go to the cathodes of the picture tube. The circuits involved for this process are discussed in Chapter 17.

SAW FILTER IF CIRCUITS

One of the problems associated with tuned circuit IF amplifiers is the ability of the circuit to process all required frequencies. In addition to this, each frequency should have equal amplification. The ideal IF response curve looks like the illustration in Figure 12-8. It is "flat" for all frequencies. This, of course, is an ideal situation. Actual response curves will vary from this ideal. The result, to the viewer, is a picture that has a loss of detail, or sharpness.

The introduction of IC technology brought about an improvement in the amplifiers used in IF systems. A further refinement was the development of a new kind of IF amplifier system. This unit is called a surface acoustical

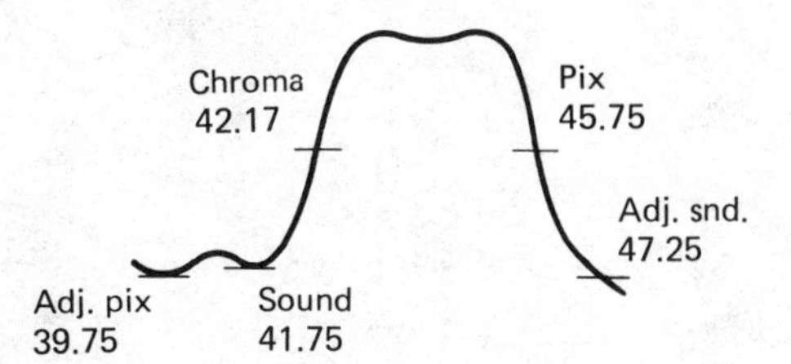

Figure 12-8 Equal amplification of all frequencies in the IF amplifier requires a flat response curve.

wave (SAW) filter. The SAW filter unit is made of a piece of piezoelectric or crystal material. This material has a property of distorting when a voltage is impressed upon it. The amount of distortion depends upon the quantity of voltage used. As the crystal material distorts, it also develops a voltage that is representative of the physical distortion.

The basic SAW filter is shown in Figure 12-9. It consists of two sets of electrodes. These electrodes are plated onto the surface of the crystal material. One of these is the input device, or transducer. The other is the output transducer. The plated filter components have varying lengths. These "fingers" are weighted during their manufacture. Weighting permits them to respond to different frequencies.

The signal voltage is connected to the input of the SAW filter. It is transmitted as a mechanical vibration at acoustical speeds to the output transducer. The signal is then coupled to other circuitry. The frequency response of this filter is determined during its manufacture. There is no way of changing this unless it becomes damaged.

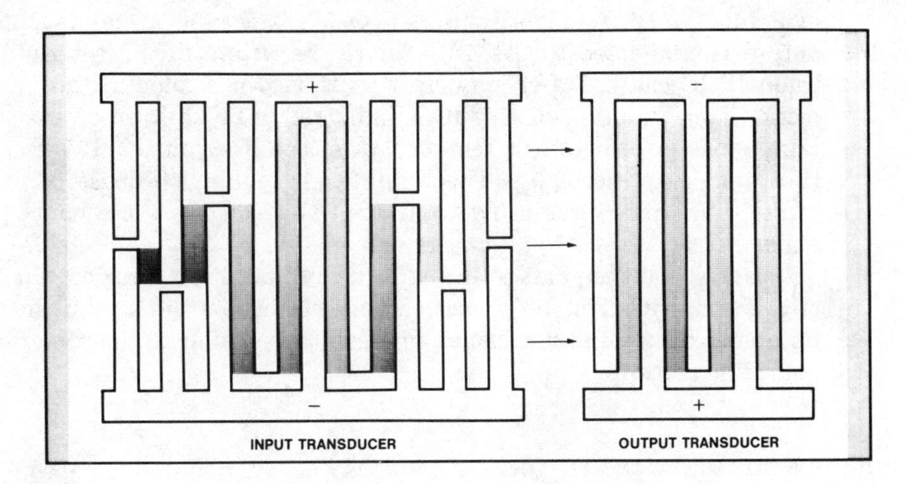

Simplified Illustration of SAW Filter Device

Figure 12-9 The SAW filter consists of plated "fingers" on a piece of crystal material. (Courtesy of RCA Consumer Electronics.)

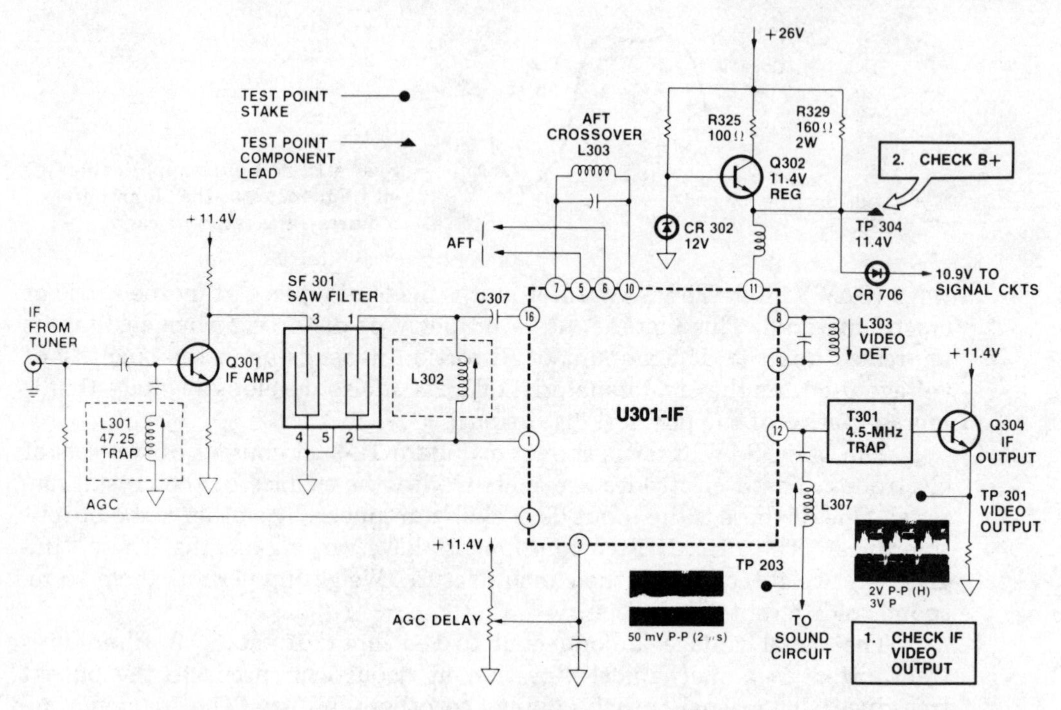

Figure 12-10 SAW filter IF amplifier system. (Courtesy of RCA Consumer Electronics.)

A partial schematic of an IF amplifier utilizing the SAW filter is shown in Figure 12-10. A single transistor, Q_{301}, is used as an input amplifier. Its output is connected to the SAW filter. The output of this filter is connected to an IF IC called the IF processor. It is shown in block form on the schematic. Each function of this IC is indicated in block form on the schematic. Luminance information is removed from the IC at pin 12. It then goes to an IF output amplifier Q_{304}. This amplifier is required because of signal losses in the SAW filter circuit. The output of this stage is a luminance, or video, signal.

The signal then passes through a delay line. From the delay line it passes into another IC. This IC processes both luminance and chroma information. Its operation for the luminance processing is similar to that described in the preceding section.

SYNCHRONOUS VIDEO DETECTORS

Diode detection systems require a relatively large signal voltage in order to operate properly. In addition, the nonlinear operational curve of the diode often allows it to operate as a mixer in addition to its demodulating function.

As a result, unwanted frequencies are developed during the detection/demodulation process. These produce undesired video interference. A solution to this is the use of a new type of detector called a synchronous video detector.

This device is a product of IC technology. It replaces almost all diode video detectors in TV receivers. It consists of two sets of identical amplifiers. A schematic for this is shown in Figure 12-11. The input signal is applied to the base of each of the two lower transistors. An 180° phase difference develops in this circuit. The outputs of the upper transistor pairs are connected in parallel. The lower transistors act as switches. They react to a modulating signal. The upper transistor pairs react to a carrier signal. The result is that the lower transistors switch on or off as the applied signal polarity changes. This is almost the same type of action accomplished by a detector diode. The result is a pulsating dc at the output of the IC. The pulsating dc is reacting to the modulating information. The waveform, therefore, is like that of a diode-detected signal. The advantage is that the negative side effects found with the diode are missing.

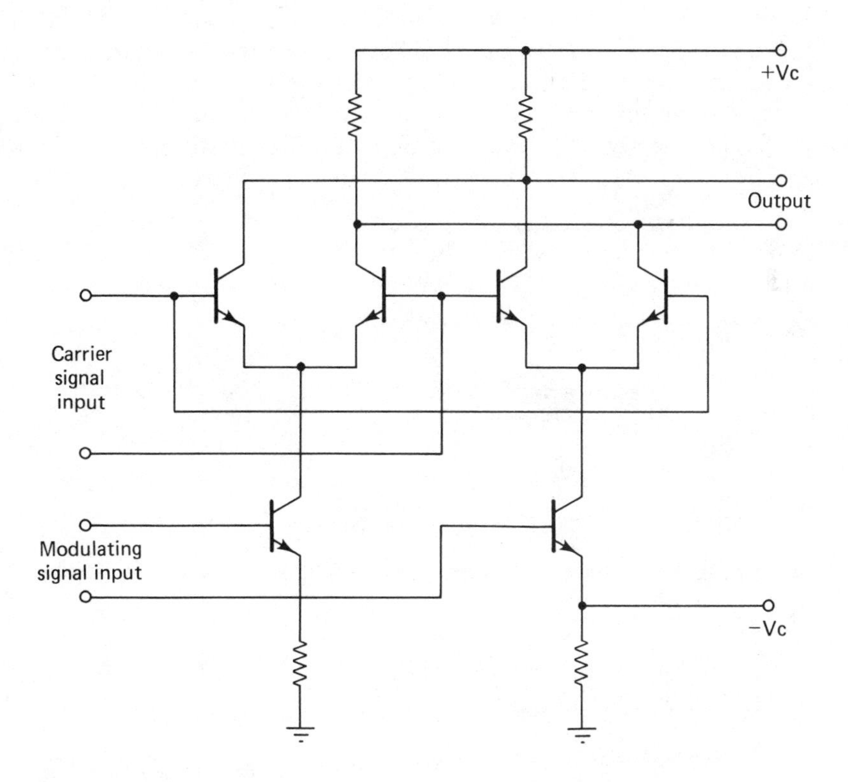

Figure 12-11 Schematic diagram of a synchronous video detector. (Courtesy of Electronic Technician/Dealer Magazine.)

COMB FILTERS

Another product of IC technology is the comb filter. This device is used to separate luminance and color information. The development of the comb filter was due to a need to overcome a form of interference. This interference looks like random colors moving on the screen of the picture tube, or CRT. High-frequency color areas, such as found when reproducing colors in vertical pin stripes or polka dots, tend to develop this interference. It is caused when some chroma signal mixes with the luminance signal. Normally, traps are utilized to minimize these signal mixes. The use of this trap tends to distort the frequency response of the luminance circuits.

Signals from the transmitter will often contain frequencies as high as 4 MHz. A response curve for a luminance channel is shown in Figure 12-12. The 3.58-Hz color signal frequency suppression will attenuate any luminence information as well. Luminance information tends to cluster around multiples of the horizontal scanning rate of 15.734 kHz. The chroma information will cluster around frequencies related to one-half of the horizontal rate. This develops an interweaving of signals, as shown in Figure 12-13. The requirement of a comb filter is to separate these two signals without introducing any distortion of either signal. The system used to separate these signals is shown in Figure 12-14. The left side shows a block diagram of the filter systems. A composite signal is presented to the input of the filter. There are three that the signal follows through the filter. One path is through a delay unit. This special delay unit will store one horizontal line of information.

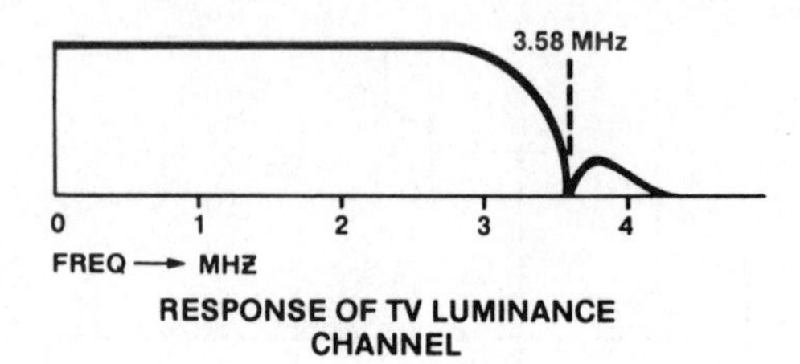

Figure 12-12 Frequency response for the luminance channel of a TV receiver.

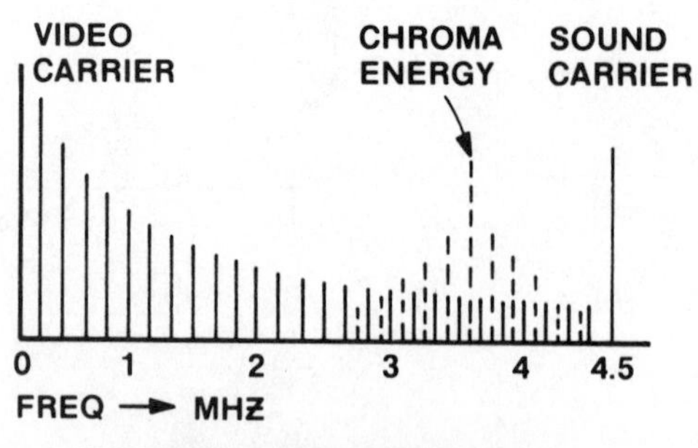

Figure 12-13 Luminance and chroma energy distribution on the TV channel.

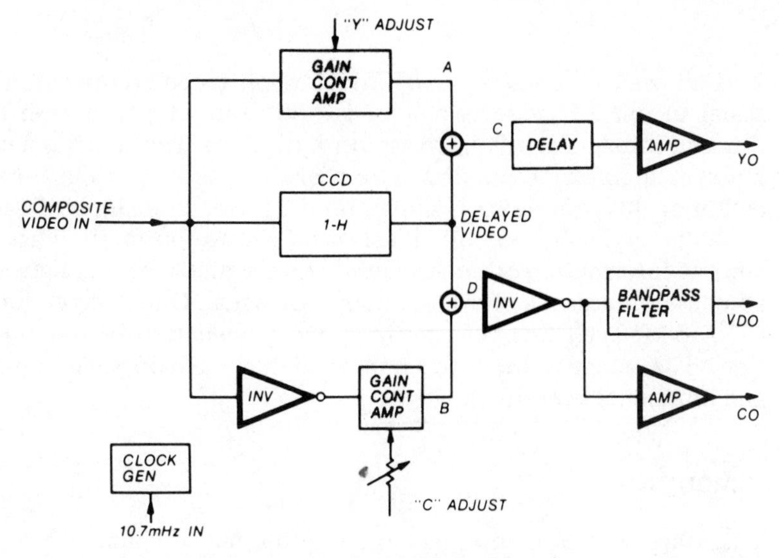

Figure 12-14 Block diagram of a comb filter. (Courtesy of RCA Consumer Electronics.)

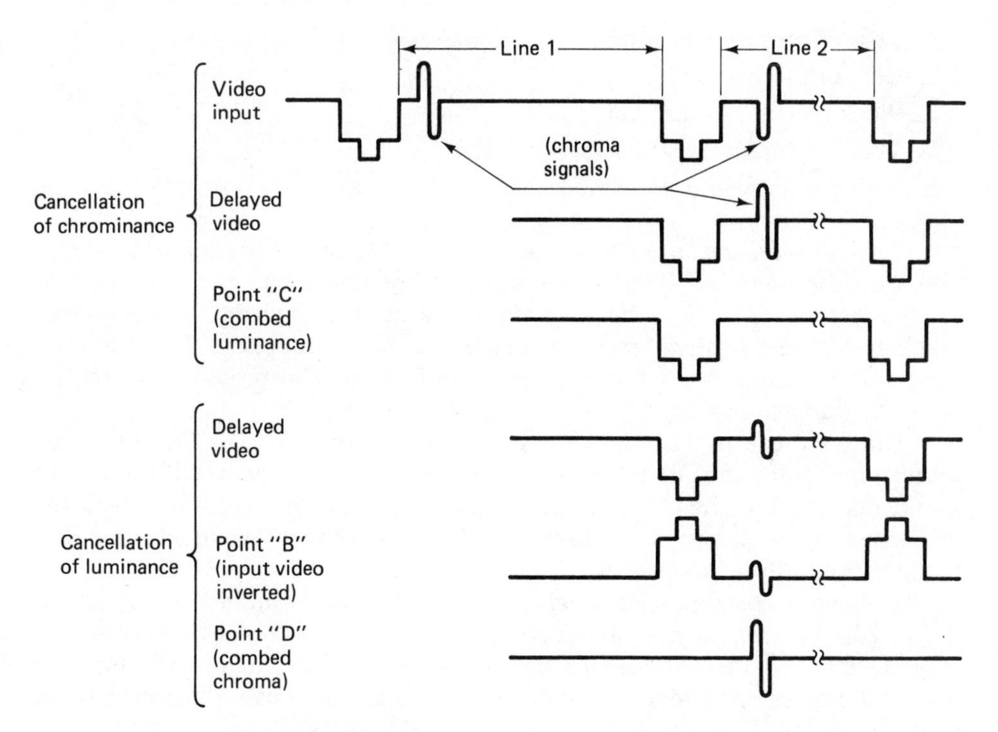

Figure 12-15 Signal processing in the comb filter. (Courtesy of RCA Consumer Electronics.)

This allows two lines of signal information to be compared. The luminance signal in one of these two lines is 180° out of phase with the other line. Chroma information is in phase on both lines. The result is a cancellation of luminance information and a reinforcement of chroma information. The action of this processing is shown on the lower three lines of Figure 12-15.

The upper half of this illustration shows the method used to comb the chroma information out of the luminance signal. Chroma signals are 180° out of phase on adjacent horizontal lines of signal. One delayed line is compared to a nondelayed line. The result is the cancellation of the chroma information. The outputs from the comb filter are fed to their respective chroma and luminance amplifiers.

TROUBLESHOOTING

There are several components to the luminance channel. These include the IF and luminance amplifiers and the detector. In addition, IF action is affected by AGC voltage. The first step in troubleshooting these systems is to determine where the area of trouble is located. There are several symptoms that are typically related to this system. The major ones are:

1. Loss of picture and sound
2. Loss of picture
3. Weak picture, sound, and color
4. Too much picture and buzzing sound
5. Sound bars seen in the video

Troubleshooting procedures described in Chapter 7 state that the trouble has to be localized. This is accomplished by eliminating working sections of the receiver. A schematic diagram and a basic knowledge of the overall block system used in any receiver makes this technique easy. There is little need for test equipment initially. Use the information provided by the receiver to start to localize the area of trouble.

Consider the signals that are processed by the blocks. When picture, sound, and color are all weak, the problem is located *before* the point of sound takeoff. On the other hand, when only picture and color are weak the problem is *after* this point. Thinking in this manner will save much time during the troubleshooting process.

It is wise to start with a valid test signal in all troubleshooting procedures. The test signal may be produced by a signal generator. Test signals may be injected at the input to the circuit. In the case of the IF amplifier one of the most convenient points is where the shielded cable from the tuner connects to the IF input.

One method used is to inject the signal at the point above. An oscilloscope is then used to trace this signal through the IF and video amplifiers. A second method is to apply the signal generator's output to each section. This requires moving the generator output cable to each section. The cable is then used as a probe in order to inject a signal at a specific point in the receiver. These methods are described in Chapter 7.

A "quick check" method for testing the IF and video amplifiers is to inject a noise signal. This does not provide accurate information. It does, however, quickly determine if a stage is functioning. The method used is very simple. The only piece of equipment required is a screwdriver or solder aid. This device is held in the technician's hand. The technician's finger is held on the metal end of the probe. The probe is then used to touch an input point in the circuit. The rest of the technician's body does not touch any part of the receiver, in order to eliminate any shock hazard. Injection of this type of noise will appear on the CRT as a form of video. This informs the technician that the signal processing from the test point to the CRT is functioning. An illustration of this procedure is shown in Figure 12-16. Keep in mind that this procedure is for quick-check purposes. It only works in signal-processing blocks of radios and TV receivers. Do not attempt to use it for scanning circuits.

The first item on the list is related to the loss of amplification in the system. Since picture, sound, and color levels are controlled by the amount of amplification, the problem could be related to a loss of AGC. AGC is a feedback type of system. The major signal path, however, is the forward path. Isolation of the problem is relatively simple. All that is required is to establish a valid AGC voltage level. This is accomplished by using a variable dc power source. The power source is adjusted to re-create the correct dc voltage normally produced by the AGC system. This voltage is inserted at the output of the AGC filter. If this procedure returns picture, sound, and color to normal levels, the problem is in the AGC circuit. If the problem is not cured, the trouble is not related to AGC. It is in the main signal path system. Normal signal tracing or injecting procedures may then be followed.

A detector/demodulator probe must be used with an oscilloscope when attempting to trace IF signals. The carrier frequency for these signals is 45.75 MHz. Almost all the oscilloscopes used in the repair of television receivers are not capable of displaying these high frequencies. The composite video signal has an upper frequency range of less than 6 MHz. This component is the modulation on the IF carrier. It may be observed by using a detector/demodulator probe. This probe acts in a manner similar to the diode video detector in the receiver. Its output is the composite video signal without any carrier.

When tracing any of the signals found in the receiver, be sure to include an observation relating to the proper polarity of the signal. Compare the ob-

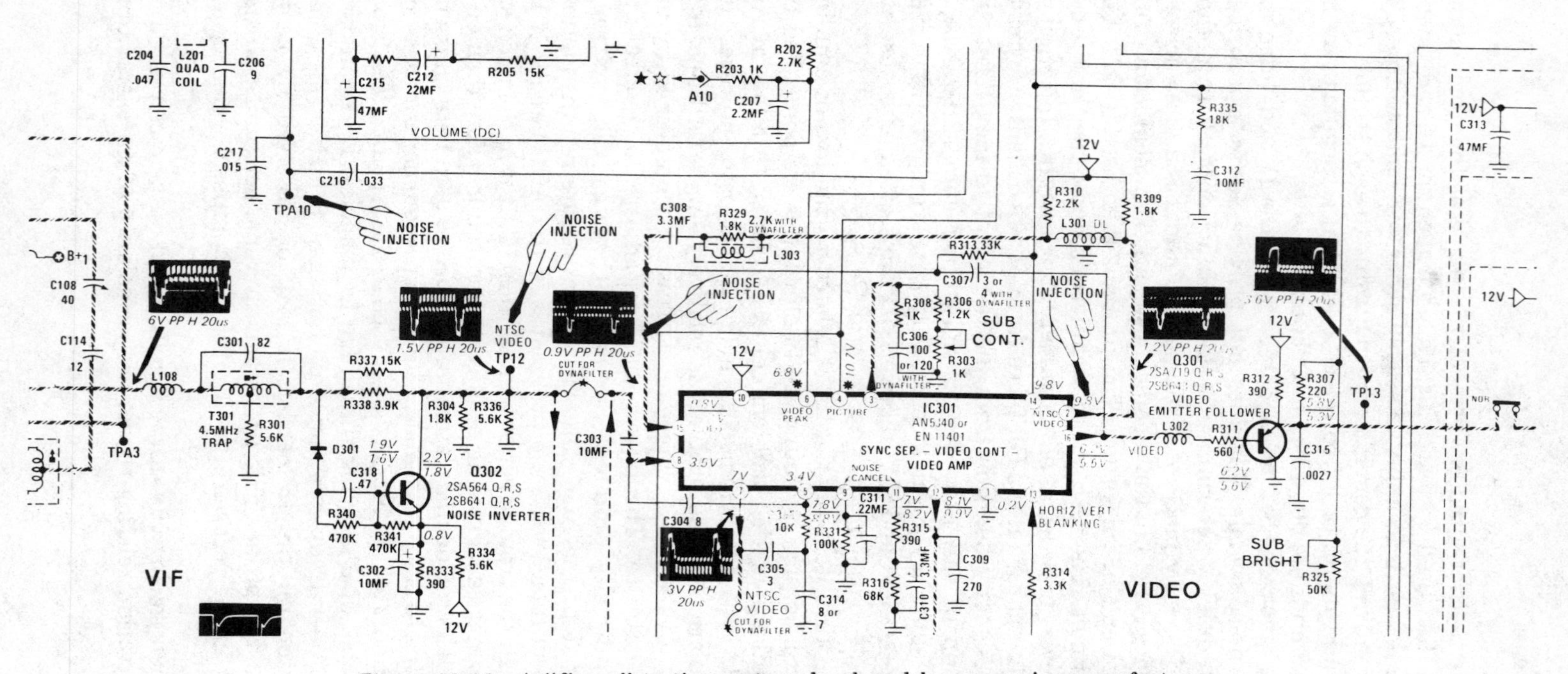

Figure 12-16 A "finger" testing system developed by one major manufacturer. (Courtesy of Quasar Electronics Corp.)

served waveform polarity with that shown in the service literature. Incorrect polarity indicates a defective amplifier stage in the system.

When the symptoms of the receiver are a loss of video at the picture tube and sound is good, the problem is located after the sound takeoff point in the receiver. A check for raster is one of the first steps to take. The presence of raster indicates that the CRT is working and that high voltage is present. Normal linear path signal flow troubleshooting with an oscilloscope is the easiest method to use in troubleshooting this section. Look for loose or broken leads at the contrast control in the early stages of troubleshooting.

The condition of too much picture and buzz in the sound may be encountered. This problem is related to excessive AGC voltage. This problem is discussed in Chapter 15.

The condition of sound bars on the video is one other relatively common problem. It is caused by a misalignment of the 4.5-MHz traps. The viewable result is lines or bars that pulse as the sound level changes. These are removed by adjustment of the 4.5-MHz traps. The traps are adjusted to null out any of these bars.

IC SERVICING

Once the problem area is localized, the technician has to identify a defective component. Signal testing will determine if the input to the IC is correct. The next step is to determine why the output is not correct. One cause for this is that the chip is defective. Before it is removed, one should check for valid operating voltages. A loss of operating voltage will also make the circuit inoperative. The schematic will provide information about these voltages. If all supply voltages are available, the problem is in the chip. It must be removed and replaced.

This procedure is typical for all IC circuits. In general, check for the presence of a proper signal at the input to the section. Also check for a valid output. If the output is missing, or low, the problem is located between the test points. Split the system in half, as is done for any linear signal system, and make another check. Repeat this procedure until one component or stage remains between the test points. Use normal test procedures to locate the bad component.

Special circuits, such as comb filters, require more servicing information. Each of these circuits is developed specifically for one manufacturer's receiver. The best method for troubleshooting these specialized circuits is described in service literature published by the manufacturer. Using this literature, which is available from the manufacturer's service or training department, is almost a necessity for these circuits.

QUESTIONS

12-1. What is the frequency of the video IF signal?

12-2. What is the bandwidth of the IF stages?

12-3. What information is contained in the video IF signal?

12-4. What are the three major sections of the video chain?

12-5. Why is a delay line necessary in a color receiver?

12-6. How does the SAW filter differ from other IF circuits?

12-7. What is the advantage of a synchronous video detector?

12-8. Why is the comb filter used in a color receiver?

12-9. Name four symptoms of video system failure.

12-10. Why is a demodulator probe required for tracing most IF signals?

Chapter 13

Picture Tubes

One of the two output devices of the television receiver is the picture tube. The electronic signal that is induced on the antenna is amplified and modified in the receiver. The result of the receiver action is a visual image on the face of the picture tube, or CRT. The image created by this action is not truly a full picture. It is, instead, made up of hundreds of individual elements of information. The scanning system in the receiver is used to develop these individual picture elements into the total picture observed on the CRT. The basic question that requires an answer is: How is this accomplished? Material in this chapter will answer this question.

VACUUM-TUBE OPERATION

One of the basic concepts that seems to be forgotten by many technicians is the principle upon which vacuum tubes operate. In general, the purpose of the vacuum tube is to create a stream of electrons. This stream strength is varied due to the operating conditions of the tube. A collector element is used to pick up these electrons and carry them to circuits in the receiver.

A schematic drawing for a vacuum tube is shown in Figure 13-1. This tube has four active elements. It is called a tetrode type of vacuum tube. Almost all vacuum tubes operate under these conditions:

1. Filament is heated
2. Control grid is most negative element

3. Plate is most positive element
4. Screen grid is second-highest positive element
5. Cathode at or near zero volts

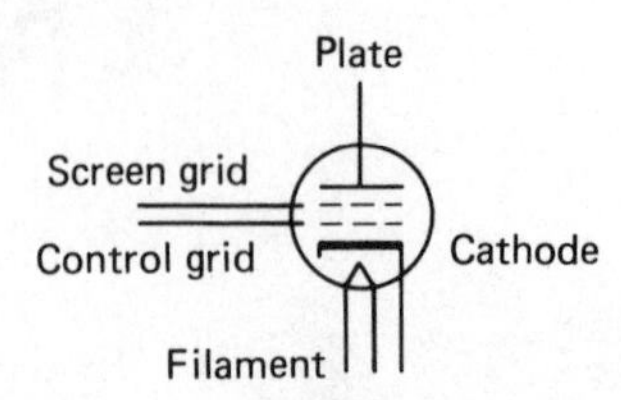

Figure 13-1 Elements of the tetrode vacuum tube.

When conditions similar to these are present, an electron current will flow from the cathode, through the tube, to the plate.

Several conditions affect the amount of current flow in the tube. First, these conditions occur only when a vacuum exists in the tube. The quantity of electrons that flow depends upon the number of electrons that are boiled off of the cathode and the amount of positive charge on the plate. The filament is used to heat a chemical coating on the cathode. This coating gives up free electrons when heated. The voltage placed on the screen grid is used to accelerate the electrons toward the plate element.

Under normal operating conditions plate, screen, and cathode heater voltages are fixed values. The element that usually has a variable voltage applied to it is the control grid. Varying control grid voltage permits a small voltage charge to control electron flow between cathode and plate. This is the origin of the name *control grid.* A vacuum tube does not really care which of its elements are varied for control purposes. The circuit shown in Figure 13-2 is probably the most common vacuum-tube circuit in use. In this circuit the signal is injected between control grid and common. The signal voltage adds to the fixed bias voltage established between control grid and common by the resistor in this circuit. The control grid resistor established an operating point for the circuit. The varying signal voltage changes the value of the charge on the control grid. This, in turn, establishes the level of electron flow in the cathode-plate circuit. As the charge in the control grid becomes more

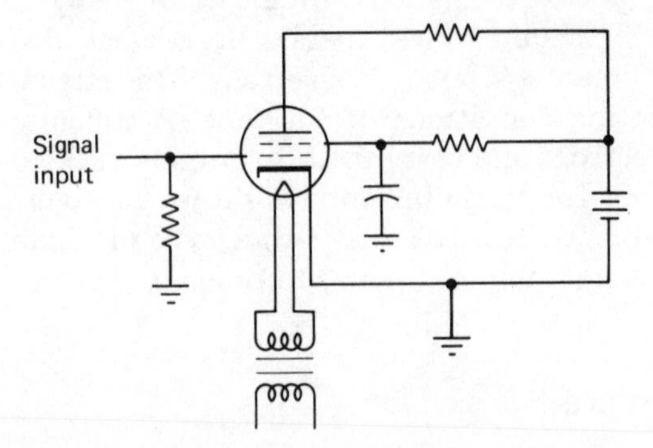

Figure 13-2 Vacuum-tube amplifier circuit with control grid signal injection.

negative, the electron flow slows to a halt. Making the charge on the control grid more positive increases the electron flow.

There is a very specific order of voltage values in a vacuum-tube circuit. The relationship of these values between cathode and control grid determine the quantity of electron flow. It is possible to vary the voltage on other elements and have similar control of electrons. Another common circuit is shown in Figure 13-3. In this circuit the signal is injected between the cathode and circuit common. The control grid voltage is held at a constant value. In this circuit it is at zero volts, or circuit common. The varying signal voltage changes the voltage at the cathode of the tube. This changes the relationship between cathode and control grid. The result is to vary the electron current in the tube. Both of these circuits are used in television receivers to inject a signal into the picture tube.

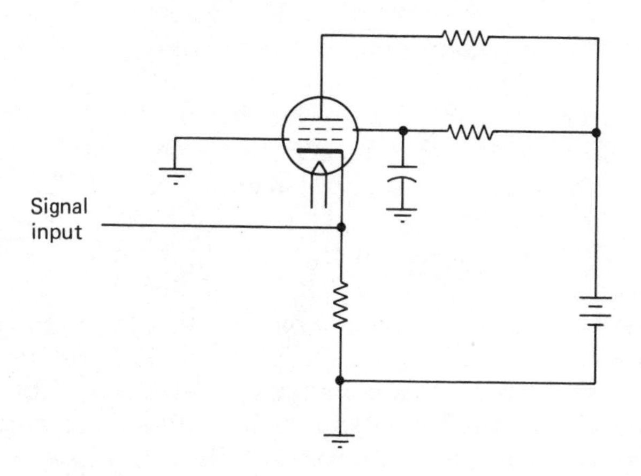

Figure 13-3 Vacuum-tube amplifier circuit with cathode signal injection.

PICTURE-TUBE OPERATION

Television picture tubes are actually one version of the cathode-ray tube (CRT). The purpose of this type of tube is to convert an electronic signal into a viewable object. Television picture tubes are produced in a variety of sizes and shapes. Fortunately, all of them operate on the same principle. The principle involved is that a stream of electrons striking a phosphorescent surface will cause the surface to glow. In other words, the electron stream is converted into a light source by means of the phosphorescent material. This principle applies to black and white as well as color CRTs.

Basic construction. The CRT used in a television receiver has a certain basic construction. An illustration of this is shown in Figure 13-4. The tube is made up of a glass envelope. The components inside the tube operate under a vacuum. The large end of the tube is slightly curved. It has a phosphorescent coating on its inner surface. In a noncolor tube this coating will glow

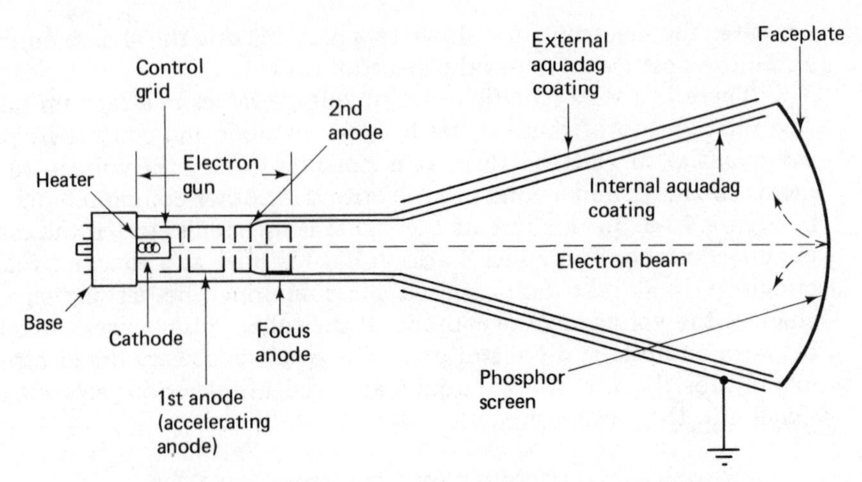

Figure 13-4 Construction of a CRT used in a TV receiver.

white when bombarded by an electron stream. The narrow end of the tube contains an electron gun. The electron gun is different from those found in receiving tubes. It does not have the usual plate element to collect electrons. The sides of the CRT are coated with a conductive paint. This coating is called *aquadag*. The inner and outer layers of this coating form a capacitor. The layer of glass that forms the body of the CRT is the dielectric material for this capacitor. This aquadag coating acts in a manner similar to the plate element of a receiving tube.

The electron gun's purpose is to create a stream of electrons. The elements of the gun that do this are the heater and the cathode. The heater, or filament, is used to raise the temperature of the cathode. The cathode is coated with a chemical that gives off free electrons when it is heated. This, by itself, is insufficient to develop the electron stream. Additional elements in the gun are required.

These additional elements serve several purposes. One of these is to create the stream of electrons. This is done by use of an accelerating anode. This anode has a high positive charge on it. It attracts the electrons from the cathode. It also tends to get the electrons moving away from the cathode so that other electrons may be given off from its coated surface.

There are two anodes associated with the CRT electron gun. These anodes are both used to accelerate the electron stream. In addition, there is a focus anode, or element. This anode is used to develop a very narrow stream of electrons. The purpose, again, is to create a stream of electrons and then direct this stream toward the coated faceplate of the CRT.

A high voltage is required to provide additional acceleration to the electron beam. A connection for the high voltage is made on the side of the glass envelope of the tube. The value of this voltage depends upon tube design. A general rule is that the level of high voltage increases with the physical dimensions of the tube. This voltage will range from a low of 600 V to a high of around 32 kV. The high voltage is developed in the horizontal output section of the receiver.

The electron beam requires some additional conditioning before a picture is created. This additional condition is the vertical and horizontal motions required for scanning. An electromagnetic device called a deflection yoke is placed around the neck of the CRT. This yoke has two pairs of windings. One pair is used for horizontal deflection of the beam. The second pair is used for vertical deflection. Positioning information for the yokes is provided by the vertical and horizontal scanning circuits in the receiver. Their output develops magnetic fields in the deflection yoke windings. The interaction of the two magnetic fields positions the electron beam in the CRT. Since the two magnetic fields have a constantly varying level, the electron beam is under constant motion. The observable result is the white raster observed on the face of the CRT. Deflection circuits are described in Chapters 18 and 19.

PICTURE-TUBE CIRCUITS

The operation of the picture tube is very similar to the operation of any other vacuum tube. Operating voltages for each element are established by the components associated with the tube. Power is obtained from the power source and horizontal output sections. A typical circuit for a black-and-white CRT is shown in Figure 13-5.

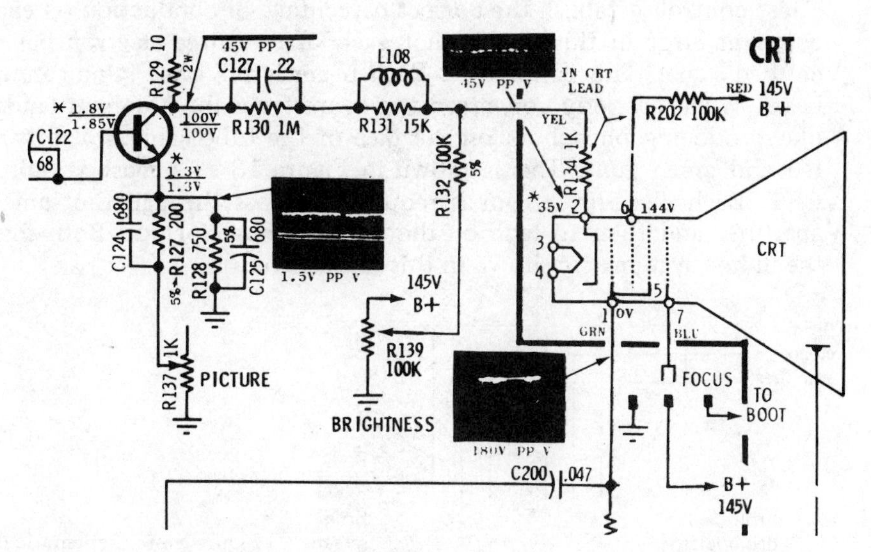

Figure 13-5 Schematic diagram of the CRT section of a TV receiver. (Courtesy of Quasar Electronics.)

The fixed operating voltages are applied to the CRT. The brightness control is used to establish a white level of raster on the CRT. This is a dc control voltage obtained from the power source. Signal level is controlled by the amount of amplification in the video output section. In this receiver a signal with a maximum level of 45 V p-p is used. When no signal is present, the cathode is positive with respect to the control grid. This permits electron-beam current and the CRT raster is seen. A signal is injected onto the cathode. This signal value is added to the dc voltage on the cathode. The result changes the level of cathode voltage. The exact value depends upon the amplitude of the signal. When no video information is present, the signal is a very high value. This keeps the CRT cathode in saturation and the raster is seen. As the signal amplitude increases, the resulting voltage at the cathode decreases. It will drop to a low value of -10 V. When the cathode is at this level, the tube is cut off. It does not conduct and a black raster is observed. The exact value of cathode voltage varies with the level of video information. It is different for each of the 525 lines of picture information. The varying level of voltage provides picture elements that range from black, though shades of gray, to white.

This same principle applies to color CRTs. The system in a color TV uses four signals. One of these is the video, or Y signal. It is usually presented to the cathodes in the tube. Most color TV picture tubes have three electron guns. Each has its own set of elements. A schematic diagram of the color TV picture tube is shown in Figure 13-6. Operating bias is established on each element in order to establish a "white" screen without any applied signal. This is usually done by adjusting "screen" controls on the back of the set. These controls establish the correct percentage of conduction on each gun. A common error in thinking is that each of the three electron beams send a colored signal. This is not true. Each beam is the same. One cannot see the beam. When properly converged the beam from the blue gun lands only on blue-producing phosphors on the face of the tube. This is also true for the red and green guns. This is shown in Figure 13-7. A mask is built into the CRT. Each electron stream is required to pass through the same hole, or aperture, and then to land on the correct phosphor dot. Both the dot and the in-line systems are shown in this illustration.

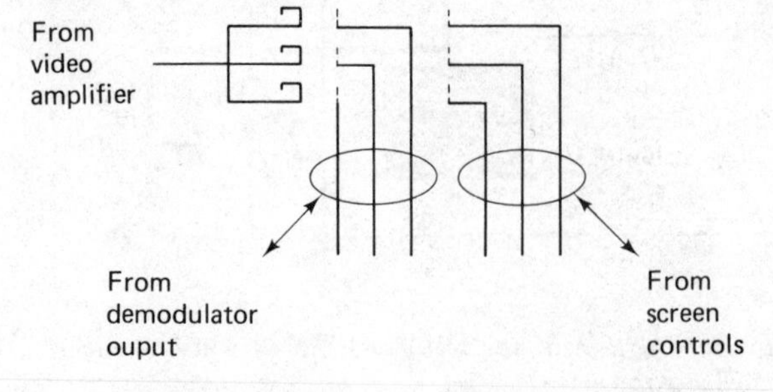

Figure 13-6 Schematic diagram of a three-gun color CRT.

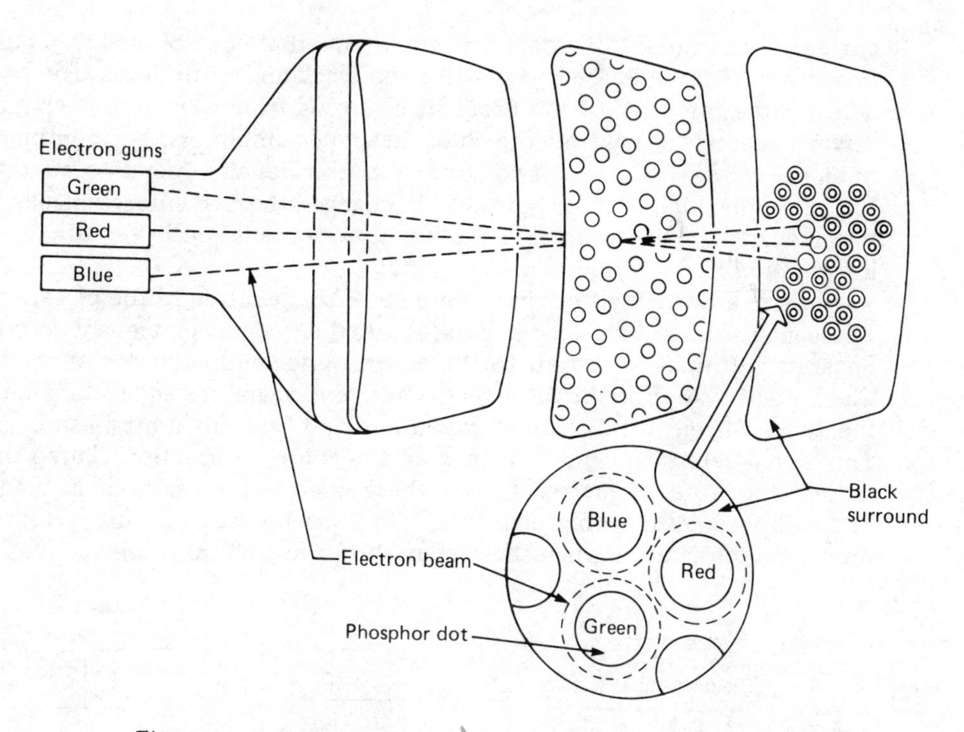

Figure 13-7 Basic shadow mask construction in the color CRT. Note the triad of color dots. (Joel Goldberg, *Radio, Television, and Sound System Repair: An Introduction*, © 1978, p. 144. Reprinted by permission of Prentice-Hall, Inc.)

The balance of the signals at the picture tube are called *color-difference signals.* These are waveforms which represent amounts of color information. One represents red, one green, and the third one blue information.

Let us look at the action of the three electron guns. Keep in mind that each gun acts independently of the other two. Picture tube setup establishes a ratio of 59% green, 30% red, and 11% blue. This ratio provides a white picture. This adjustment is related to the screen controls of the tube. Changing any one, or all, of these percentages will produce a color. The specific color depends upon the amount of electrons from the three guns. The initial setup of the tube establishes specific operating bias on the guns. This bias allows an electron flow from each gun to the face of the tube. Video information presented at the cathodes will cause a variation in the amount of electrons that flow. All three cathodes receive the same amount of Y signal at exactly the same time. This produces the black-and-white picture. The addition of color information is done at the control grid element. The color signal voltage varies the voltage on the control grid. This, then, also controls electron-beam current in the tube. Now there are two factors that control the beam current flow. One of these is used to control the black-and-white level. The second controls the intensity of the beam of electrons from each gun. Each gun receives a separate signal from the color-difference amplifiers. The combination of signal voltages from the two sources establishes the operating conditions for the individual electron guns. This changes the percentages of electron-beam current. The result is a picture that contains both video and color information.

There are two methods in common use for operation of the picture tube. These are shown in Figure 13-8. Part (a) uses the systems previously described. Separate information is sent to the control grids and cathodes of the tube. The second system is shown in part (b). Here, signal mixing is done outside the tube. An electronic circuit called a *matrix* is used to establish a signal. This signal represents both video and one-color information. Three signals are developed in the matrix. One of these is fed to each cathode in the tube. The cathodes are the only elements with a varying signal in this system. The results are the same as when the system shown in part (a) is used.

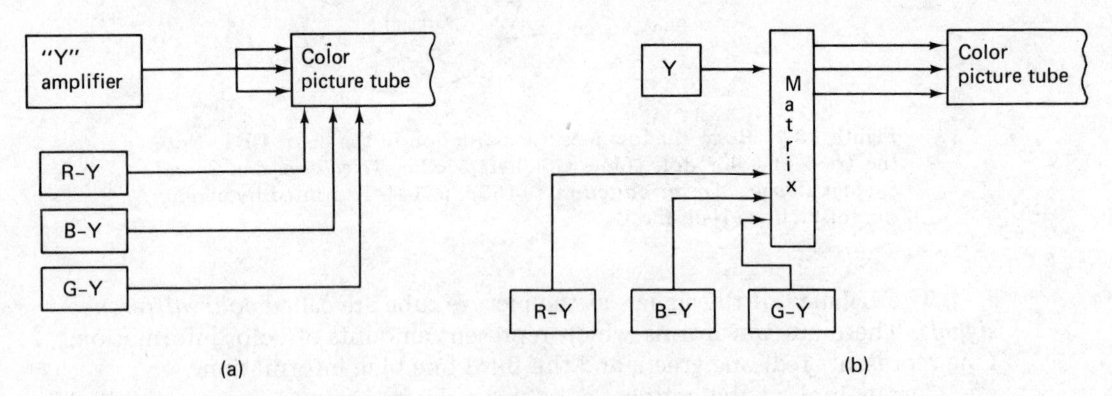

Figure 13-8 Two methods of introducing signal and color information in the CRT.

PICTURE-TUBE TROUBLES

CRT problems can be itemized under four headings. These are, in no order:

1. Shorts between elements
2. Open filaments
3. Low emission
4. Poor color tracking (color tubes only)

Let us examine each of these problems.

Shorts between elements. The elements of a CRT are very close to each other. The electron gun is constructed as a precision device. There are times when the elements may change position in the gun. This is usually the result of warpage of the gun structure due to heating and cooling during operation. A second cause for a short between elements is due to flaking of the conductive material in the tube lodging between two elements. In either case the normal voltages required for correct CRT operation are changed. Conduction, if it still occurs, is incorrect.

A black-and-white CRT will exhibit this problem with either a raster that is too bright or no raster at all. A color CRT will usually have the same types of problems. In the color CRT the problem is usually seen with only one color. The specific color depends upon which of the three guns was affected by the problem.

It is possible to repair some tubes that have shorted elements. A picture tube checker/restorer is used. One of these is shown in Figure 13-9. The procedure for repair of the CRT is described by the manufacturer of the test unit. Keep in mind that not every tube is repairable. In many cases the tube will have to be replaced. In essence, the repair attempts to "burn off" the short. This is done by applying a voltage to the shorted elements. Current flowing in the circuit will often melt the shorting particles. If no other damage is done to the tube, it may be returned to service.

Open elements. Another internal problem occurs when an element of the CRT opens. The open element will cause a dim picture. In cases where the

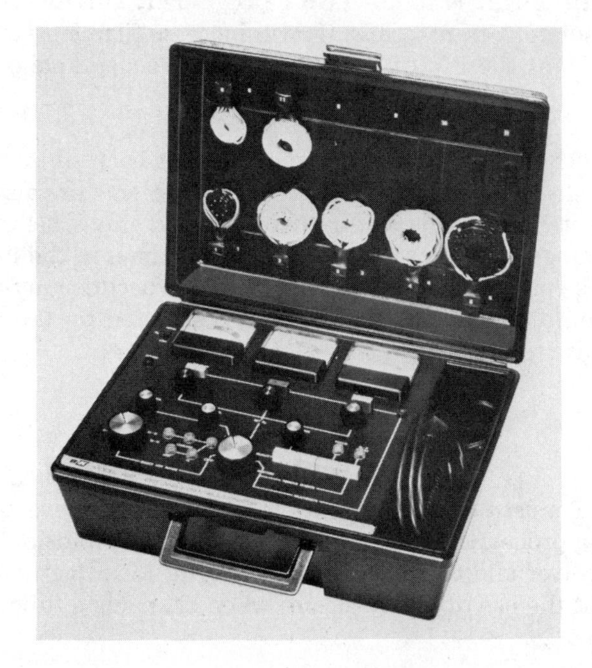

Figure 13-9 CRT tester-reactivator. (Courtesy of Dynascan Corporation.)

cathode element is open, there will be no picture at all. It is possible to weld open elements together in some situations. A fairly high voltage is connected across the open elements and the cathode. Gently tapping the neck of the tube will set up a vibration of the electron gun. This may permit the open element to touch its lead. The momentary contact will often weld them together.

Open filaments. There is no good cure for an open filament in a CRT. The fact that it is open is confirmed by measuring the resistance across the socket terminals. This should be a very low value if the filament is good. The only solution to this problem is to replace the CRT. Before this is done, check to be sure that the correct voltage for filament operation is present at the CRT socket. If the operating voltage is missing, the filament will not light. The result is exactly like that of an open filament.

Low emission. This condition is usually due to a failure of the cathode to give off sufficient electrons. The viewable effect of this is a weak picture with low brightness. In extreme cases the low brightness is accompanied by silver highlighting of objects.

There is a school of thought that says that low-emission tubes may be reactivated. The reactivation is accomplished with a CRT tester/rejuvenator. The filament is operated at an above-normal temperature and a high-voltage charge is applied to the tube. This may provide temporary relief for the problem.

A second school of thought is to use a CRT brightener. This device is a filament boost and/or isolation transformer. It will increase filament voltage by about 10%. This shortens the life of the CRT. A more acceptable procedure is to replace the tube.

Poor color tracking. The color CRT is designed so that each of its three guns will provide equal electron emission. The control elements are also designed with similar characteristics. This feature permits good gray-scale tracking as brightness levels range from low to high. There are times in the life of the CRT when one gun's emission level changes. When this occurs, gray-scale tracking cannot be accomplished. This feature may be tested on the CRT tester. Usually, the tube is replaced when this condition is found.

CRT REPLACEMENT

Replacement of the CRT must be done very carefully. The tube is evacuated during the manufacturing process. If it is cracked, the resulting implosion may cause damage to the receiver and to anyone who is near it. Manufacturers of picture tubes recommend the use of safety glasses when changing a tube. The procedure to follow is normally provided by the receiver manufacturer. In general, these steps are taken:

1. All interconnecting wiring is removed. Note where each wire should be reconnected before removing any of them!
2. Any obstructions, such as the receiver chassis, are removed and set aside.
3. The receiver is placed with its tube face down on a protected surface.
4. All mounting hardware is removed from around the tube. Note how each connection is made.
5. The old tube is carefully removed and placed upon a padded surface. Any mounting hardware is now removed from the tube.
6. The mounting hardware is placed on the new tube. It is then carefully placed in the cabinet and fastened in place.
7. All interconnecting wires and components are replaced in the same positions as they were found on the original tube.
8. The cabinet is turned upright and the chassis replaced. All parts are fastened securely.
9. Power is applied to verify that the new CRT works properly.
10. Degaussing and color setup are done. This is accomplished by following the directions in the service literature.

This should complete the installation of the replacement CRT. The receiver is checked to be certain that all parts are correctly mounted. All should appear as it did before the CRT was changed.

QUESTIONS

13-1. List the voltage ranking of a vacuum tube's elements.

13-2. Which element is usually used to control electron flow?

13-3. What is the purpose of the high voltage required for the CRT?

13-4. What is the function of the deflection yoke?

13-5. What is a raster?

13-6. What three color phosphors are used to create a white picture?

13-7. What percentages of illumination for each phosphor are required for a white raster?

13-8. How are the C and Y signals combined in the CRT?

13-9. Why is it necessary to degauss a new CRT?

13-10. What is the cause of low emission in a CRT?

Chapter 14

Sound Section

One of the two outputs of a television receiver is the audio, or sound, signal. This signal is developed at the transmitter. It is broadcast on a frequency-modulated carrier to the receiver. The sound signal carrier is established at a frequency that is 4.5 MHz higher than the picture carrier frequency. The sound signal is processed through the receiver tuner section together with picture information. Both picture and sound carriers are mixed with the local oscillator in the receiver. The result is a frequency-modulated sound carrier in the IF amplifier. This carrier is processed through the IF amplifier section until it reaches a take-off point in the receiver. This takeoff point will be located near the video detector block. The specific location depends upon whether the receiver is a color or a noncolor unit. Each system for sound takeoff is discussed in this chapter. In addition, the processing of the sound carrier, its ultimate demodulation, and the amplification of the remaining audio signal are discussed in this chapter.

BLACK-AND-WHITE SOUND SECTIONS

A block diagram for the sound section in a black-and-white receiver is shown in Figure 14-1. The picture carrier and the sound carrier in the video IF section of the receiver have a separation of 4.5 MHz. The two carriers are mixed in the video detector block of the receiver. The result is a carrier with a frequency of 4.5 MHz. This carrier is identified as the sound IF signal. It contains the frequency-modulated sound information. This carrier also contains

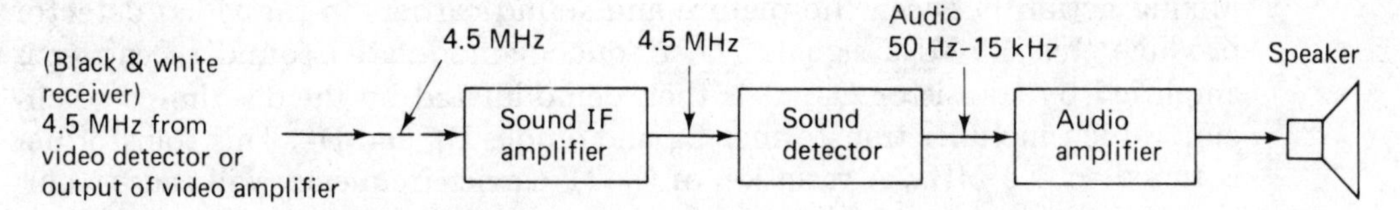

Figure 14-1 Block diagram of the sound system in a TV receiver.

the amplitude-modulated video information. This video information is not able to be demodulated in the FM sound system. It is usually ignored when discussing the sound section.

The 4.5 MHz sound signal is called the sound IF signal. It is coupled to the sound IF amplifier by a sound takeoff coil. This coil is resonant at 4.5 MHz. It picks up the sound IF signal from the video amplifier circuits. A schematic diagram for a black-and-white receiver using discrete components is shown in Figure 14-2. This is a partial schematic. It shows the section related to the sound processing.

The sound signal is taken from the collector of the video amplifier, Q_9. It is coupled to the sound takeoff coil L_4. This circuit is tuned to 4.5 MHz.

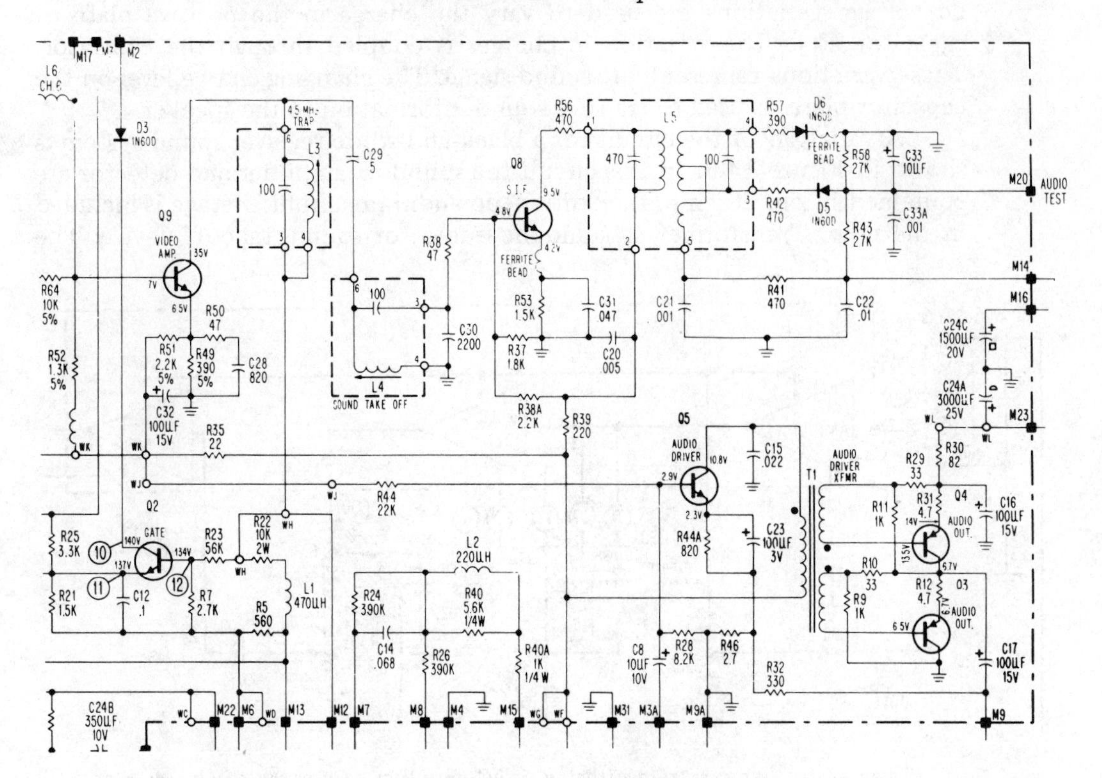

Figure 14-2 Partial schematic diagram showing the sound system in a discrete-component receiver.

Mixing action between the picture and sound carriers in the video detector produce this 4.5-MHz signal. The frequency-modulated sound IF carrier is amplified by transistor Q_8. It is then demodulated by the discriminator circuit, which includes transformer L_5 and diodes D_6 and D_5. This transformer is tuned to 4.5 MHz. A variation of the IF carrier frequency will cause a current flow through one of the two diodes. A frequency increase forward biases one diode. A frequency decrease will forward bias the other diode. The result of these frequency variations develops a current flow that resembles the original sound waveform.

The sound signal is connected to the volume control VR_{100}. It, in turn, is connected to the base of the first audio stage, Q_5. The audio signal is then transformer coupled through T_1 to the bases of the audio output transistors Q_4 and Q_3. These operate in a push-pull configuration. The audio signal developed in these power amplifiers is then coupled to the speaker LS_1 through capacitor C_{17}.

Signal is developed at the collector of output transistor Q_4 by the alternating conductions of the two output transistors. This is actually a varying dc voltage. It varies due to the changing resistance of each transistor. The dc voltage variations are used to vary the charge in the positive plate of capacitor C_{17}. The variation in charges is coupled through the capacitor. These variations represent the sound signal. The changing charge level on the capacitor plates is used to transfer sound information to the speaker.

A variation of the circuit for a black-and-white receiver sound system is shown in Figure 14-3. In this circuit the sound IF amplifier and detector are contained in one IC chip. In addition, an audio preamplifier stage is included in the chip. Transformer L_{109} is the audio, or sound, takeoff device. The

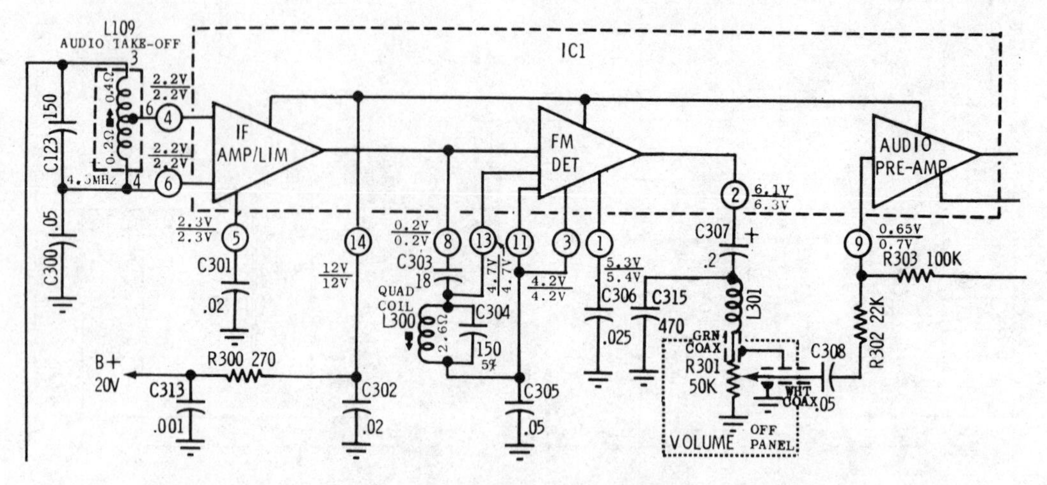

Figure 14-3 Sound system utilizing an IC amplifier. (Courtesy of Quasar Electronics.)

4.5-MHz signal is then injected into the IC. The first stage is a combination sound IF amplifier and limiter. Limiting action will remove any amplitude-modulated information. The remaining signal contains useful frequency-modulated information.

The FM detector uses a quadrature coil. This is tuned to 4.5 MHz. It reacts to changes in frequency by developing a current flow and signal voltage in the IC components. The resulting wave is a reproduction of the modulating signal. It is coupled through C_{307} to the volume control R_{301}. The audio signal is then returned to th IC audio preamplifier stage. It exits from the IC at pin 10. From this point it is connected to three stages of audio amplification. The sound IF, detector, and preamplifier are all contained in one IC. The only components that are not included on the IC are those that are too large or that cannot be manufactured in the IC.

The balance of this circuit is shown in Figure 14-4. It contains three stages of audio amplificiation. These stages are quite typical of those used in discrete component circuits. The audio amp and audio driver are common-emitter amplifier stages. Signals enter at the base of each transistor and leave at the collector element. The output stage is called a complementary symmetry output stage. It consists of two power transistors. Each has similar electrical characteristics. One transistor is an NPN and the other is a PNP type. This arrangement eliminates the need for a driver transformer. Since the two transistors are of opposite polarity, one turns on with the positive half of the incoming signal. The second transistor turns on only during the negative half-cycle of the input signal. This, then, is a push-pull type of amplifier.

The output signal is taken from the emitters of the two transistors. It is coupled to the speaker through capacitor C_{312}. Voltage variations across each of the two collector-emitter connections of these transistors develop a waveshape that is similar to that of the audio-modulating signal. The capaci-

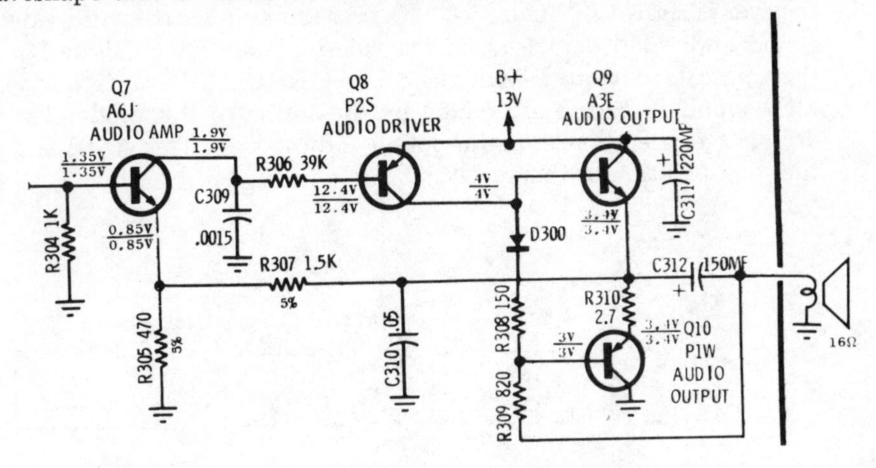

Figure 14-4 The balance of the sound system shown in Figure 14-3.

tor is used to couple these variations to the speaker. In some receivers these audio stages, including the output, are contained in one integrated circuit. This permits the development of physically smaller and more compact receivers.

COLOR SOUND SECTIONS

The audio output sections of color receivers are essentially the same as the circuits used in black-and-white receivers. This is also true for the sound IF amplifier and the sound detector circuits. The major differences in the sound sections of color receivers is in the method of developing the 4.5-MHz sound IF signal. The sound carrier developed at the transmitter is very close to the color subcarrier frequency. The difference between these two frequencies is 910 kHz. When these two signals heterodyne, the result is a viewable herring-bone pattern in the picture. To eliminate this interference the sound takeoff point in a color receiver is placed ahead of the video detector circuit. Any resulting 4.5-MHz signal is attenuated by a trap tuned to 4.5 MHz. A 4.5-MHz trap is located on either side of the video detector in order to minimize this interference.

A block diagram for a color receiver sound section is shown in Figure 14-5. The sound IF amplifier, sound detector, and audio amplifier blocks are the same as those found in a noncolor receiver. The difference is shown as an additional block. This block is called the sound IF detector. Its function is to detect the 4.5-MHz sound signal before the composite signal enters the video detector section. The takeoff point for the sound IF detector is usually in the third video IF amplifier. A schematic diagram of this section of a color receiver is shown in Figure 14-6. This circuit shows the third video IF, both sound and video detectors, and a video output stage. Diode D_1 is used as the sound IF detector. Both the 45.75-MHz video IF carrier and the 41.25-MHz sound IF carrier are mixed by the action of this diode. The result is a 4.5-MHz carrier that contains sound information. This signal is then fed to the sound section of this receiver.

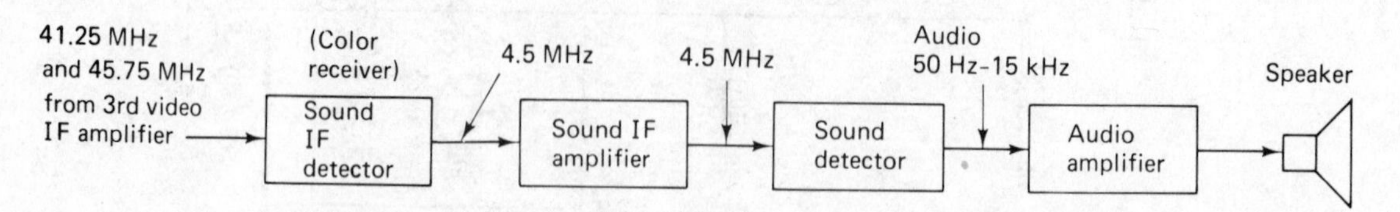

Figure 14-5 Block diagram of a color receiver sound system.

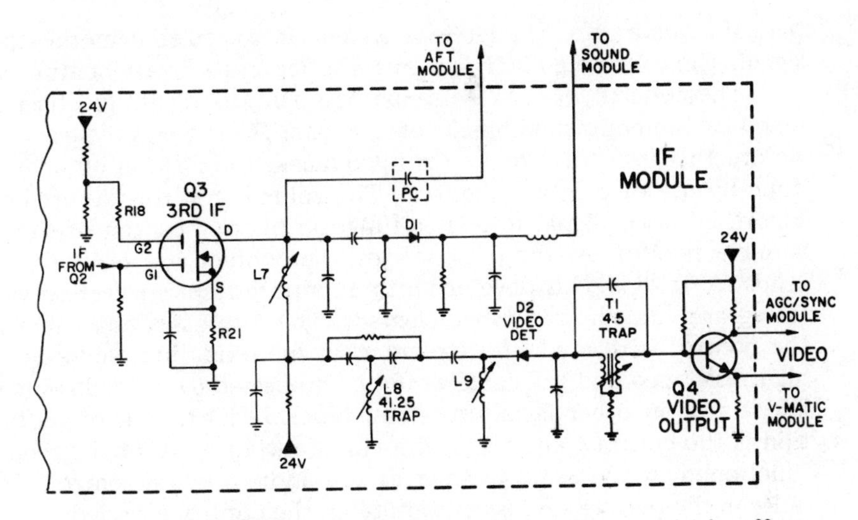

Figure 14-6 Schematic diagram of a color receiver sound take-off system. (Courtesy of Magnavox Consumer Electronics.)

The sound section is contained in one integrated circuit. This is shown in Figure 14-7. The functions of this IC are shown in block form in this illustration. It is common to show the functional section of an IC in this manner. It certainly does make signal tracing much easier for the technician. The input to the IC has a 4.5-MHz resonant circuit. This circuit, L_1, is used to couple the 4.5-MHz sound IF carrier to the IC. The other resonant circuit, L_2, is

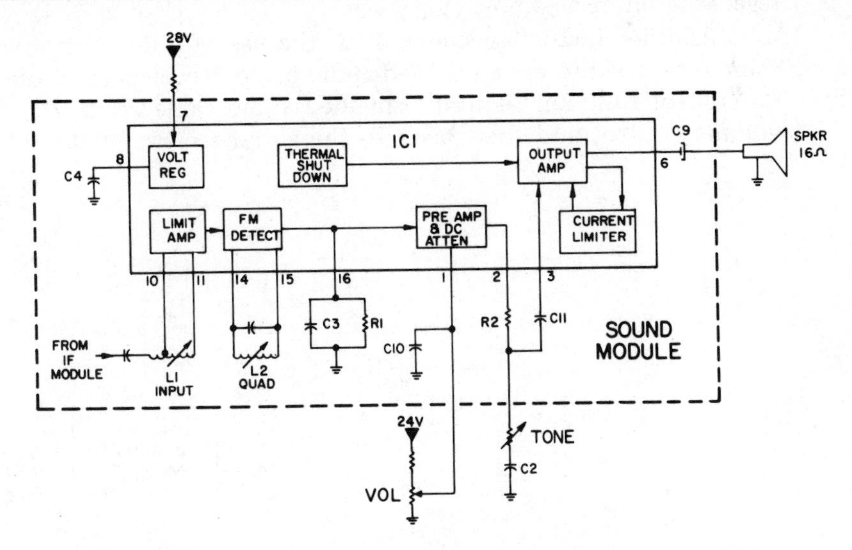

Figure 14-7 Block diagram of an IC sound system. (Courtesy of Magnavox Consumer Electronics.)

part of a quadrature FM detector system. It is used to demodulate the carrier signal. The following block is a preamplifier and a dc attenuator.

The concept of the dc attenuator is being used with greater frequency in many of the controlled blocks of receivers. The reason for this is very logical. Before this type of circuit was introduced, the control of signal levels was done by use of a volume control. The volume control was used as shown in Figure 14-8(a). Signal level is a function of the position of the arm of the variable resistor. As the resistor ages, the contact arm on the rotating shaft tends to oxidize. This develops into an intermittent contact between the rotating arm and the resistance element. The result is a noise that is heard as the control is rotated. In extreme cases the oxidation could develop into a high resistance and no signal would be processed to the following stage. This is also true in other signal-processing blocks. An example of another application is the contrast control in the video output stage of the receiver. It has functioned in the same manner as the audio volume control. The result is noise in the picture, or loss of picture, as the control is varied.

The solution to this problem is shown in Figure 14-8(b). A variable control is used in this circuit, also. The major difference is that the control is used to vary the dc operating voltage of the stage, or block. The use of dc minimizes the problem of noise getting into the circuit. The dc voltage applied to the variable control does not permit a rapid development of oxides. Also, since the variable voltage is applied as an operating voltage, the action of amplifying electrical noises is not present. This eliminates the annoying noise problems associated with the circuits that control signal level by varying the level as it enters an amplifying stage.

Another factor associated with the use of a dc control voltage is that there is no need to use a shielded audio cable. Receivers that do not have this dc control function require a shielded audio cable from the chassis to the volume control and then back to the audio section of the receiver. This is

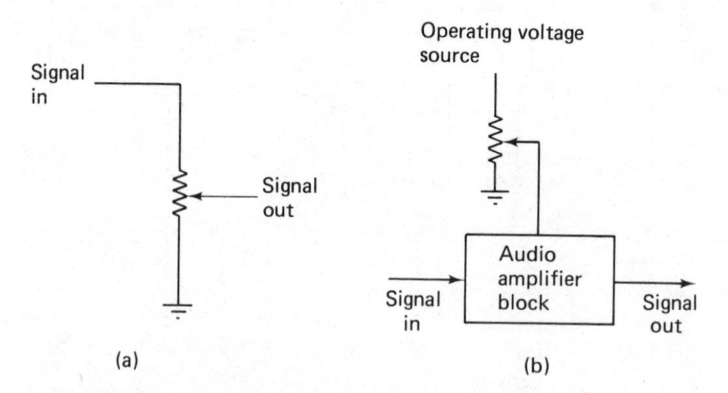

Figure 14-8 (a) Typical signal level volume control system, and (b) newer voltage-controlled amplifier system.

necessary in order to minimize the pickup of unwanted signals. The dc control system only requires an unshielded wire. The final factor for use of this dc control system is that it is readily adapted to microprocessor control of the receiver functions. Microprocessor control uses a varying level of dc voltage in order to change levels of sound, picture, or channels.

The final block in this section is the output amplifier. It develops the power required to operate the speaker. There is a current-limiting block that is used with the output stage. It contains a shutdown circuit. In the event that current flow in the output stage should exceed design values, the current limiter will turn off the output stage. A second control block connected to the output is the thermal shutdown. This circuit reacts to temperature. It is another fail-safe type of circuit. Excess current will raise the temperature of the IC. This block will cut off current flow when the IC temperature goes above design values. A heat sink is used to keep operating temperature below this critical value.

TROUBLESHOOTING THE SOUND SECTION

There are certain common troubles in a television receiver that can be isolated to the sound section. The fact that one observes normal picture and raster usually eliminates the video, color, and sweep sections as problem areas. Sound problems are classified as either weak sound or absence of any sound. In addition, there could be distorted sound or a buzz in the sound.

All the symptoms above are related to the sound section with the exception of the last one. Normally, a buzz in the sound is caused by an overloaded tuner or IF section. An overloaded tuner is usually caused by improper AGC voltage levels. This topic is discussed in Chapter 15. Another noise type of malfunction is when an open filter capacitor develops in the receiver power supply. If the filter capacitor is used with either a 60- or an 120-Hz source, the result is a low-frequency hum. This may be observed with an oscilloscope.

Discrete component system. These systems use a linear signal flow path. The oscilloscope is utilized for initial checks. A test generator that can deliver a 4.5-MHz FM signal is required. One could use a video generator that also produced a sound signal. The video generator may be connected to either the antenna terminals of the receiver or the IF input connection from the tuner. When a 4.5-MHz sound carrier is used, the generator is connected to the sound takeoff point in the receiver.

A wise initial test for any system is to check both input and output points on the receiver. The input point test will determine if the proper signal is present at that point. A check at the output point of the module or chassis will indicate if the output transducer is working. A 4.5-MHz test signal is usually modulated by either a 400-Hz or a 1-kHz audio tone. This sine-wave form should be observed once the signal is processed by the demodulator.

Speakers do fail. When they do fail, any of the symptoms described at the beginning of this section can be present. When the presence of a valid input signal has been determined together with a distorted or absent output signal, the next steps involve applying the rules for troubleshooting. This is a linear signal flow circuit. The rule to apply at this point is to make the next test at or near the midpoint of the circuit. A good place to make this test is at the volume control. An audio signal should be observed at this point. The presence of a nondistorted signal usually indicates that the sound IF and demodulator are functioning.

A distorted waveform at this point indicates that the problem is in the sound IF or demodulator stages. The problem may require adjustment of the IF or demodulator transformers. These are tuned with an alignment tool that adjusts the position of the core of the transformer. Follow the manufacturer's service literature when making any adjustments of this type. If this alignment does not repair the problem, circuit analysis is required. Make another test, again splitting the system in half. This will localize the problem to either the sound IF or the demodulator. Follow this test with voltage and/or resistance measurements to determine the component that has failed.

The same procedure is used when checking the audio-processing sections that are between the volume control and the speaker. Tests are made that split the area of suspicion in half until only one stage remains. The problem is in this stage. Use of a meter to make voltage and resistance measurements will determine the component that failed.

Integrated-circuit servicing. The servicing of an IC system is usually much easier than servicing a discrete component system. Here, too, the input and the output are checked as an initial step. The next step is to check for proper operating voltages. An instruction sheet for making tests of this type is shown in Figure 14-9. The sequence of the tests is listed in each instruction box. There are two voltage tests to be made. One of these is the source voltage. It should be 25 V in this circuit. The B+ voltage may be checked with the oscilloscope to be certain that the filter capacitor is working properly.

A second voltage test is used to determine that the volume control is working. This is a dc control in this circuit. Rotating the control shaft should vary the dc voltage at the resistor or the IC. This voltage should range as indicated on the sheet. Major manufacturers will provide this type of information as a part of an ongoing service training program. The information is devoted to specific models produced by the receiver manufacturer. Information and procedures will often vary from one model to another.

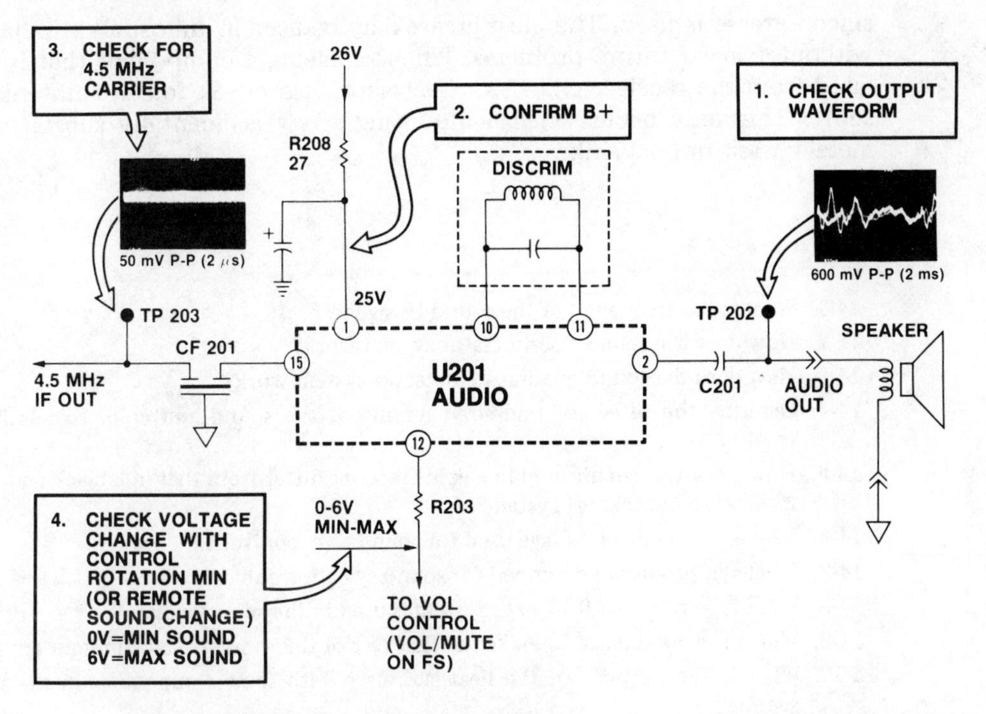

Figure 14-9 Service data for troubleshooting the sound system. (Courtesy of RCA Consumer Electronics.)

HEAT TRANSFER

One by-product associated with electronic devices is heat. This may be a serious problem if it is not controlled. Solid-state devices often use a heat radiator, or sink, to dissipate excess heat. The heat sink is usually mounted on the transistor of IC. In some cases the heat sink serves as a mounting device. In either situation there must be a good transfer of heat away from the device. A silicon grease is used to aid in the transfer of heat. Often, the collector of a power transistor is connected to the case of the transistor. The case has to be insulated from the heat sink. A mica washer is used to insulate the case/collector from its mounting device. When replacing components that require a heat sink, check or replace any insulating hardware. Also be sure to apply a light coating of silicon grease.

One word of caution. In some situations silicon grease may attack the cases of plastic "power tab" transistors. This has occurred when a white silicon grease is used. Check with a local supplier to be certain that the proper silicon grease is used. The clear grease may be used in almost all installations without fear of future problems. When replacing a component that is insulated from the receiver chassis or heat sink, also check for an unintentional short. This may occur when a mounting screw accidentally shorts to the metal chassis or heat sink.

QUESTIONS

14-1. What is the frequency of the sound IF signal?

14-2. How does the sound discriminator system work?

14-3. How does the sound quadrature detector system work?

14-4. Describe the effect of undesired mixing of the sound and color signals in the receiver.

14-5. How does the circuit used in a color receiver differ from that in a black-and-white receiver sound takeoff system?

14-6. Why is a dc control voltage used for sound level control?

14-7. What symptoms are described for sound system troubles?

14-8. What type of signal flow path system is used in the sound section?

14-9. Where is a good place to make the first test of this system? Explain your answer.

14-10. What is the purpose of the heat sink used with some components in the sound section?

Chapter **15**

Automatic Gain Control

The amplitude of the video signal at the cathode of the picture tube determines the quality of the image developed in the screen. The signal voltage level will range between 30 and 150 V p-p. The receiver RF and IF amplifier are designed so that signal amplification produces this range of signal for the CRT. Signal levels at the receiver antenna will range from a few microvolts to well over 50,000 μV. The low end of this range will produce a very weak picture. Sync pulses may not be strong enough to stabilize the picture. Additional amplification in the receiver is required to develop the required signal level at the CRT. The other extreme situation will produce too much signal for the amplifiers and CRT to process. The viewable result is a negative and distorted picture.

Receivers are designed to operate under a wide range of signal conditions. In some locations one channel may have a very strong signal and a second channel may be very weak. The viewer would be required to adjust signal amplification levels to compensate for these different signal strengths. It is not convenient to have the viewer adjust signal levels for different channels. In fact, the trend at the present time is to require less and less viewer control. Many of the circuits that had required viewer (or technician) adjustment have now been replaced with internal circuits that make automatic adjustments.

A feature that is in common use for almost all types of receivers is called automatic gain control (AGC). This circuit is used to adjust receiver amplification automatically. It is used to compensate for a wide range of received signal levels. The AGC section of the receiver is basically a type of feedback circuit. A block diagram for it is shown in Figure 15-1. A signal is

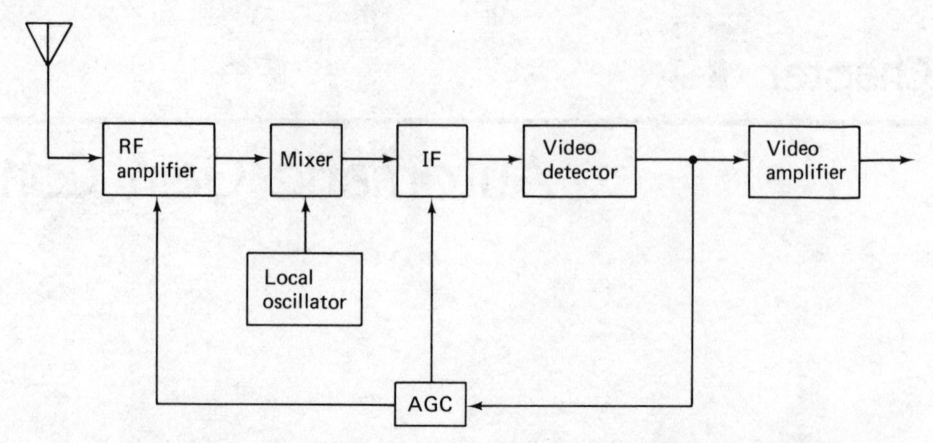

Figure 15-1 Block diagram of an AGC-controlled IF amplifier system.

taken from the output of the video detector. This signal is processed in the AGC block. The output of the AGC block is a dc control voltage. This dc control voltage is applied to the IF amplifier of the receiver. Application is usually made at the base of the IF transistors. This dc voltage is added to the fixed bias on the transistor base. The result is a change in base bias and a resulting change in the amount of amplification occurring in the IF amplifier. A second control system connects the AGC output to the receiver RF amplifier. This voltage is used to control the gain in the IF amplifier stage.

The ideal AGC system would be designed to either increase or decrease the amount of amplification in the IF and RF stages. When a very strong signal is received, the AGC action would reduce the amplification. When a weak signal is received, AGC action would increase amplification. Under usual operating conditions, the receiver AGC is adjusted during the installation process. This adjustment is made using the strongest local channel signal as a reference. AGC is adjusted by rotating a variable control located at the rear of the receiver. This control may also be located on a module. The adjustment is made to provide minimum amplification for the strongest signal. Weaker signals will then be given additional amplification in the receiver RF and IF sections.

BASIC AGC

A basic AGC circuit is shown in Figure 15-2. A signal is taken from the output of the detector. This signal is filtered by the *RC* network on the AGC line. The output of the filter is a dc voltage. This voltage is applied to the bases of the IF and RF amplifier transistors. A system of this nature has certain limitations. One such limitation is the amount of voltage that may be

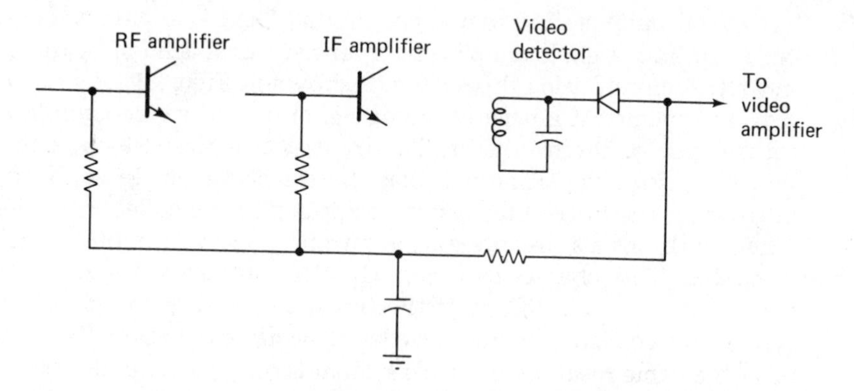

Figure 15-2 Partial schematic diagram of the IF system.

developed in this simple system. Often, this voltage is not sufficient to compensate for extremes in picture levels. In some receivers an amplifier stage is included in the AGC system. The amplifier will correct this limitation. Another problem created by this basic system is related to the charge time for the filter capacitor. Compensation for a rapidly changing signal level cannot be accomplished with a large value of capacitance. Too small a value of capacitance will not allow enough dc control voltage to develop. Often a compromise value is used. When a moving airplane is in the signal path, a flutter in signal level occurs. AGC action tries to correct for this. It is not able to do so completely. The flutter in signal level is observed on the screen of the CRT.

KEYED AGC

The disadvantages of the basic AGC system are overcome when a keyed AGC system is utilized. The term "keyed" refers to a pulse that is developed in the horizontal output transformer circuit. This pulse is synchronized with the sync and blanking pulses from the transmitter. A sample block diagram for this system is shown in Figure 15-3. Both the composite video signal and the

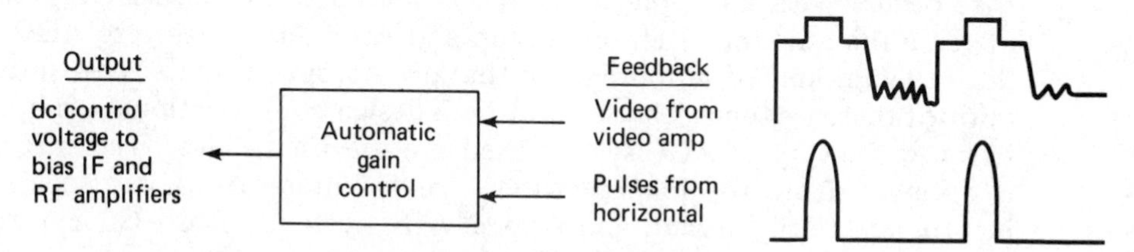

Figure 15-3 Two signal inputs are required in the keyed AGC system.

horizontal pulse are fed to a keyer circuit. The keyer circuit is designed to be on when both signals are present. If either signal is missing, there is no keyer output. A circuit using this system is shown in Figure 15-4.

In this circuit a negative-going signal from the video amplifier is applied to the base of the transistor. The transistor is biased to be cut off with no signal applied. The signal will bias the transistor on. In addition, a negative pulse from the horizontal system is applied to the collector of the transistor. When both signals are present, a current flows in the output circuit of the transistor. This charges capacitor C_1. The output of the capacitor is fed to the IF and RF amplifiers. If the transistor used in the circuit was an NPN type, the two signals would have to be positive in nature. This would accomplish the same results. Since this system is only partially dependent upon the level of the incoming signal, it will overcome the limitations described earlier in this chapter.

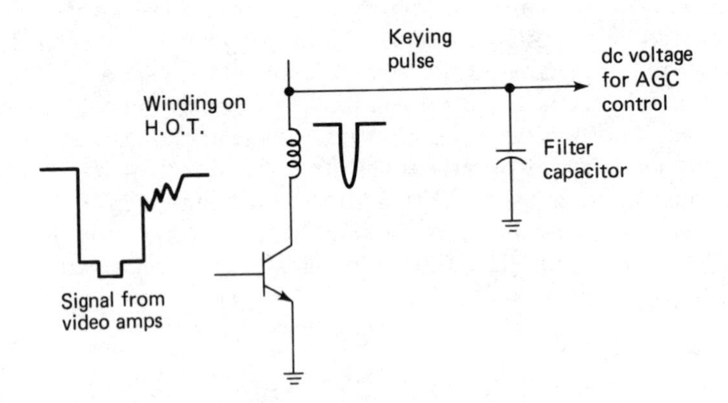

Figure 15-4 Schematic diagram of a keyed AGC system.

FORWARD AND REVERSE AGC

These terms refer to the effect on the conduction of the controlled stages that occurs with an application of AGC voltage. Forward AGC action will increase the amount of amplification and conduction. Reverse AGC will reduce the amount of amplification that occurs. A solid-state receiver may use both of these systems. The selection is a design consideration. It is important to know the type of AGC system used in a specific receiver. This information is obtained from the manufacturer's service literature. Some receivers will incorporate both forward and reverse AGC systems. The AGC control voltages are provided in the service literature.

NOISE CANCELLATION

A problem in the television receiver is that the electronic circuits are not able to think. This problem is particularly evident when noise pulses are present in the signal. Noise appears as an amplitude-modulated wave. It rides along with the other amplitude-modulated information on the signal. When the noise pulses are the same amplitude as the sync pulses, the receiver circuits cannot tell them apart. The result is a receiver that is often out of sync. A section of the receiver called a noise gate or noise blanker is used to remove the noise pulses. A simplified block diagram for a noise gate circuit is shown in Figure 15-5. In this circuit the noise pulse is separated from the video signal. It is then amplified and inverted. The inverted noise pulse is added to the composite video signal. The inverted noise pulse, being opposite in polarity of the original pulse riding on the video signal, will cancel itself. The result is noise-free video. The noise-free video is then processed in its normal manner.

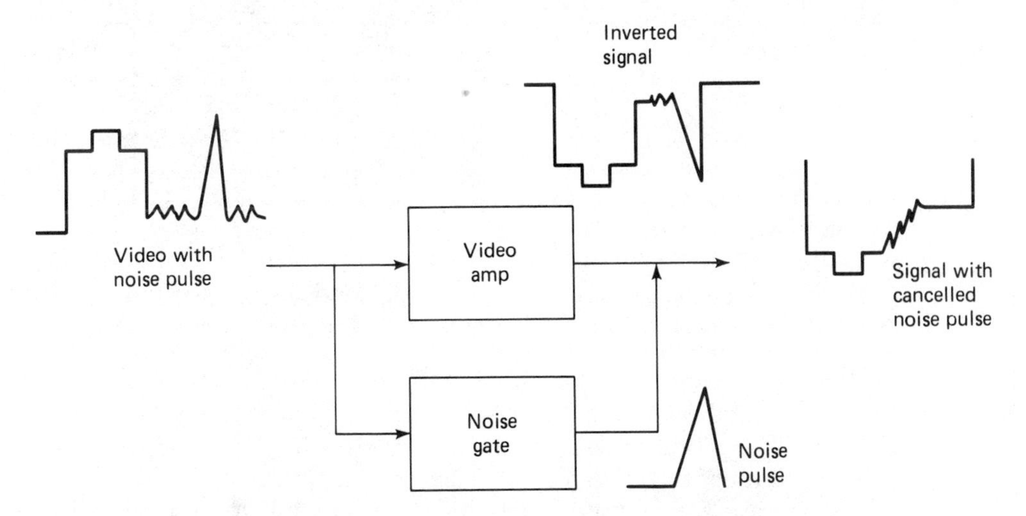

Figure 15-5 Block diagram of a noise gate system.

INTEGRATED-CIRCUIT AGC

Use of IC technology in the television receiver has resulted in the incorporation of AGC and noise separators in the same chip. A block diagram for a typical IC is shown in Figure 15-6. This IC performs several functions. These include three stages of IF amplification, video detection, a noise gate, AGC, and sync separation. The signals required for this IC are the composite video IF from the tuner and a keying pulse from the horizontal output stage. In

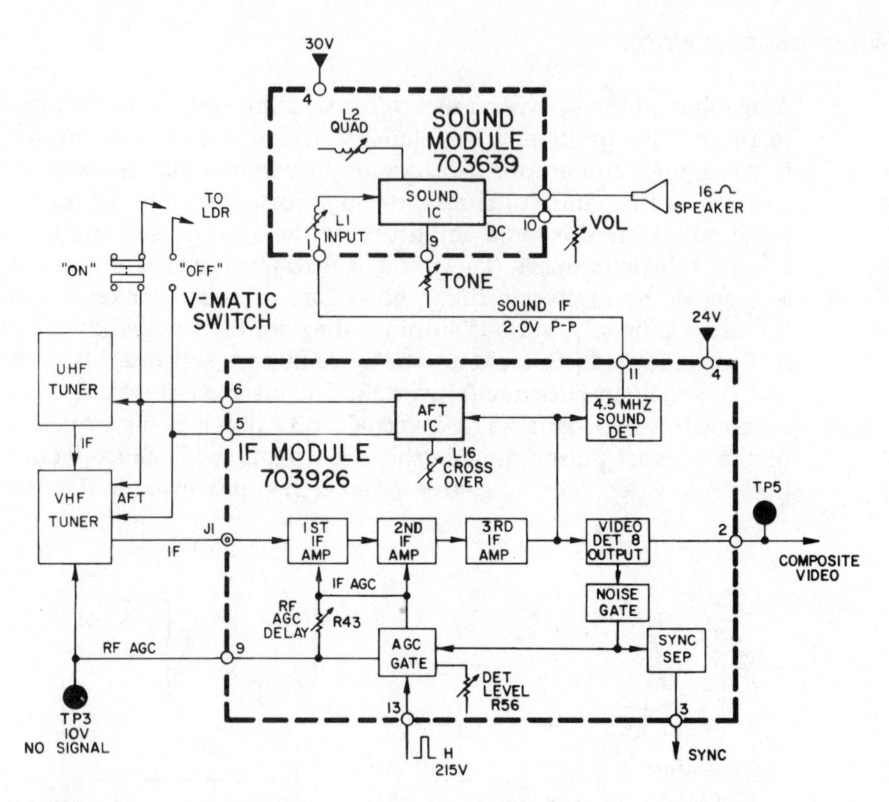

Figure 15-6 Block diagram of an IC IF and AGC system. All blocks for IF are contained in one IC. (Courtesy of Magnavox Consumer Electronics.)

addition to these functions, this IC includes a 4.5-MHz sound detector. The function of the sync separator is discussed in Chapter 16.

There are five outputs from this IC. These are the composite video, a sync pulse, the 4.5-MHz sound, automatic fine tuning (AFT), and an RF AGC voltage. The RF AGC provides the required level of AGC voltage to the RF amplifier in the tuner. In many receivers this system is called *delayed AGC.* This term refers to the application of the AGC voltage. In a delayed AGC system the delay voltage develops only after the IF amplifier AGC action reaches its maximum level. Then, if additional amplification is required, the delayed AGC voltage is used to increase the amplification in the tuner RF amplifier. The system is used to reduce the level of atmospheric noise that would be amplified in the tuner section if the RF amplifier operated at maximum gain. Using the IF amplifier first will tend to minimize this effect.

SERVICING AGC SYSTEMS

There are two conditions that occur with a failure in the AGC system. One of these results in a weak picture and sound. The extreme condition is complete loss of both picture and sound. The second condition results in excessive picture information. The picture may become negative. This condition is usually accompanied with a buzz in the sound.

The first order of business is to determine if the problem is a forward (IF) problem or a feedback (AGC) problem. This may be determined by making a series of checks. These checks are similar in both discrete component and IC systems. There are some differences, however. Because of this each system is discussed separately.

Discrete component systems. The initial check is to be sure that a signal is present at the video detector output. This indicates that the IF amplifiers are working. The next test is to apply a fixed dc voltage at the output of the AGC system. The exact value and polarity of this voltage is determined from the manufacturer's service literature. This is called *clamping* the AGC line. What it does is to restore the AGC system to its design value. If this clamp restores the system to its normal operation, the problem is isolated to the AGC system. If clamping the AGC line does not restore the correct level of video, the problem is in the IF amplifier system.

The clamping voltage is usually obtained from a variable dc supply. The supply output is adjusted to obtain the proper voltage value. The leads from the power supply are then connected between circuit common and the appropriate point in the circuit. This system may be used to clamp any dc voltage in the system. When a transistor is used as an AGC amplifier, such as the one illustrated in Figure 15-7, the clamp voltage is applied to each point to localize the problem area. Do not overlook the possibility of an open AGC filter capacitor.

Integrated-circuit systems. The utilization of the IC has enabled receiver manufacturers to develop more stable circuits. This, then, has made the service technician's work easier. When testing an IC circuit one has first to determine that the proper input signal and operating voltages are present. If a variable control voltage is used, this is checked to be certain that it is able to swing through its entire range. If there is a filter capacitor connected to the IC, the voltage and waveform are measured.

Manufacturers' literature is used as a reference when making these tests. Be certain that any keying pulses are present. These must have the proper polarity as well as the correct amplitude. One receiver manufacturer (RCA) has eliminated the need for keying pulses. Service literature for a specific

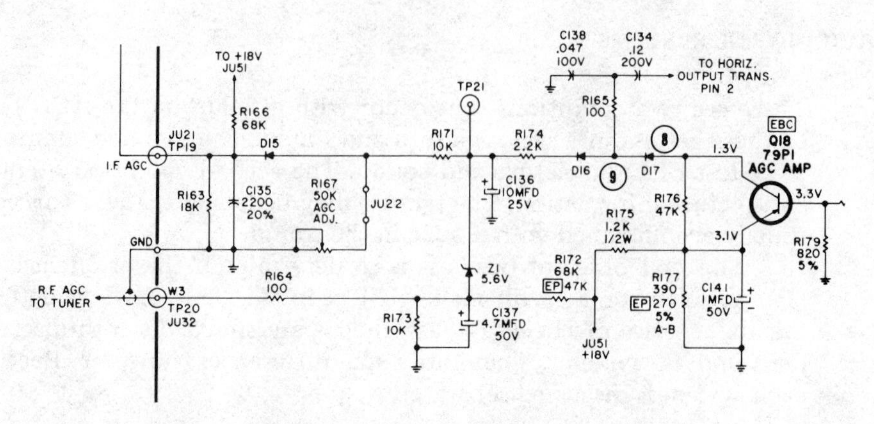

Figure 15-7 Solid state AGC amplifier circuit. Applying a fixed dc voltage to its output connections determines if it is functioning properly.

model or chassis provides this information. Since not all receivers use this system, it is wise to check carefully. The methods described in the service literature use a "positive elimination" type of system. This term describes the method by which all working circuits are eliminated from a problem area. When this is done properly, the only area left is the one that is malfunctioning. It really is as simple as it sounds. The main ingredients are a knowledge of how the receiver works and a trust in the correctness of the measurements made in the circuit.

QUESTIONS

15-1. What is the purpose of AGC?
15-2. Which blocks are affected by AGC action?
15-3. Describe the action of the AGC system.
15-4. Why is keyed AGC used?
15-5. Describe forward and reverse AGC.
15-6. Describe AGC failure symptoms.
15-7. What is meant by "clamping" the AGC?
15-8. What is delayed AGC?
15-9. What signal path system is used for the AGC system?
15-10. Where is a good place to make the first test of the AGC system?

Chapter 16

Synchronization Systems

The television receiver system uses two free-running oscillators to create a raster. The raster is then displayed as an illuminated screen. It does not contain luminance, or video, information. A vertical deflection system develops vertical sweep. The horizontal deflection system develops horizontal sweep. The two systems produce magnetic fields in the proper windings of the deflection yoke. The result is a raster that contains 525 lines.

The two deflection oscillators, being free running, need to be synchronized with the transmitted signal. This synchronization is required to recreate the proper image that is developed at the camera. If this is not done, the scanning systems could easily reproduce a picture similar to that shown in Figure 16-1. This picture shows the result of loss of both horizontal and vertical synchronization. A pulse is added to the transmitted signal. This pulse is called the *sync pulse*. Its purpose is to develop a timing signal in the receiver. The timing, or sync, signal is used to ensure the reproduction of the correct picture. It has nothing to do with the development of the raster.

The sync pulse is transmitted as a part of the composite video signal. Each of the 525 lines that make up the picture contain a sync pulse. The waveform of the signal for a line of picture information is shown in Figure 16-2. The sync pulse is transmitted during that period of time when the raster is blanked, or at the black level. It rides on the blanking pulse. There is additional sync information transmitted during the vertical blanking interval. This interval is made up of the first 21 lines of each field. This information is used to develop the correct timing information for the vertical deflection system.

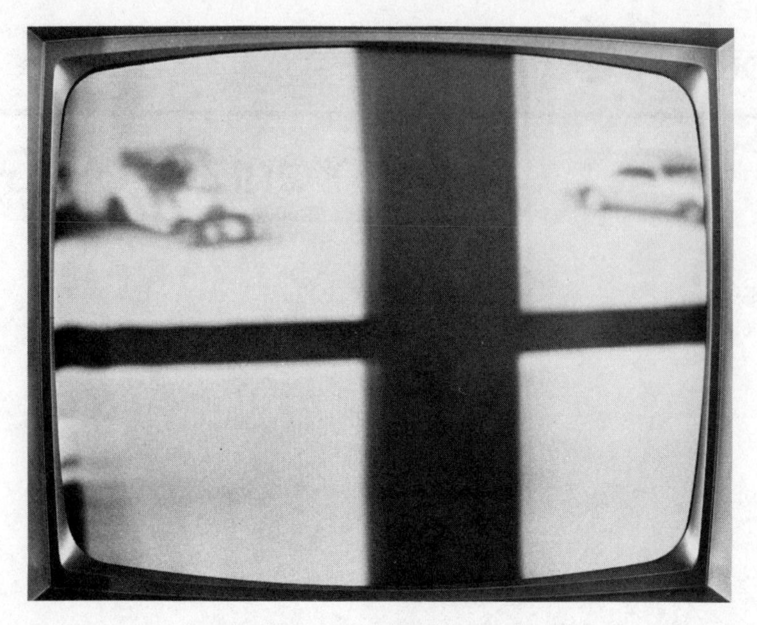

Figure 16-1 Nonsynchronized picture.

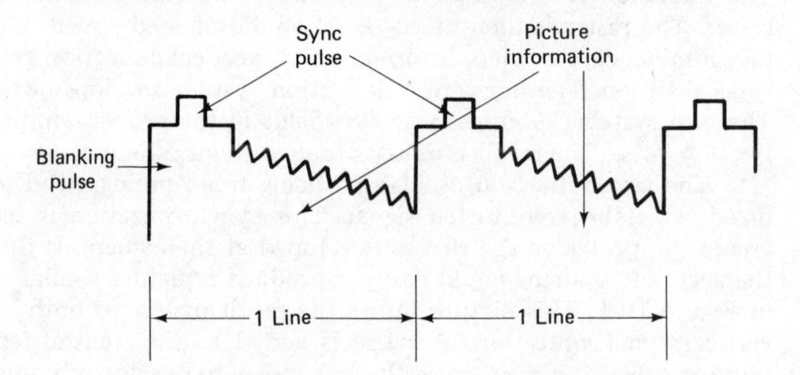

Figure 16-2 Composite luminance signal waveform.

Before the sync information can be used, it must be separated from the composite video signal. After it is separated, it receives additional processing. This additional processing changes the sync pulses into a form that is used for timing of the deflection oscillators. A block diagram of the portion of the receiver that processes the sync signal is shown in Figure 16-3. The sync separator is designed to conduct only when the amplitude of the video signal is above the level of the blanking pulse. The only information transmitted at this level is the sync pulses. This block is labeled the sync separator because it separates sync pulses from the balance of the composite video signal.

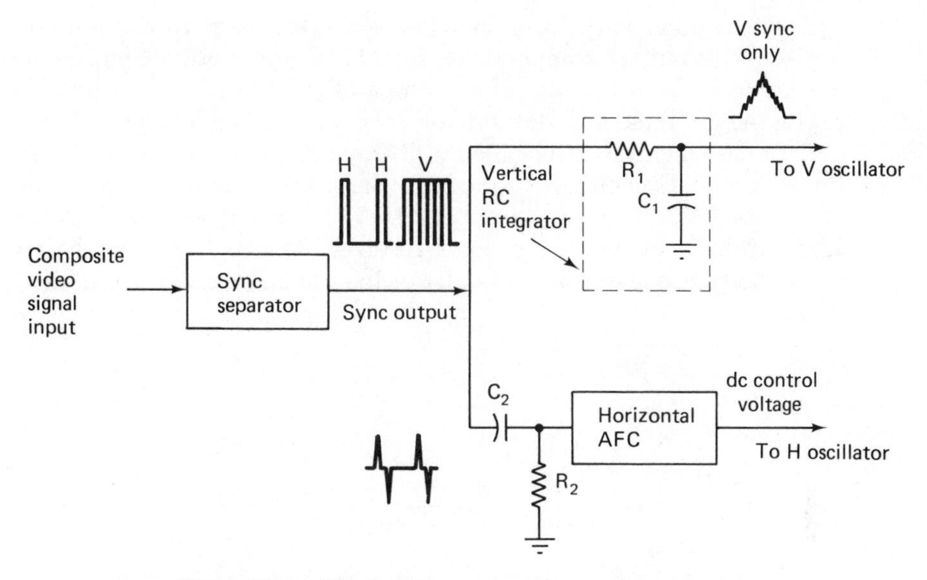

Figure 16-3 Block diagram of a sync separator system.

After leaving the sync separator, the signal may be amplified. This depends upon the requirements of the receiver. From this point the sync pulse path divides. One branch leads to the vertical deflection oscillator. The sync pulses have to be processed before they can be useful at this point. In addition, the higher-frequency horizontal pulses must be removed from the path. This is accomplished by use of an integrator circuit. This is basically a low-pass filter circuit. It passes the 60-Hz vertical sync information and stops the 15,734-Hz horizontal frequency pulses. The resulting waveform is shown in Figure 16-4. This increasing voltage is used to trigger the vertical oscillator circuit. In a color receiver this oscillator operates at a frequency of 59.94 Hz.

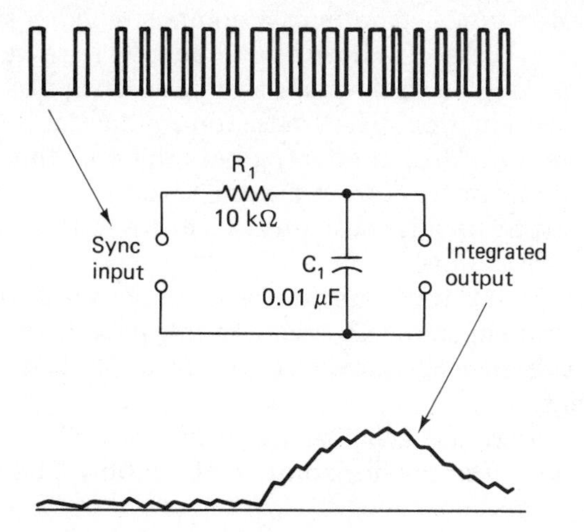

Figure 16-4 Vertical integrator schematic diagram.

The second path from the sync separator leads to the horizontal AFC section. This section compares the sync pulse and a pulse from the horizontal output stage. It is a basic phase comparator circuit. The output of the horizontal AFC block is a dc control voltage. The wave-shaping circuit in the horizontal sync system is called a differentiator. It is effectively a high-pass filter circuit. This circuit produces a sharp pulse. It is shown on the bottom of Figure 16-5. These spikes occur 15,734 times a second. This is the horizontal frequency rate of a color receiver. This is the general theory as it relates to sync separation. Now, let us look at some specific circuits.

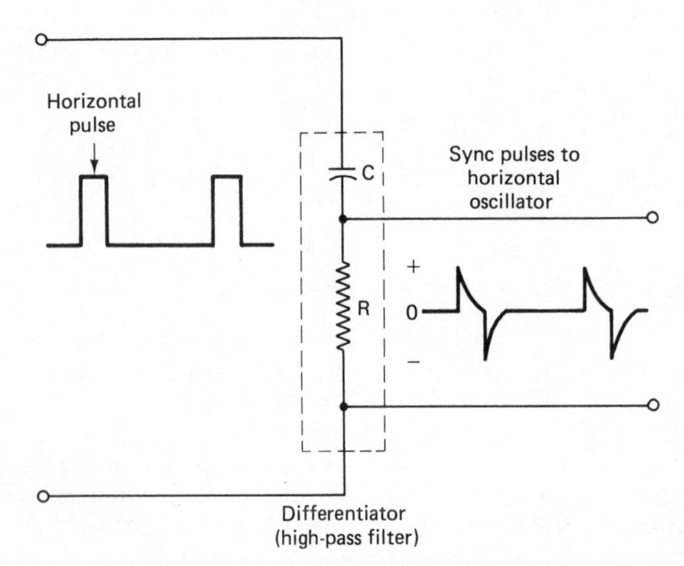

Figure 16-5 Horizontal differentator schematic diagram.

Solid-state circuits. The schematic shown in Figure 16-6 is typical of many transistorized sync circuits. This system uses a single transistor, Q_1. The input to the transistor sync separator is a composite video signal. The sync separator transistor is biased so that it is in cutoff. A pulse from the horizontal output stage is coupled to the transistor base. The pulse is used to drive the transistor into conduction. The transistor conduction period lasts as long as the pulse is present. This, in effect, allows the sync tips to be processed and stops all video information from passing through this stage. Waveforms observed at the output of this stage are also shown in this illustration. These are negative-going waveforms.

One of the outputs from the sync separator is connected to the vertical oscillator. Resistor R_{59} and capacitor C_{34} act as a low-pass filter. The result is an integrated signal. This signal is shown as waveform 13. It is coupled to the vertical oscillator stage.

The second output from the sync separator is passed through a high-pass filter. It is then connected to the horizontal AFC section. The horizontal

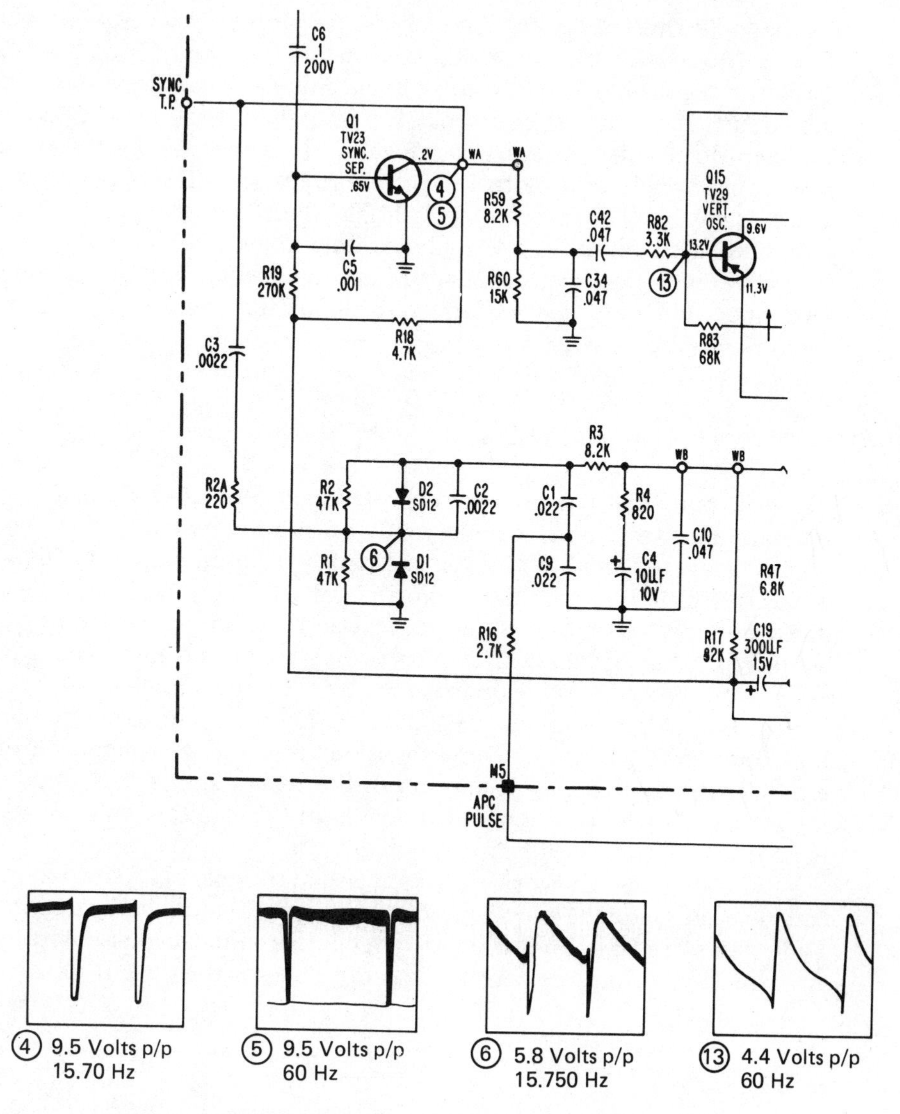

Figure 16-6 Schematic diagram of a sync separator system.

AFC is a phase comparator circuit. It uses a pair of diodes to produce a dc control voltage. The AFC section has two inputs. One of these is the differentiated sync pulse. The second input is a pulse from the horizontal output transformer. The sync signal is fed to the cathodes of diodes D_1 and D_2. These are the horizontal AFC diodes. These diodes are wired in opposite polarity. They produce a positive and a negative voltage. The horizontal pulse is fed to the center of the series capacitors C_1 and C_9. These also act as a voltage-dividing network. The circuits develop a dc voltage at resistor R_3. This is filtered by capacitor C_4. It then goes to the base of the horizontal oscillator transistor. The control voltage value is established in the design of the circuit. If the horizontal oscillator frequency rises above 15,734 Hz, the control voltage shifts from its design value. This changes the bias on the oscillator transistor. It returns to the proper frequency. The opposite is also true if the frequency drops below 15,734 Hz. This process occurs 15,734 times a second. The result is an oscillator that maintains its design frequency. Discrete-component sync separators are also built into integrated circuits. The input and output waveforms are essentially the same as when discrete components are used.

COUNTDOWN AND DIVIDER CIRCUITS

The efforts on the part of receiver manufacturers to produce a "foolproof" set have resulted in the development of some very stable frequency references. These are used in the vertical and horizontal sections. Their purpose is to stabilize the vertical and horizontal oscillator circuits. The need for consumer-adjusted controls for vertical and horizontal hold purposes is also eliminated. An explanation of phase comparator and phase-locked-loop circuits must precede any explanation of these new circuits. A phase-locked-loop (PLL) circuit is shown in Figure 16-7. This circuit is very similar to the horizontal AFC circuit used in television receivers. The names of the blocks are different from those usually used. For example, the horizontal oscillator is now called a voltage-controlled oscillator. The letters "VCO" are used to label this block.

The PLL circuit works in this manner to lock in the horizontal oscillator. The VCO operates at a frequency near that of the horizontal circuit. Sync pulses and horizontal pulses are compared in the phase detector block. A correction voltage of zero volts is obtained at the output of the phase de-

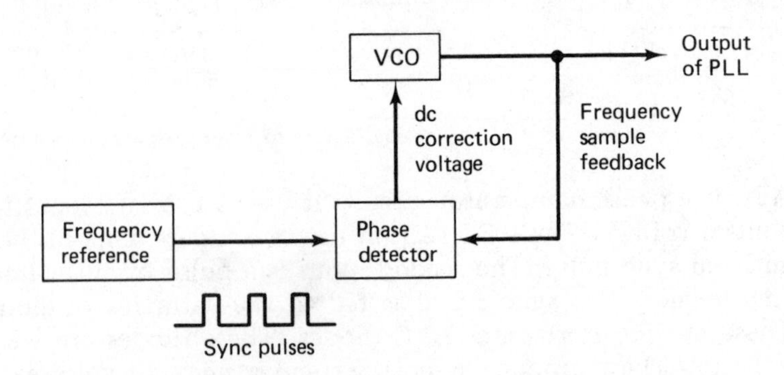

Figure 16-7 Block diagram of a basic phase comparator system.

tector when both frequencies are equal. If the two frequencies are unequal, the dc correction voltage will have either a positive or negative value. The correction voltage is fed to the VCO. Its frequency will raise or lower, depending on the polarity of the correction voltage. Should the VCO change its frequency for any reason, the dc correction voltage from the phase detector will correct this change.

There are certain circuits in which the frequency of the VCO is not the same as that of the reference signal. It is still possible for a circuit of this type to work as a PLL. An IC called a digital divider is added to the system, as shown in Figure 16-8. This divider will divide the VCO frequency by some predetermined number. The result is that the divided frequency must be made equal to the reference frequency by PLL action. The balance of the PLL circuit functions in the same manner as described earlier. It is possible to use more than one divider in this system. A circuit that uses two dividers is shown in Figure 16-9. The addition of a programmable divider allows the selection of any frequency that is a multiple of the reference oscillator. When the reference oscillator is operated at a basic frequency, such as 1 MHz, the combination of VCO frequencies is almost unlimited. The programmable divider is adjusted to the desired output frequency "divider." The VCO then tunes to the correct frequency because a changed correction voltage is applied to it. When the phase detector can no longer detect any frequency

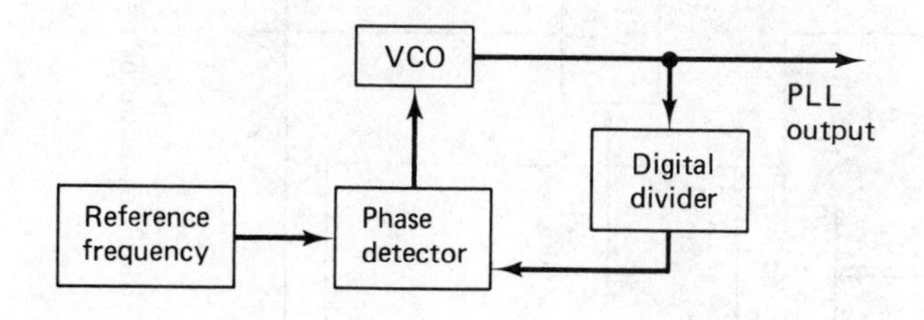

Figure 16-8 Block diagram of a phase comparator with a digital divider circuit.

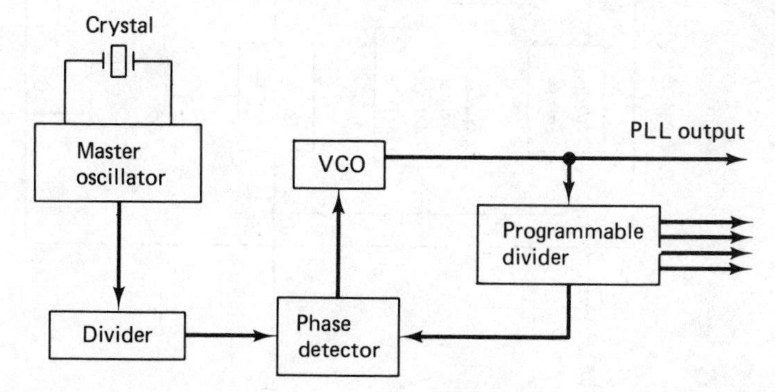

Figure 16-9 Block diagram of a programmable phase comparator system.

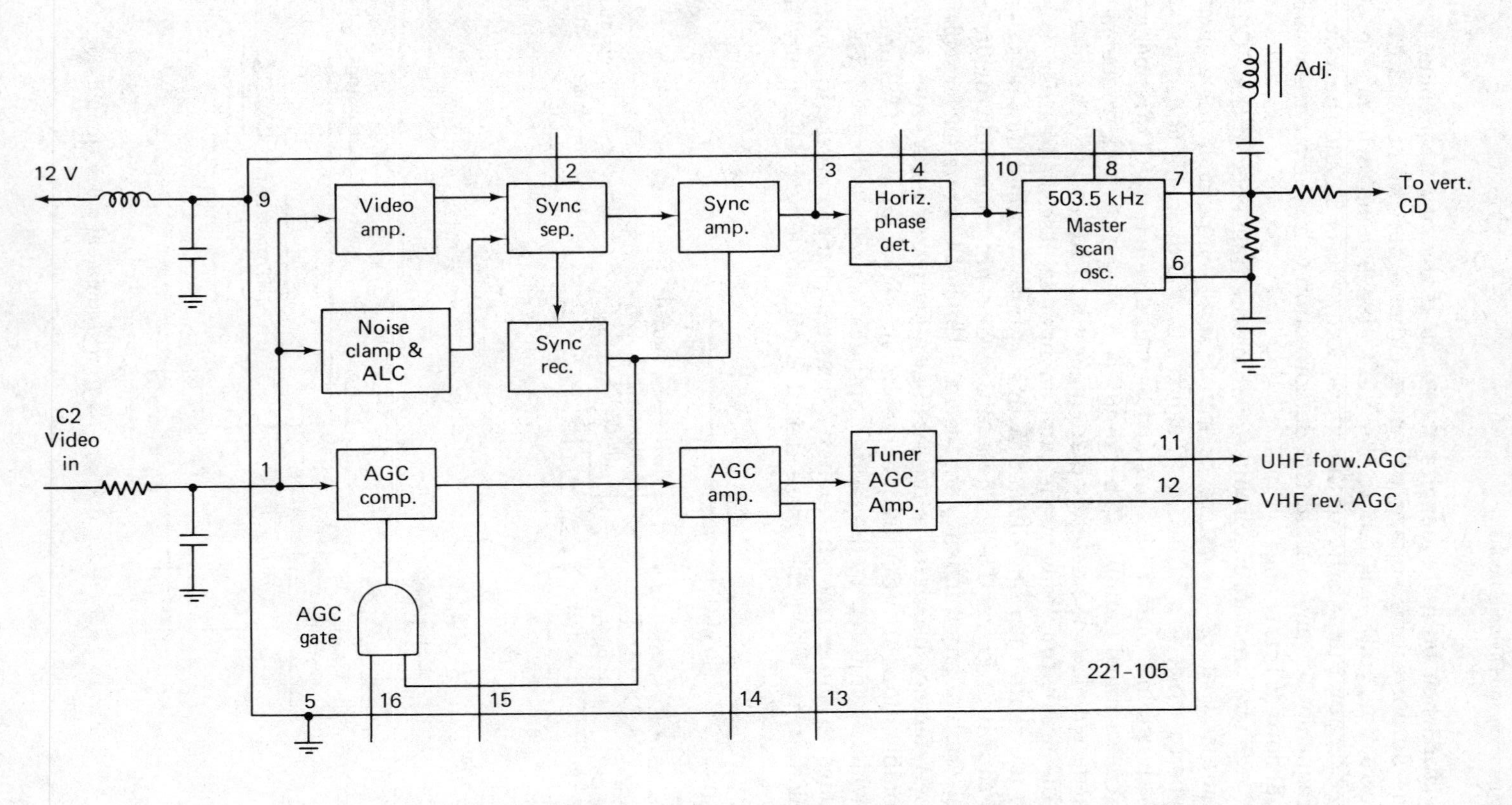

Figure 16-10 Block diagram of a horizontal system and sync amplifier IC.

differences, the output is a dc voltage whose magnitude and polarity are proportional to the initial frequency and phase difference.

Television applications. The PLL system is used extensively in citizens' band (CB) radio-frequency synthesizers. In these receivers the desired frequency is "programmed" when the channel selector knob is rotated. The system used in a television is preprogrammed. It operates on only one frequency. In the TV receiver one IC is used to replace both the horizontal and vertical oscillator circuits. One PLL circuit is used to create both the 15,734- and 59.94-Hz signals required for sweep.

A block diagram of an IC used for, among other functions, a master scanning oscillator is shown in Figure 16-10. This IC also incorporates video amplification, sync, horizontal AFC, and AGC functions. The master scan oscillator operates at a frequency of 503.5 kHz. This frequency is 32 times the horizontal frequency rate of 15,734 Hz. This frequency is used because solid-state technology makes it easier to stabilize the higher rate. A ceramic resonator is used for frequency stability instead of a crystal. The reason for this is a lower cost factor. Very little frequency stability is lost with the resonator. Synchronization of this master oscillator is done by use of horizontal sync pulses. The output from this IC is a 503.5-kHz signal. This signal is fed to another IC in the system. This second IC is called the vertical countdown IC.

The vertical countdown IC is shown in block form in Figure 16-11. The 503.5-kHz signal is divided by 16 and 2 to develop the 15,734-Hz horizontal

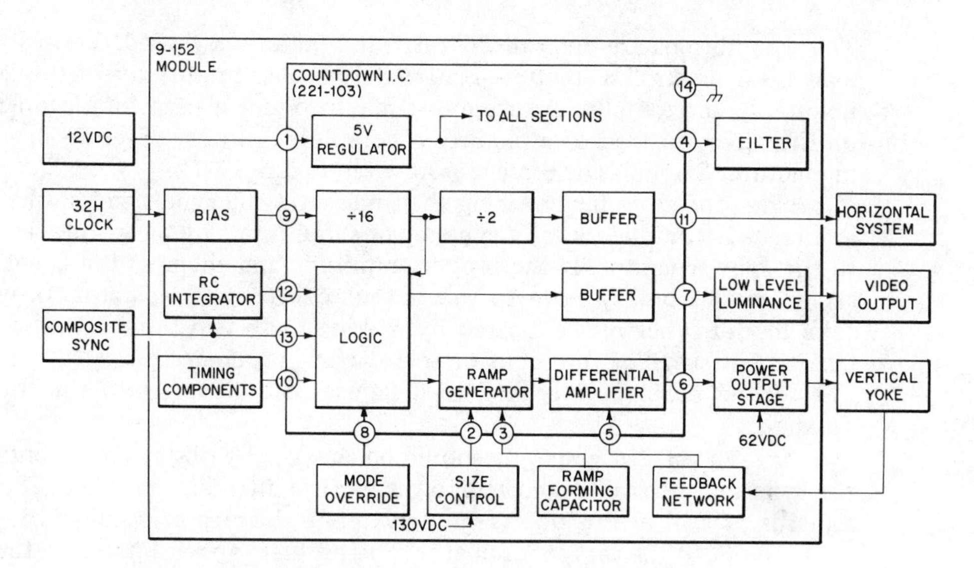

Figure 16-11 Block diagram of a countdown system IC. (Courtesy of Zenith Radio Corporation.)

rate. This signal is in the form of a sawtooth wave. It is fed to the horizontal section of the receiver, where it is amplified. In addition, the signal after it is divided by 16 is fed to a vertical countdown section. Its frequency at this moment is 31,468 Hz. It is now divided by 525 in order to develop the 59.94-Hz vertical frequency rate. The reason for using the 31-kHz frequency is that digital dividers do not easily divide by half numbers. The 15,734-Hz signal would have to be divided by 262.5 in order to develop the 59.94-Hz frequency rate. The correct 59.94-Hz signal is fed from this IC to the vertical amplifier section of the receiver.

Circuits of this type eliminate the need for consumer-adjusted vertical and horizontal frequency controls. In addition, they are more compatible with video-tape recorders. The reason for this is that the two basic vertical and horizontal rates are standardized. This is desirable for sync requirements between machines and receivers.

TROUBLESHOOTING THE SYNC SYSTEM

The initial step in troubleshooting the sync system is to determine the extent of the failure. Three basic conditions may occur. These are:

1. Loss of vertical sync
2. Loss of horizontal sync
3. Loss of vertical and horizontal sync

These conditions are determined when the picture is viewed. Vertical problems result in a rolling picture. Horizontal problems result in a picture that is not stable horizontally. A severe loss may produce a series of diagonal lines on the screen instead of a picture. A complete loss of sync results in a floating picture. Examples of these are shown in Figure 16-12.

The procedure for checking then measures the sync signal in the problem area. A complete loss of sync repair should start by measuring the input to the sync separator. If the proper amplitude and shape of the signal is observed at this point, move to the output of the sync separator. Usually, a total loss of sync will be located by making these two tests. Either there is insufficient signal at the input or there is an improper signal at the output. The second problem is related to a failure of the components in the sync separator.

Loss of vertical sync only should be checked by observing the output of the sync separator as it connects to the vertical integrator network. If proper signal is present at this point, the next test is the measurement of the signal at the input to the vertical oscillator. The problem area is limited to the integrator circuit if one adjusts the receiver vertical hold control and the picture fails to stabilize. It should "lock in." The hold control usually can be rotated

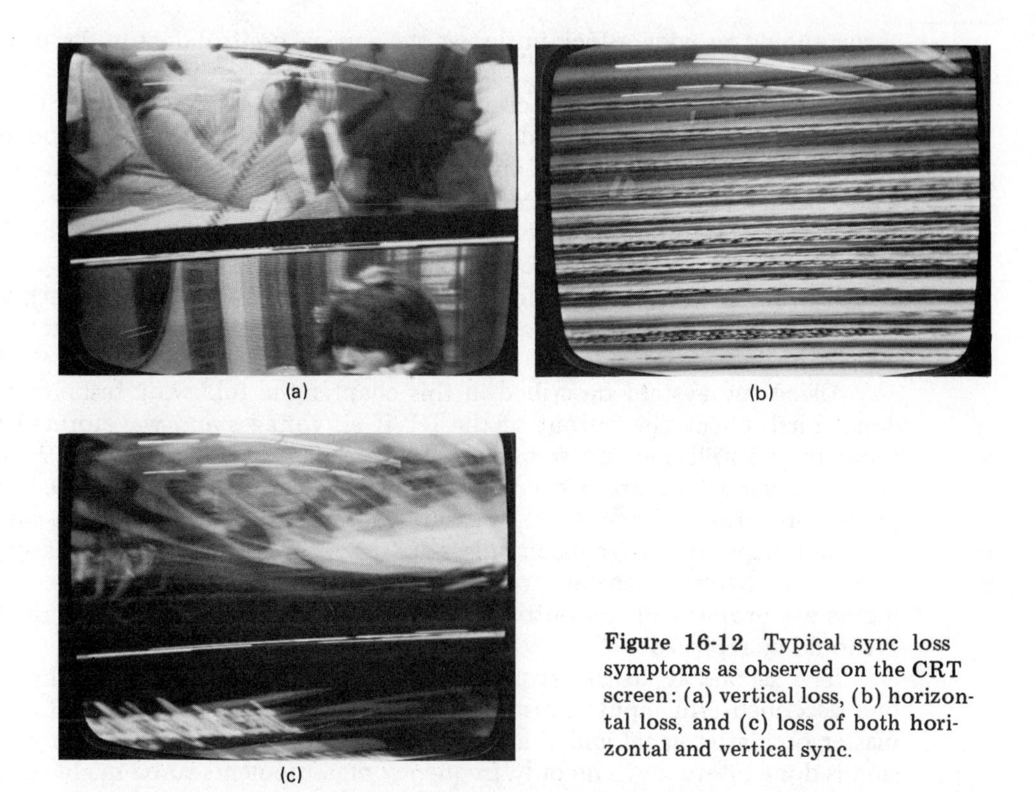

Figure 16-12 Typical sync loss symptoms as observed on the CRT screen: (a) vertical loss, (b) horizontal loss, and (c) loss of both horizontal and vertical sync.

a small amount in either direction before rolling occurs. Failure to lock in indicates a lack of proper sync signal.

Service of the horizontal sync system is a little more complicated. This is because of the addition of the horizontal AFC block. A diagram of this section is shown in Figure 16-13. Two signals must be measured at the input to the AFC block. One of these is the differentiated sync pulse from the sync separator. The second is a feedback pulse from the horizontal output circuit. Both of these must be measured for proper shape and amplitude. The output of the AFC block is a dc control voltage. Its value must be measured if both input signals are correct. Rotation of the horizontal hold control in the re-

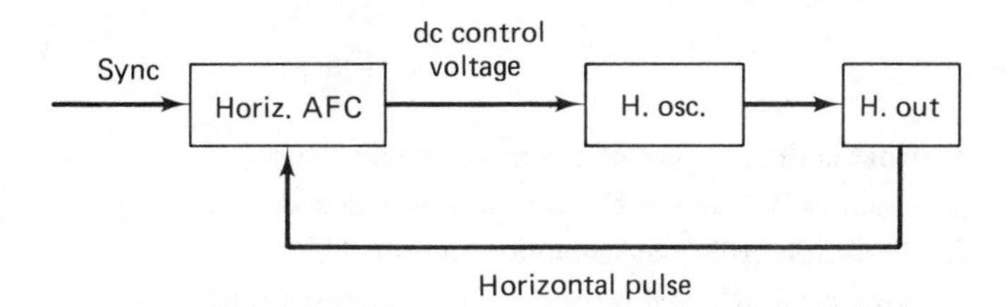

Figure 16-13 Block diagram of a horizontal AFC system.

ceiver should provide a lock-in just as the vertical control does under proper operation.

Countdown and divider circuits require a different approach. This is because a lot of the blocks used are found inside one or two ICs. A frequency counter is a required addition on the test bench when working on these circuits. Once again the literature developed by the receiver manufacturer is required. Service literature from other sources may also be used. The first step is to study the schematic and/or block diagram. Look for test points. Decide what to measure and determine what should be observed at each test point. Also decide which piece of test equipment to use before the measurement is made.

Using the system described in this chapter the following tests may be done. First, check the output of the IC. If all voltages and waveforms have the correct amplitude and frequency, the problem is *after* the IC in the circuit. If the outputs are not correct, go to the input of the IC. Check for proper operating voltage. Next, measure the input signals for correct amplitude and shape. If any feedback pulses are used, these must also be measured. Be sure to check for proper polarity as well as the correct amplitude. If the inputs are proper and the output is not, the problem is probably in the IC. It should be replaced.

The second IC in this system is tested in the same manner as the first one described. All inputs must be observed. These include the 503.5-kHz master oscillator signal and the sync signal. In this IC all the frequency division is done internally. The only frequency measurements to be made are the two output frequencies. There is also a luminance, or video, signal processed by the IC. This, too, must be checked. A failure of any of the IC outputs usually indicates the need for replacement.

Probably the most difficult part of troubleshooting this type of system is believing that the results of the tests are correct. For some reason technicians who have gone through the transition of vacuum tubes, transistors, and now integrated circuits have difficulty in accepting the concept of a bad IC. It seems too easy. In reality, it is easy. The main ingredient for successful servicing is the understanding of how the circuit functions. Once this is understood, the next phase is knowing which test is to be made and where to make it. The final part of the process is to analyze the results of the tests.

QUESTIONS

16-1. What is the purpose of the sync signal?
16-2. Where is it found in the transmitted video signal?
16-3. Briefly describe sync separator action.
16-4. What level of signal is used for sync information?

16-5. Which blocks are affected by sync signals?
16-6. What is the countdown system in the receiver?
16-7. What outputs are available from the countdown system?
16-8. What symptoms are observed when sync is not correct?
16-9. What signal flow system is used in the sync system?
16-10. Where should the first check for the sync system be made?

Chapter 17

Color Processing

One of the basic requirements for a color television system was that it had to be compatible with the present black-and-white broadcast system. The black-and-white TV receiver processes video information. This information is converted into luminance information in the CRT. The signal is used to control the intensity of the CRT electron beam. The beam sweeps across the face of the CRT and develops an image when it energizes a segment of the phosphor coating on the inside of the tube. Timing of the position of the beam is done by the sync system in the receiver. To summarize, this system produces an image on the CRT in terms of luminance, or brightness, values.

A color signal requires two additional components. One of these is related to the tint, or hue, of the color. The other is related to the quantity, or saturation. of the amount of color information. Three primary light colors are used to create the color image in the CRT. These are red, green, and blue. Most of us have mixed paint colors. The result of mixing these three colors of paint will produce a brown-black color. That is because we are mixing pigments. When light is mixed, we obtain a different result. The correct values of red, green, and blue light will produce a white color. By definition, white is a lack of color.

Color is transmitted on a 3.579545-MHz suppressed subcarrier. This signal contains the two components required to create the color image in the receiver. One of these is phase modulated. This signal contains hue information. The second is the amplitude-modulated saturation information. The sidebands of the two signals are transmitted together with the luminance in-

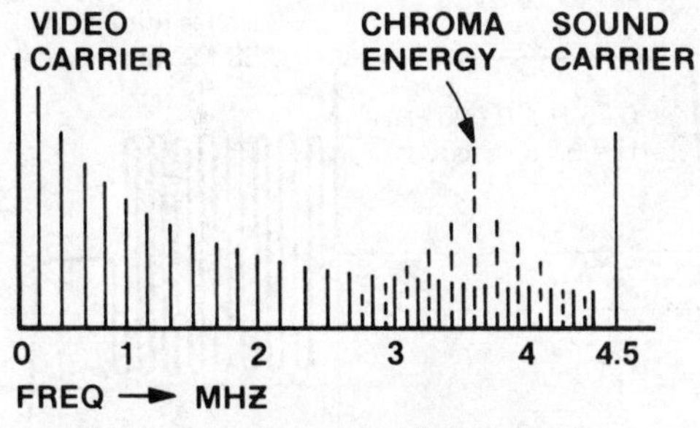

Figure 17-1 Placement and distribution of signals used in the receiver.

formation in the station carrier. This information is placed 3.579545 MHz above the station carrier frequency. Figure 17-1 illustrates the placement of this subcarrier.

Because the color subcarrier is not transmitted, it must be re-created in the receiver. This is necessary to demodulate the carrier. The reinsertion of the carrier is done by use of a 3.579545-MHz oscillator in the receiver. This oscillator's frequency has to be exactly that of the station's subcarrier generator. This is done by the addition of another signal on the station carrier. This signal is called the *color burst*. It consists of a minimum of eight cycles of a 3.579545-MHz signal. The color burst signal rides on the "back porch" of the horizontal blanking pulse. It serves two purposes in the receiver. One is to synchronize the two color carriers. The second is to turn on the color-processing blocks in the receiver.

In the receiver a pulse from the horizontal sweep section is used to key the burst amplifier section. This is shown in Figure 17-2. Once the burst is processed, it is applied to the color oscillator section as a reference for oscillator phase correction. The amplified burst is also applied to the color killer block, where it is used to turn on the color-processing section of the receiver.

The process of adding color information to the black-and-white signal transmission was a complex one. The color information had to be added in such a manner that it would not interfere with black-and-white information. Studies of the distribution of the luminance signal information showed that this information had a tendency to cluster at multiples of the horizontal scanning rate. In other words, a carrier that required a band width of 4 MHz actually occupied spaces that are multiples of 15,734 Hz. The development of color uses a frequency that is half of a multiple of the luminance signal information. This allows the color information to cluster midway between clusters of luminance information. This is illustrated in Figure 17-3. This principle is used to separate luminance and chrominance information. An explanation of this is given later in the chapter.

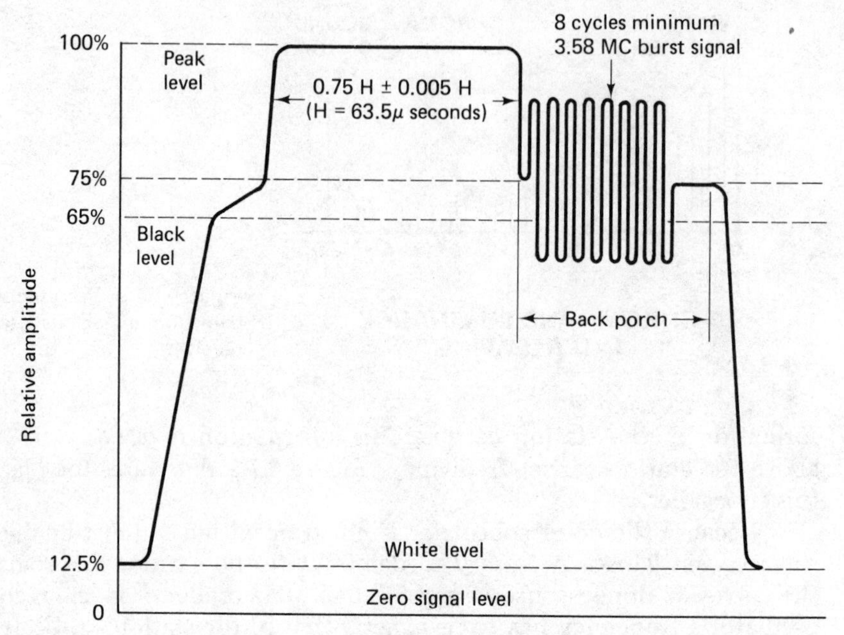

Figure 17-2 Signals used to key the color burst system in the receiver.

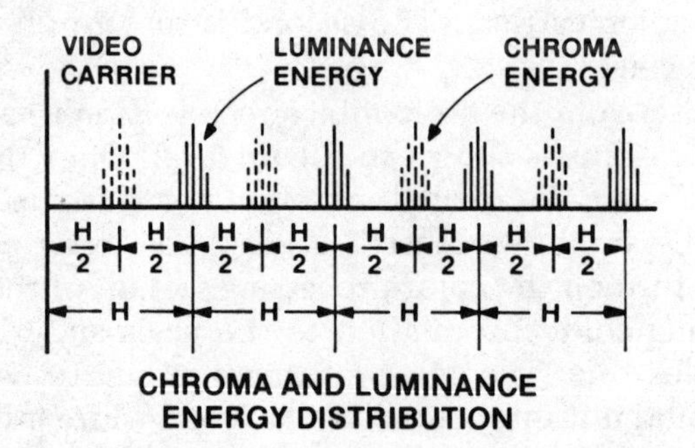

Figure 17-3 Chroma and luminance energy distribution. (Courtesy of Magnavox Consumer Electronics.)

BASIC COLOR PROCESSING

There are several ways of processing the chroma portion of the signal. One system is shown in Figure 17-4. A composite video signal is fed to the first chroma amplifier from the video section of the receiver. It passes through a high-pass filter which eliminates the luminance portion of the signal. This signal contains both luminance and burst information. The chroma signal is amplified and coupled to the chroma output stage. A color-level control is placed between the bandpass and output amplifier stages. In addition, the signal is fed to the burst amplifier from this point. Two signals are required in order to turn on the burst amplifier. One is the burst signal. The second is a keying signal from the horizontal sweep section of the receiver. The output of the burst amplifier is fed to the killer block. It is used to turn off the killer circuit. This permits the color section controlled by the killer to operate.

In addition to the control of the killer, the burst signal is used as a reference for the color oscillator in the receiver. The color oscillator's phase must be identical if the correct colors are to be reproduced. When the color oscillator signal in the receiver is in phase with the color oscillator signal from the transmitter, the true colors are reproduced in the receiver. The color information is transmitted as two signals. These signals have a 90° phase shift. This must also be reproduced on the receiver.

The color carrier signal is separated into two carriers that have the 90° phase difference in the phase-splitter block. These signals are amplified and sent to two demodulators. These are identified as the R-Y and the B-Y demodulators. This reference is to red less Y or luminance information and to blue less luminance information. In some receivers these are labeled X and Z demodulators.

A second signal is required for demodulation. This is the luminance signal. Both are injected into the demodulators. The output of the demodulators is comprised of the two color-difference signals. These have a 90° phase shift. The color difference signals are applied to two amplifiers. These amplifiers serve two purposes. One is to amplify the chrominance signal. The other is to develop the third color signal. Portions of the R-Y and B-Y signals are fed to the G-Y amplifier. These signals are mixed. The result is the development of the G-Y signal information.

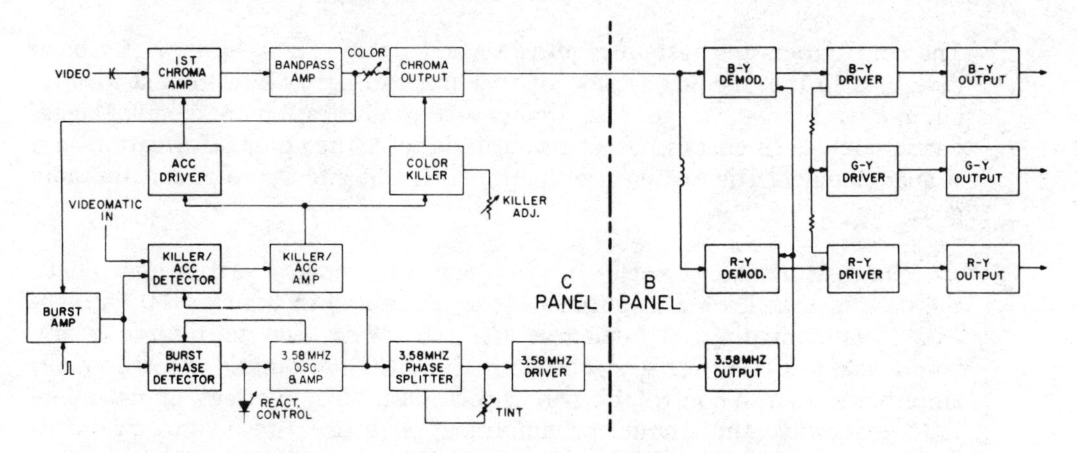

Figure 17-4 Block diagram of the color-processing blocks in a receiver. (Courtesy of Magnavox Consumer Electronics.)

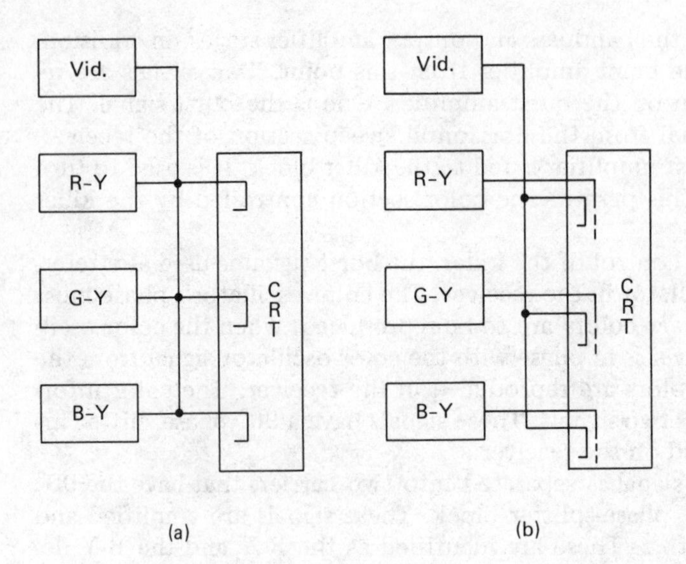

Figure 17-5 Methods of mixing luminance and color information for the CRT system.

The output of the three color-difference signals is mixed with the luminance signal. This is done to re-create the composite signal required for CRT operation. Two methods of doing this are shown in Figure 17-5. One method combines the two signals before they enter the CRT. The recombined color and luminance signal is then fed to the cathodes of the CRT. A second method injects the color signal at the control grid of the CRT. The luminance signal is fed to the cathode. The levels of the two signals then control emission from each of the three electron guns in the CRT.

DISCRETE COMPONENT COLOR PROCESSING

The color-processing system requires several stages. These have briefly been described in the opening section of this chapter. Let us now look at specific circuits used in each stage. The circuits used as illustration are from a Magnavox Model T979 chassis. Other receivers process the color information in a similar manner. They often use slightly different circuits to reach the same goal.

Bandpass and chroma amplifiers. A composite video signal is injected to the base of the chroma amplifier, Q_1, as illustrated in Figure 17-6. A high-pass filter consisting of capacitors C_1 and C_2 is used to filter out any luminance information. A signal from the ACC (automatic color control) amplifier is used to control the gain of this stage. This, in effect, operates like AGC to control the amount of amplification in this stage. Chroma information is amplified by the bandpass amplifier Q_2 and the chroma output amplifier Q_3. It is then coupled to the demodulators.

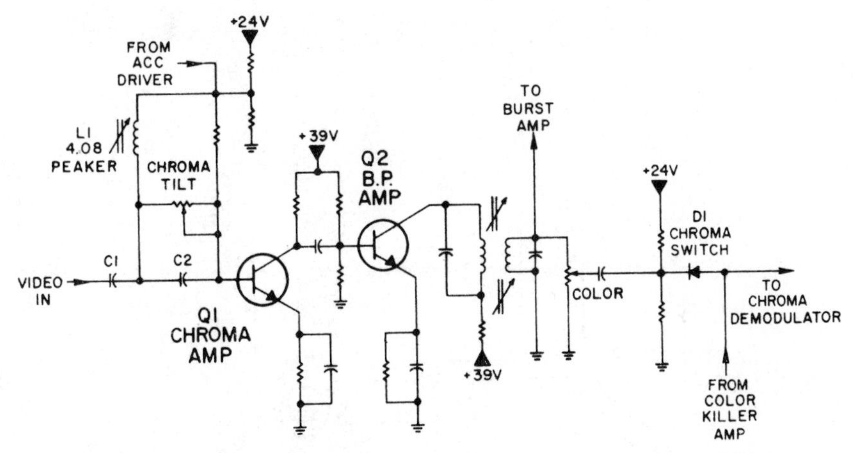

Figure 17-6 Chroma bandpass amplifier circuit. (Courtesy of Magnavox Consumer Electronics.)

The burst signal has to be kept out of the chroma amplifier circuits. This is done by use of a diode, D_1. This diode is normally forward biased. When a pulse from the horizontal output section is applied to the cathode of the diode, it becomes reverse biased. This shuts off signal flow during the period of time when burst is transmitted. The result of this action keeps the burst signal from entering the chroma output section.

Color killer control is accomplished by turning the chroma output stage on or off. A bias voltage is applied to the base of Q_3. This voltage will bias the output stage in either an on or an off state. The color killer action to develop this bias voltage follows.

Burst, ACC, and color killer. A chroma signal is coupled from the secondary of the bandpass transformer (Figure 17-6) to the base of the burst amplifier Q_7. The burst amplifier is shown in Figure 17-7. A pulse from the horizontal sweep is also applied to the base of this transistor. When both signals are present, the transistor is biased on. It conducts only during the period when the burst is being transmitted. Any chroma information is blocked from this stage because it is off during that portion of the line transmission.

The output of the burst amplifier is connected to two different circuits. One of these is out of phase with the other. Each is compared to the 3.579545-MHz color oscillator signal. The circuit used for comparison is similar to that used for horizontal AFC. The output from each circuit is a dc control voltage. This correction voltage is connected to a varactor diode in the oscillator circuit. The correction voltage from the phase-locked-loop circuit adjusts the frequency of the color oscillator.

The second output from the burst amplifier is used as an AGC control for color level. This is done by use of a second PLL circuit. This circuit is

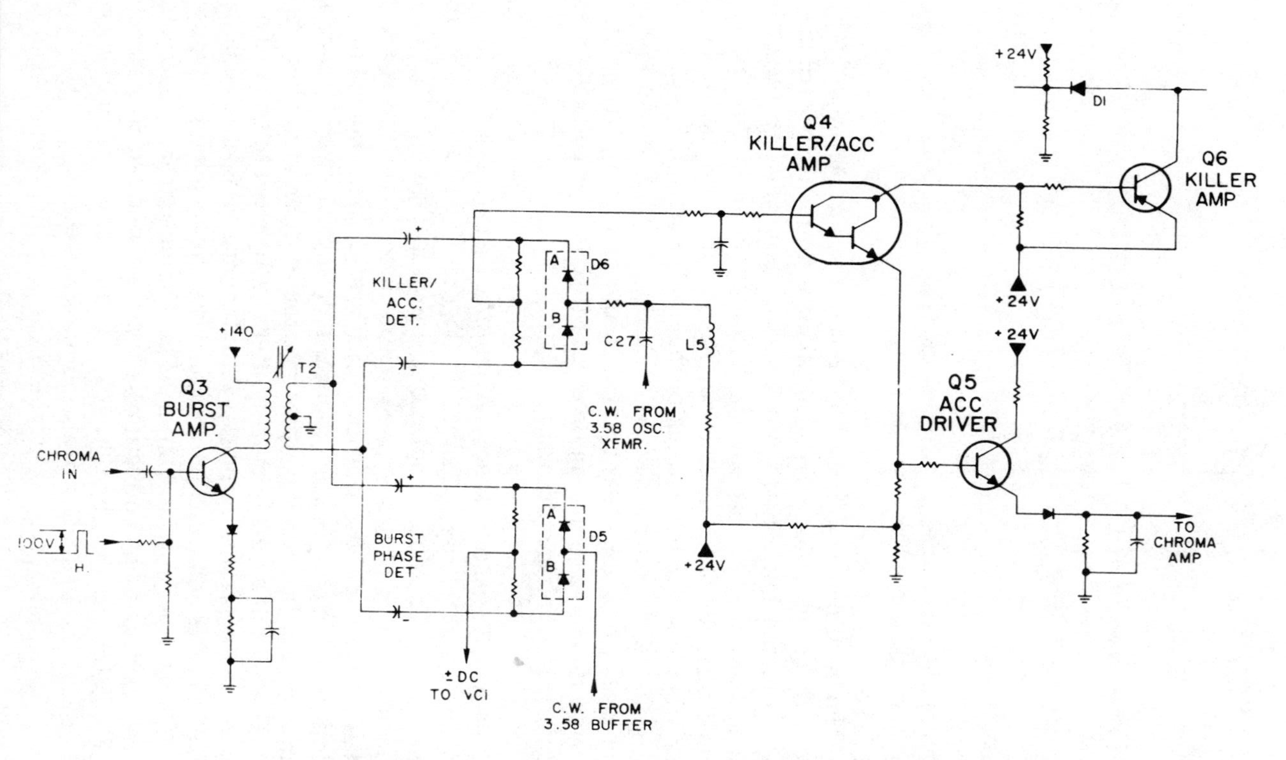

Figure 17-7 Burst amplifier circuit. (Courtesy of Magnavox Consumer Electronics.)

connected to Q_6, which is the color killer and automatic color control (ACC) amplifier. The output of this transistor is used to establish a bias on the driver transistor, Q_5. This transistor will decrease its gain as its bias is increased. The result is a fairly constant level of color in the receiver.

The color killer section is used to bias the chroma output, Q_3. The output of the killer circuit is dependent upon the level of burst present at the base of Q_6. This is used to control the bias on the killer transistor Q_4. The output of Q_4 will bias the output transistor Q_3 into conduction when a burst signal is present. When there is no burst signal, this transistor is biased into cutoff. The result is a shutdown of the color output blocks.

Color oscillator. This circuit is shown in Figure 17-8. A crystal oscillator is used to establish the carrier frequency of 3.579545 MHz. A varactor diode, VC, is used to adjust the oscillator to the exact frequency. Minor adjustments of this frequency are accomplished by the dc correction voltage from the burst amplifier. The carrier signal is amplified by transistor Q_9. Outputs of this circuit are coupled to the ACC and burst detectors. These are feedback signals used to control respective stages. Another output is connected to a phase splitter. This circuit inverts one signal and shifts the signal by 90°. Varactor diode VC_2 is used together with the tint control to shift the output signal to a maximum of 55° each side of the 90° point. The shift in frequency permits a range of tints for the color picture.

Color demodulators. The color demodulator circuit is shown in Figure 17-9. It consists of one transistor and two phase-detector networks. The output from the transistor is coupled to a phase-splitting transformer. This transformer has two output connections. Each is connected to a demodulator network. In addition, the 3.579545-MHz color carrier is reinserted by applying it to the transformer windings. A chroma signal is applied to the diodes in the demodulator network. The output of the demodulators becomes the R-Y and B-Y color signal. The two signals are then fed to color-difference amplifiers.

Color-difference amplifiers. Three color signals are required for correct CRT operation. Two of these are developed in the demodulator section of the receiver. The color difference amplifiers shown in Figure 17-10 are used to amplify existing signals and to develop the G-Y signal. The G-Y signal is developed by taking a portion of the R-Y and B-Y signals from the emitters of their amplifiers. These two signals are mixed and the result becomes the G-Y signal. This is also amplified in the system. All three color-difference signals are given additional amplification by the output stages. The three signals are then coupled to the proper element of the CRT. They recombine with the luminance signal to create the color images viewed on the screen of the CRT.

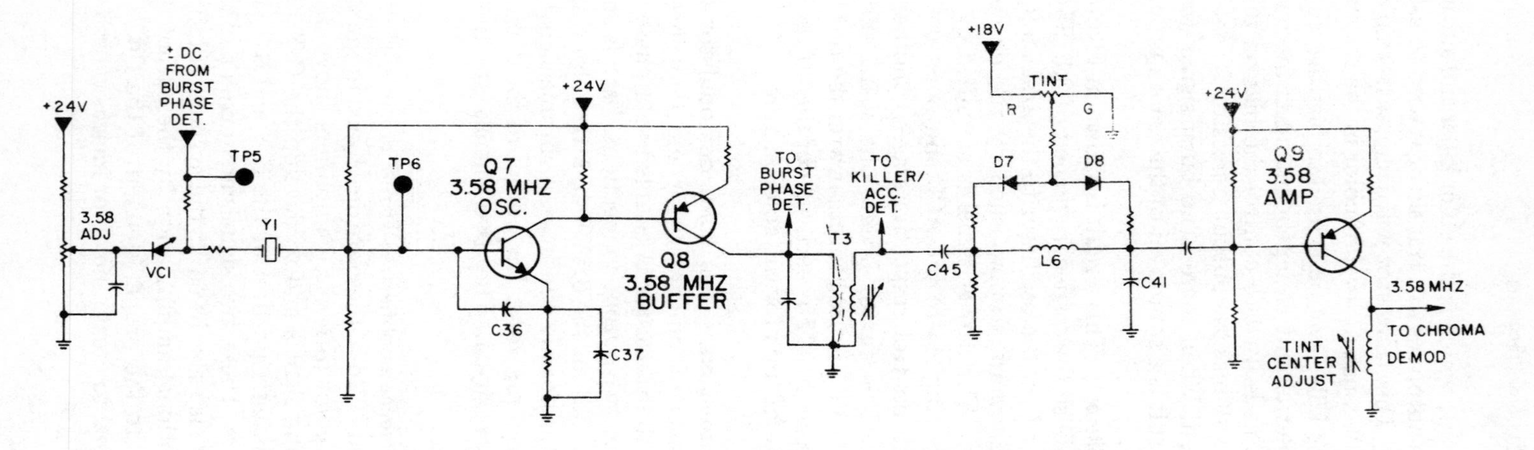

Figure 17-8 The color oscillator and phase control circuits found in a color receiver. (Courtesy of Magnavox Consumer Electronics.)

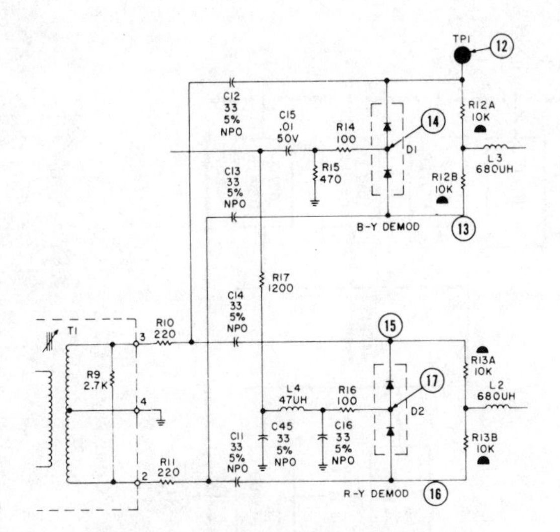

Figure 17-9 Typical chroma demodulator circuit. (Courtesy of Magnavox Consumer Electronics.)

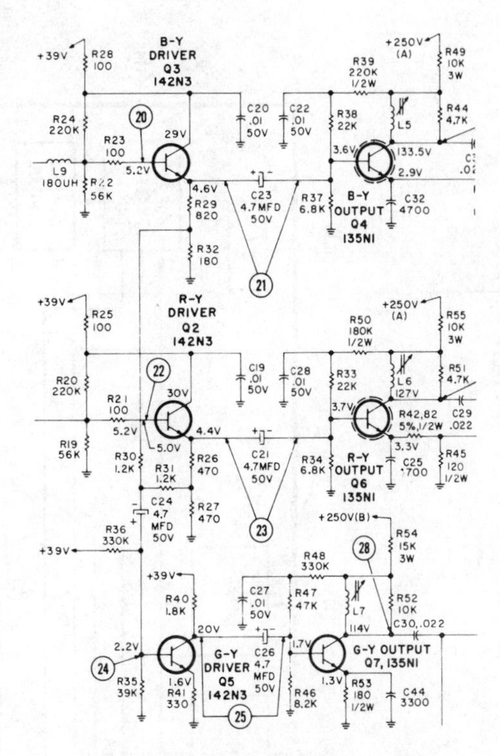

Figure 17-10 Three color-difference amplifiers are required for proper color display. (Courtesy of Magnavox Consumer Electronics.)

IC COLOR PROCESSING

Hopefully, the message that IC technology has simplified servicing is reaching you by this time. Figure 17-11 shows a block diagram for *one* color-processing IC. This single IC replaces almost all of the discrete components described in the beginning of this chapter. Some of the terms used in earlier sections of this chapter and the preceding chapter are also used to describe the functions of this IC. The 3.579545-MHz oscillator is simply called a VCO, or voltage-controlled oscillator. The other blocks have names that are also similar to those used to describe discrete component blocks. A partial schematic for a receiver using this IC is shown in Figure 17-12. Almost all the functions are obtained with a few inputs to the IC. There are three basic outputs. These are the three color difference signals. They are coupled to

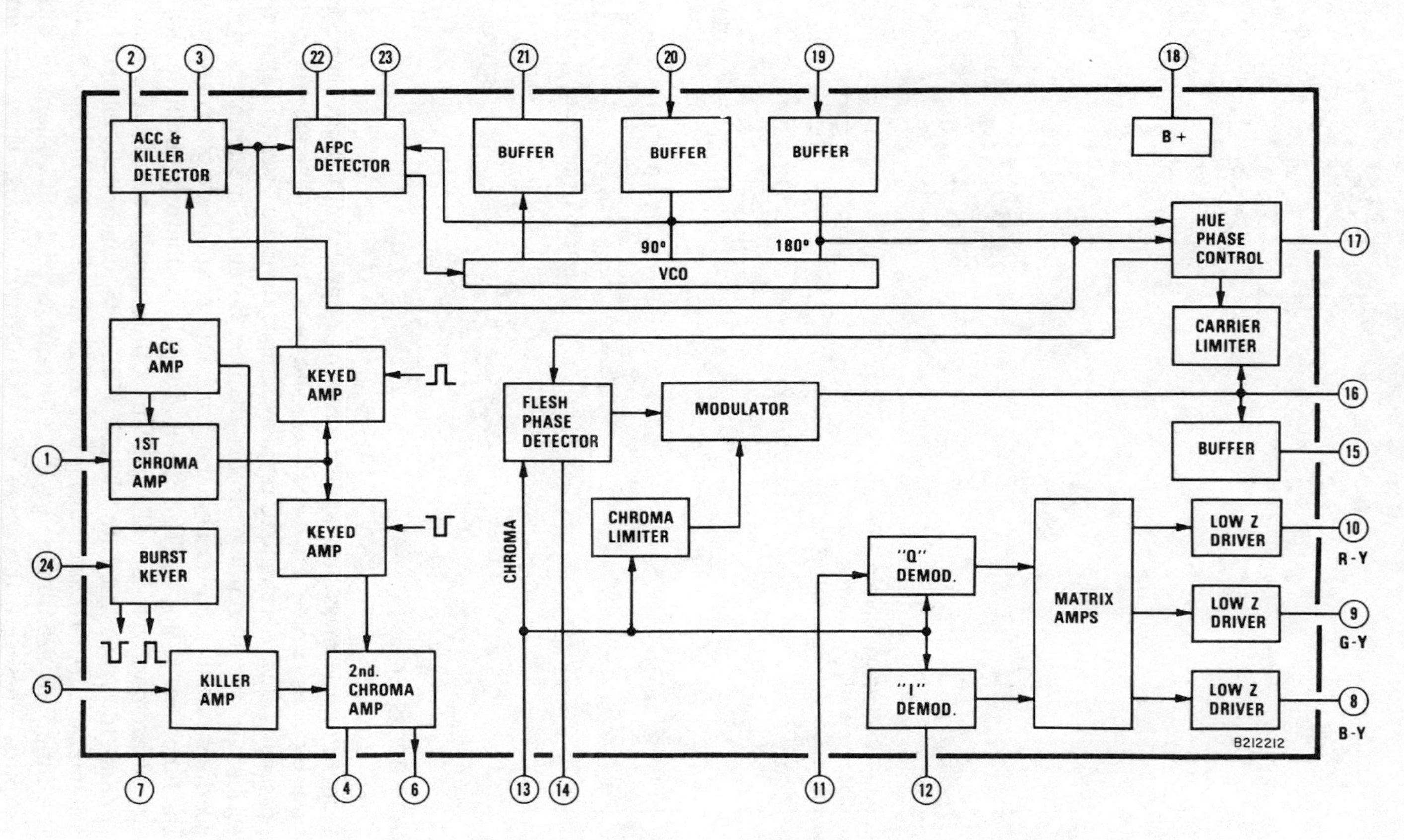

Figure 17-11 Block diagram of an IC color processing block. (Courtesy of RCA Consumer Electronics.)

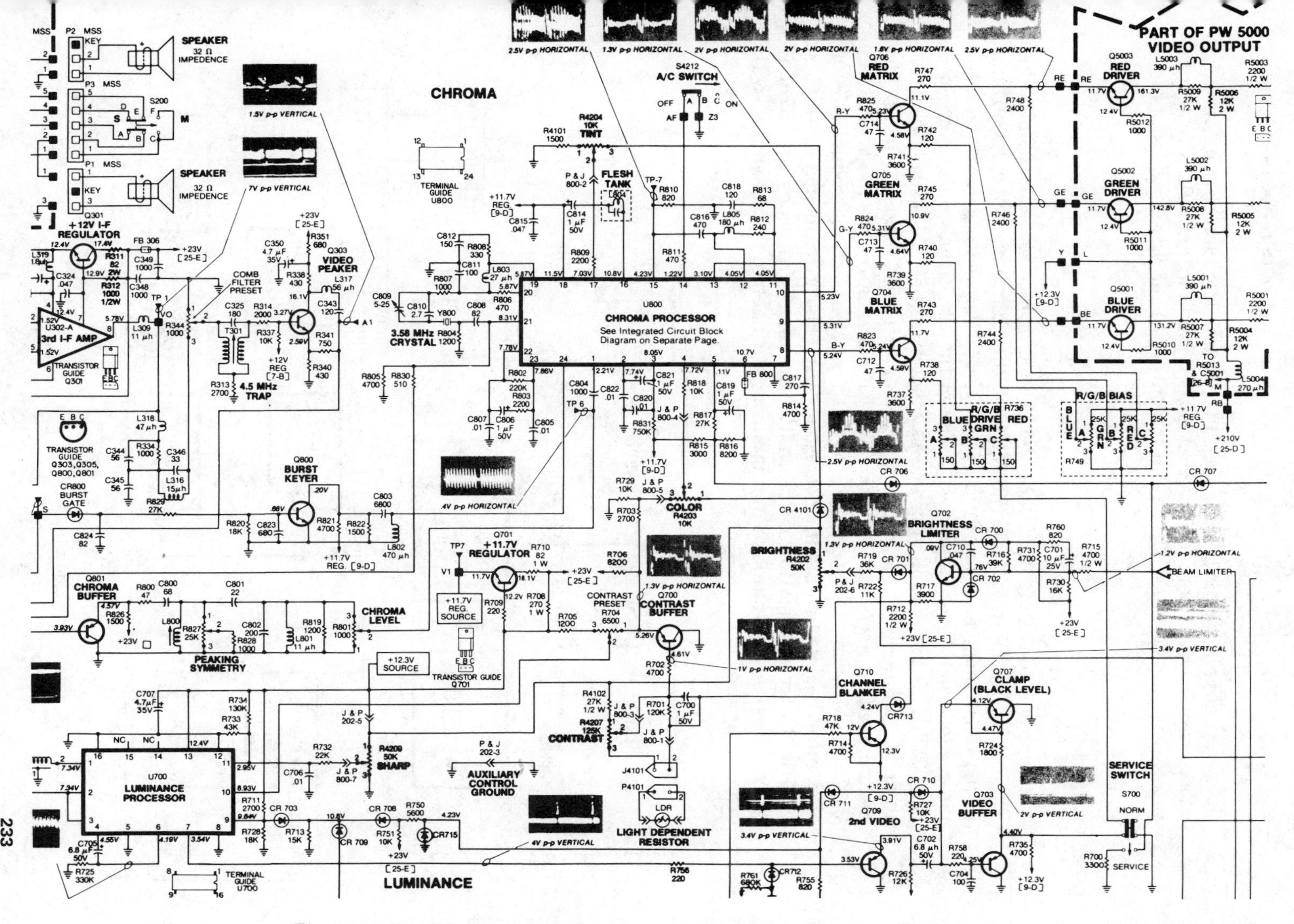

Figure 17-12 Schematic diagram for an IC color-processing section. (Courtesy of RCA Consumer Electronics.)

their respective matrices. In this section the luminance and chrominance signals recombine. The composite signal is then fed to an amplifier/driver transistor. It is coupled to the CRT from the output of the driver transistors.

COMB FILTERS

A recent entry into the circuits found in a color receiver is the comb filter. This circuit is used to separate chroma from the luminance signal. Its position in the block diagram is shown in Figure 17-13. Before this circuit was introduced, one could see interaction between the two signals on the screen of the CRT. This interaction appeared as random colors. It is usually observed in areas of fine detail, such as men's ties or patterned jackets. Normal procedure used to keep chroma out of the luminance system was to install a 3.579545-MHz trap. The addition of this trap circuit also introduced distortion into the picture. Luminance signals often have frequencies as high as 4 MHz. Color information is transmitted around a frequency of 3.579545 MHz. The use of a trap kept any luminance signal frequencies above 3 MHz from being used.

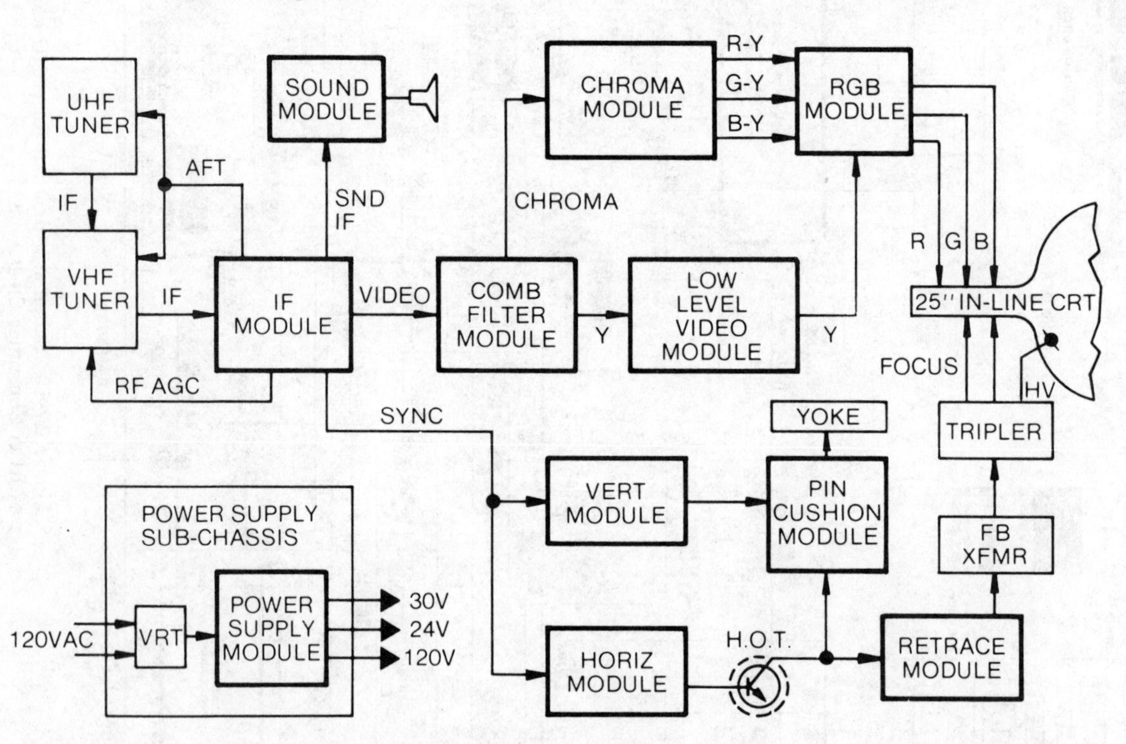

Figure 17-13 Position of the comb filter in the color receiver. (Courtesy of Magnavox Consumer Electronics.)

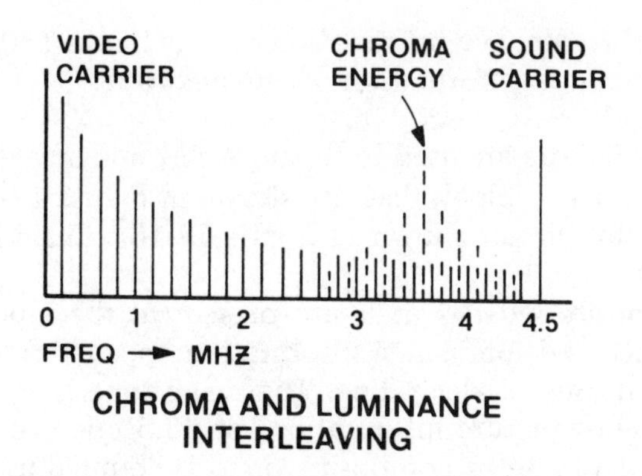

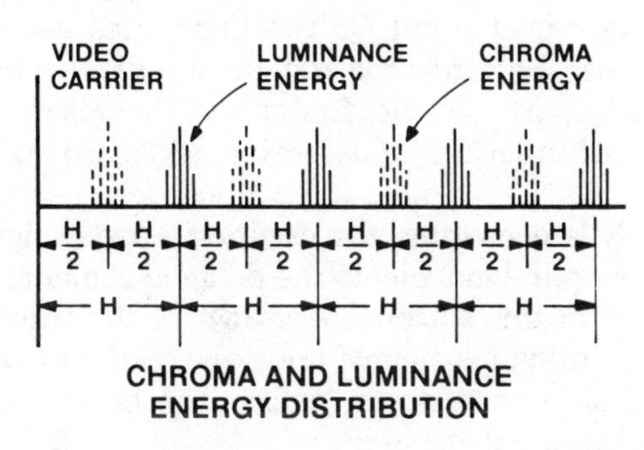

Figure 17-14 Clustering effect of chroma and luminance signals on the signal. Chroma and luminance energy distribution. (Courtesy of Magnavox Consumer Electronics.)

The comb filter takes advantage of the fact that chrominance and luminance signals tend to cluster around different harmonics of the horizontal frequency. This clustering effect is shown in Figure 17-14. The clustered frequencies appear as teeth of a comb, hence the name of this filter. When one observes the signal present in each line of a field, some conclusions can be made. These are:

1. The information on one line is not very different from the information on the next line.

2. The chrominance information on one line is 180° out of phase with the chrominance information on the next line.

These two factors are used to separate chrominance and luminance signals in the receiver. The block diagram shown in Figure 17-15 and the set of waveforms for this circuit shown in Figure 17-16 will aid in the understanding of this circuit.

Composite video signals are present at the input of this system. Both chrominance and luminance information appears on this signal. The signal is processed through a delay line. The delay time is equal to the time required for one line of picture information, or 63.5 microseconds (μs). At the same time, a second line of composite video is coupled around the delay system. Both signals appear at point C. The luminance information is almost identical. The chrominance information on the second line is 180° out of phase with its counterpart on the first line. The result is a cancellation of the chrominance information. This action is shown in the upper three lines of the diagram.

While this is occurring, the composite video signal passes through an inverter. The inverted signal meets the delayed signal at point D. The two waveforms resulting in this action are shown as the fourth and fifth lines of the drawing. The luminance signals are now 180° out of phase with each other. The chrominance signals are in phase. The chrominance signals add and the luminance signals are canceled.

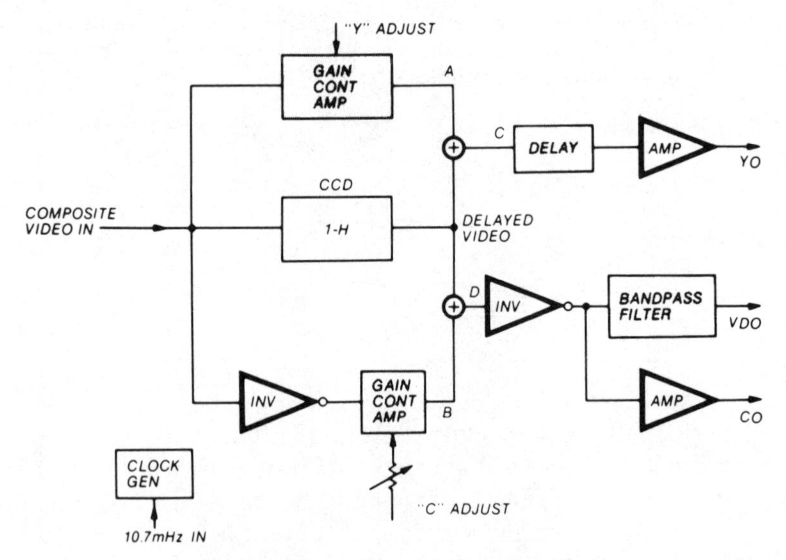

Figure 17-15 Block diagram of a comb filter system. (Courtesy of Magnavox Consumer Electronics.)

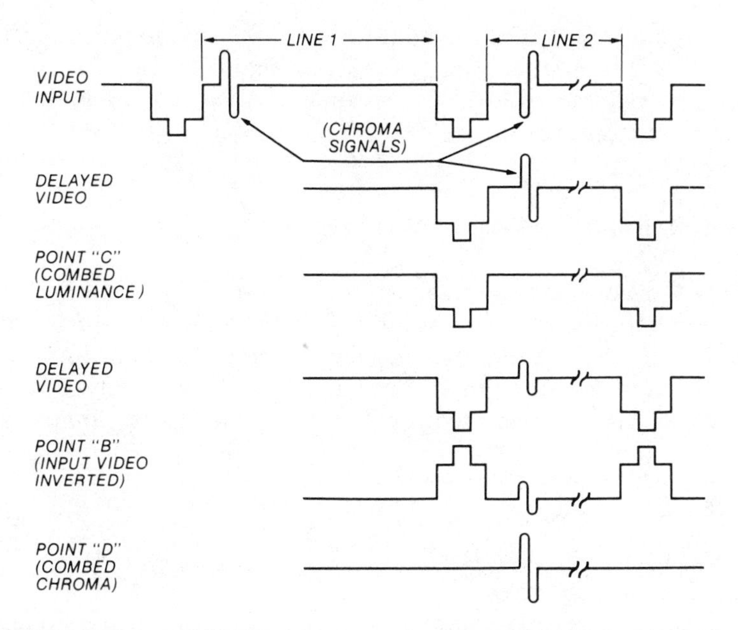

Figure 17-16 Signal processing in the comb filter system. Phase reversal of signals cancels chroma (lines 1, 2, and 3) or video (lines 4, 5, and 6). (Courtesy of Magnavox Consumer Electronics.)

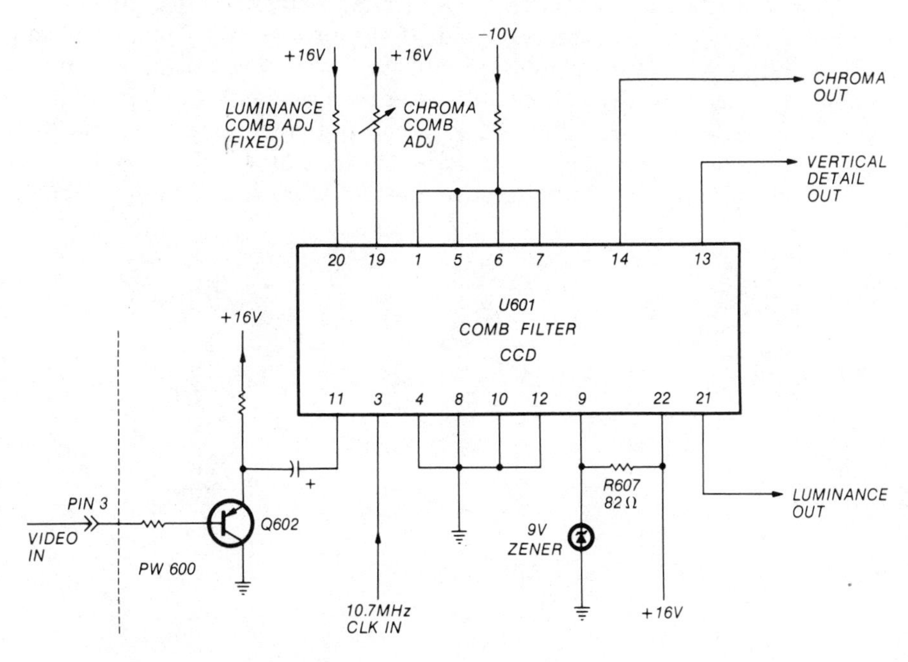

Figure 17-17 Comb filter IC and its connections in the receiver. (Courtesy of RCA Consumer Electronics.)

All of this circuit is contained in one integrated circuit. A diagram of the IC and its connections is shown in Figure 17-17. The 10.7-MHz clock signal is developed as a third harmonic of the color oscillator frequency. The signal is used in the operation of the components in the IC.

TROUBLESHOOTING COLOR-PROCESSING SECTIONS

Consideration has to be given to the types of signals that are processed by this section before any troubleshooting can be done. Each of the input signals has to be checked. These are the burst, luminance, chrominance, and keying pulses from the horizontal section. In addition to these, the 3.579545-MHz oscillator signal is also required. Problems in this section can be classified as:

1. Lack of any color
2. One color missing
3. Two colors missing
4. Barber-pole color

Some of these problems can be checked by operation of the receiver controls. Rotation of the color control will quickly indicate if any color is present. The color killer control may require adjustment. Also, the hue, or tint, control may have been misadjusted. If the proper adjustment of these controls does not restore proper color, additional troubleshooting is required.

A lack of any color could be caused by a color killer circuit defect. The color killer circuit failure could keep the color circuits biased off. Another reason for loss of all color could be the lack of a 3.579545-MHz color carrier. This will stop the demodulation system from functioning properly. Another

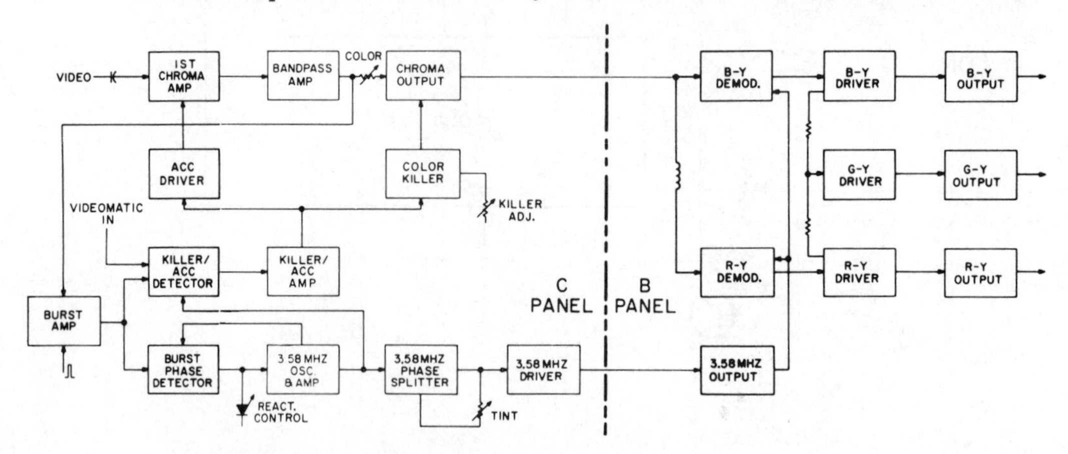

Figure 17-18 Block diagram of a color receiver color-processing section. (Courtesy of Magnavox Consumer Electronics.)

cause for a lack of color is a failure in the bandpass circuit. If a stage failed, the chrominance signal could not reach the demodulators.

The best tool to use in troubleshooting a color section is the oscilloscope. Each of the circuits described earlier in this chapter has typical waveforms displayed. The service literature for the receiver also has waveform information. Test points are shown on schematics. The service technician needs to refer to these to be effective. A block diagram for a receiver color processing section in shown in Figure 17-18. A systematic approach for troubleshooting this section follows.

The first thing to do is to measure the input waveforms. If these are correct in amplitude and shape, move on to another point. This system was a combination of signal paths. An excellent place to make a second set of tests is at or near the midpoint. A convenient place to do this is at the demodulator. Two signals are required at the input of the demodulators. One signal is measured at the output point. A test point that would quickly tell if both inputs are present is at the output of the demodulators. If correct signal is present, all sections between the input and the test points are good. The drivers and output blocks are wired as a series signal path system. These are tested as linear paths by splitting them at the midpoint. This quickly establishes the area of trouble. Once the area of trouble is located voltage and resistance measurements are made to determine the defective component.

A lack of one or two colors usually indicates a failure in either one demodulator or in the color amplifiers. Waveform analysis will quickly determine if these blocks are working properly. The barber-pole effect is produced when the color oscillator in the receiver is off frequency. The viewable result appears as stripes of colors. These color stripes run diagonally across the face of the CRT. The adjustment for this is described as an AFPC (automatic frequency phase control) adjustment in most service literature.

QUESTIONS

17-1. What components are found in the color signal that are not found in a noncolor signal?

17-2. How is the color information transmitted?

17-3. What is "burst"?

17-4. How is burst used in the receiver?

17-5. What is the purpose of the color killer system?

17-6. Why is a 3.579545-MHz oscillator used in the color receiver?

17-7. What does the color demodulator do in the receiver?

17-8. What is meant by the terms "B-Y," "R-Y," and "G-Y"?

17-9. Why is the comb filter used in a color receiver?

17-10. Describe typical color problems.

Chapter **18**

Vertical Sweep

Movement of the electron beam that is created in the CRT is accomplished by the sweep systems in the receiver. There are two of these systems. One of them provides horizontal movement of the beam. This system creates each of the 525 lines of the raster. The second system provides vertical movement to the electron beam. Both of these systems create the white raster on the face of the CRT. The combination of the effects of the output of both the vertical and the horizontal sweep systems moves the beam around the face of the CRT in an orderly fashion. Pulses generated at the transmitter provide synchronizing signals. These are required so that the camera tube electron beam and the CRT electron beam are in the same positions at all times.

Each of the two scanning systems develop a sawtooth wave. This wave is amplified as it passes through the stages of the sweep system. An output transducer, called a deflection yoke, is used to create a magnetic field. The deflection yoke is placed around the neck of the CRT. It is used to bend the electron beam so that it will travel to all points on the face of the tube. This chapter covers material related to the vertical deflection system of the receiver.

A block diagram of a typical vertical deflection system is shown in Figure 18-1. Until the introduction of vertical countdown circuits, all receivers had a vertical oscillator block. Its function is to develop a sawtooth waveform. The frequency of this wave is 60 Hz for a black-and-white receiver and 59.94 Hz for a color receiver. These are the basic frequencies for the vertical system of the receiver. A sync signal from the sync separator is integrated by a low-pass filter network. The output of the integrator net-

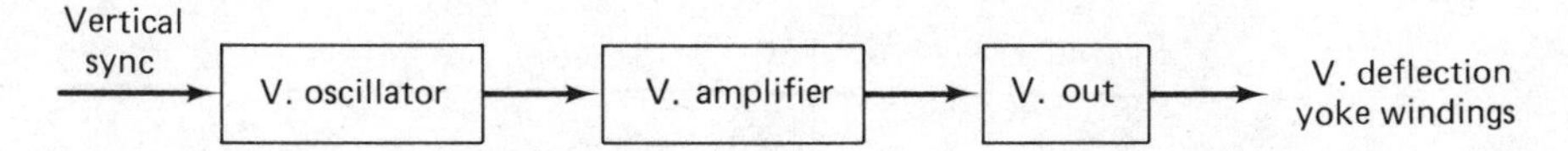

Figure 18-1 Block diagram of a vertical deflection system.

work is a control voltage. This voltage is used to sync the vertical oscillator. The vertical oscillator is synchronized with the camera signal in this manner.

After the sawtooth wave is developed, it is amplified. The number of stages of amplification depends upon the design of the receiver. Amplitude of the signal is controlled by a vertical height control. The shape of the wave may be modified to some degree by a linearity control. A third control, called vertical hold, is used to put the oscillator on its fundamental frequency.

After being amplified, the signal is fed to an output stage. Further amplification occurs in this stage. In addition, the output of the stage is coupled by an impedance-matching device to the vertical windings of the deflection yoke. The current passing through this yoke is used to develop a magnetic field in the yoke windings. The magnetic field provides vertical positioning of the electron beam. The field intensity changes in the windings. This is due to the polarity and strength of the output current. The varying strength of the magnetic field is used to give movement to the electron beam.

The beam movement is controlled so that it covers the full vertical field of the CRT 60 times a second. Two fields are produced. One of these develops the odd-numbered lines of the picture. At the completion of its cycle, a second field is created. This field develops the even-numbered lines of the picture. Together the total number of lines create the visual image on the screen.

VERTICAL OSCILLATORS

A circuit for a solid-state vertical oscillator is shown in Figure 18-2. The operation of this circuit is dependent upon the charging time of capacitor C_{17}. Operating power is applied to the circuit from the horizontal output transformer. The pulses are rectified by diode D_1 and filtered by C_{11}. The output of this scan-derived supply is 25 V dc. A voltage-dividing network consisting of R_1 and R_9 applies a positive voltage to the base of the oscillator transistor Q_1. This positive voltage holds the transistor in a cutoff mode. Capacitor C_{17} is charged by the positive voltage source. This capacitor is connected to the emitter of the transistor. The emitter voltage raises due to the charging action of the capacitor. When the emitter voltage increases by about 0.7 V the transistor switches on and C_{17} discharges. This causes the oscillator transistor to shut down. This cycle is repeated. Its output is the sawtooth wave required for vertical sweep.

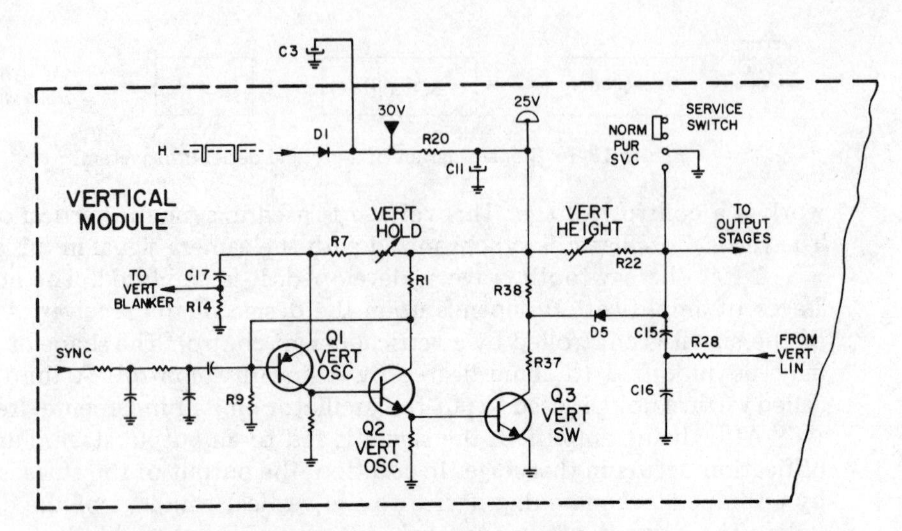

Figure 18-2 Schematic diagram of a solid-state vertical oscillator. (Courtesy of Magnavox Consumer Electronics.)

Synchronization of the oscillator is accomplished by use of the integrated sync signal. It is injected to the base of the oscillator, Q_1. Basic frequency control is accomplished by adjusting the charging rate of capacitor C_{17}. Turning the vertical hold control will vary the voltage on this capacitor. This will control the charging rate.

Transistors Q_1, Q_2, and Q_3 are wired in a Darlington circuit. When Q_1 is turned on, it will turn on the other two transistors. The varying voltage out of transistor Q_3 is used to charge capacitors C_{15} and C_{16}. Transistor Q_3 is used as a switch. It is controlled by the output of Q_2. The collector voltage of transistor Q_3 is used as the sawtooth output voltage of the oscillator section. This sawtooth waveform is fed to the vertical amplifier stages. The height control controls the amount of amplification in this stage. This is done by varying the dc level of the output sawtooth waveform. A "service" switch is helpful in a color setup procedure. When it is in the "service" position, the vertical sweep collapses. The result is a horizontal line on the face of the CRT. Collapse of the sweep is done by connecting the vertical signal output to common through the switch.

VERTICAL OUTPUT

The circuit for the vertical output section is shown in Figure 18-3. The sawtooth waveform is fed to the base of transistor Q_6. This, and transistors Q_7 and Q_8, are connected in Darlington configuration in order to obtain high gain. The collector of the output transistor is connected to the vertical windings of the deflection yoke and to inductance L_1. When Q_8 is at minimum

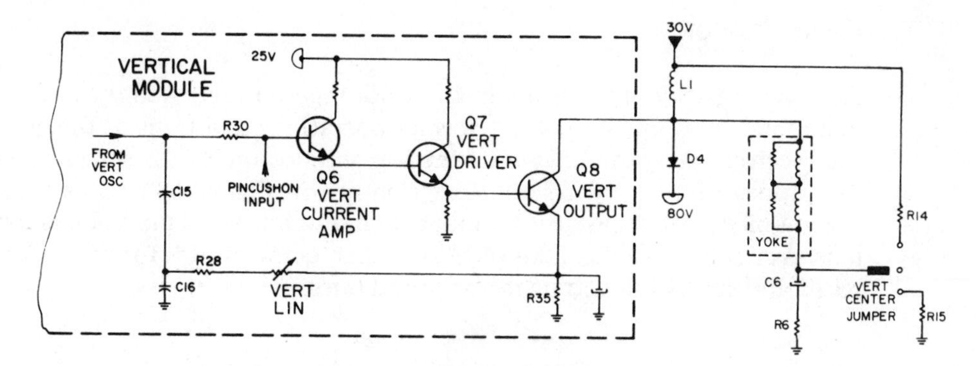

Figure 18-3 Schematic diagram of a vertical output circuit. (Courtesy of Magnavox Consumer Electronics.)

conduction, the output current flows through L_1, the yoke, and R6. This positions the electron beam at the top of the screen. When Q_8 increases its conduction, the current in shunted away from the yoke. This changes the magnetic field in the yoke and forces the beam to move down. It continues the downward motion until the retrace period. During the retrace period the transistor shuts off and the current passes through the yoke. This returns the beam to the top of the screen. The cycle repeats at the vertical rate.

VERTICAL BLANKING

The electron beam is not shown during the first 21 horizontal lines of each field. The result is a blanked screen during the time these lines are transmitted. Information is transmitted to the receiver during this period. The information relates to sync and color information. A vertical blanking circuit schematic is shown in Figure 18-4. Transistor Q_9 is the vertical blanking transistor. It is biased to either the on or the off condition by the charge on capacitor C_{17}. The result is a square-wave output. This signal is fed to the chroma demodulators. The signal blanks, or shuts off, the demodulators during the period it is present.

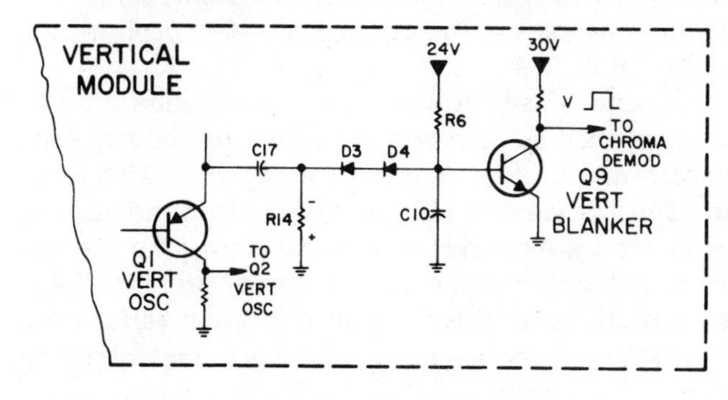

Figure 18-4 Vertical blanking circuit used in a solid-state receiver. (Courtesy of Magnavox Consumer Electronics.)

VERTICAL PINCUSHION

A problem that only appears in the color receiver relates to the shape of the horizontal lines on the raster. Normal operating conditions of the circuits in the receiver cause the lines near the top and bottom of the screen to pull in. The effect of this creates the image shown in Figure 18-5. This is called a "pincushion" effect because it looks a lot like commercial pin holders, or cushions. A circuit included in the vertical section compensates for this effect. The result appears as fully straight horizontal lines.

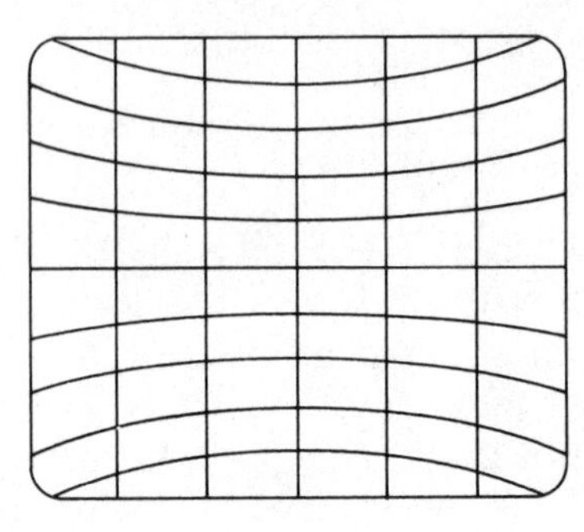

Figure 18-5 A pincushion effect in the receiver is corrected by adjusting a pincushion-effect transformer.

CONVERGENCE

The correct operation of the color CRT requires that the electron beam from each gun hit the correct phosphor dot. During the scanning period the three beams must land on the proper set of phosphor dots. A problem that occurs is that the physical position of the three electron guns does not lend itself to doing this easily. The drawing in Figure 18-6 will aid in understanding this. When the deflection system centers the beams on the face of the CRT, the problem is minimized. During the sweep period the beams have different lengths. This requires an adjustment in the travel time and position. A section of the receiver called the *convergence board* is included in a color receiver. This board has inductive and resistive circuits. These circuits are used to modify the true sawtooth wave in the sweep sections. The result is that the three electron-gun beams will converge on the correct set of phosphor dots at all places on the face of the CRT.

Most receivers have two sets of adjustments for convergence. One of these is called the *static convergence adjustment.* It is used to converge the beams at the center of the screen. This adjustment is usually done with three electromagnets. The static convergence assembly is mounted on the neck of the CRT behind the yoke. A separate convergence board is used for the rest of the adjustments. These are called *dynamic convergence adjustments.* Each receiver manufacturer has instructions for this procedure. It often varies from one brand to another. It is best to follow the procedure. Each control of the

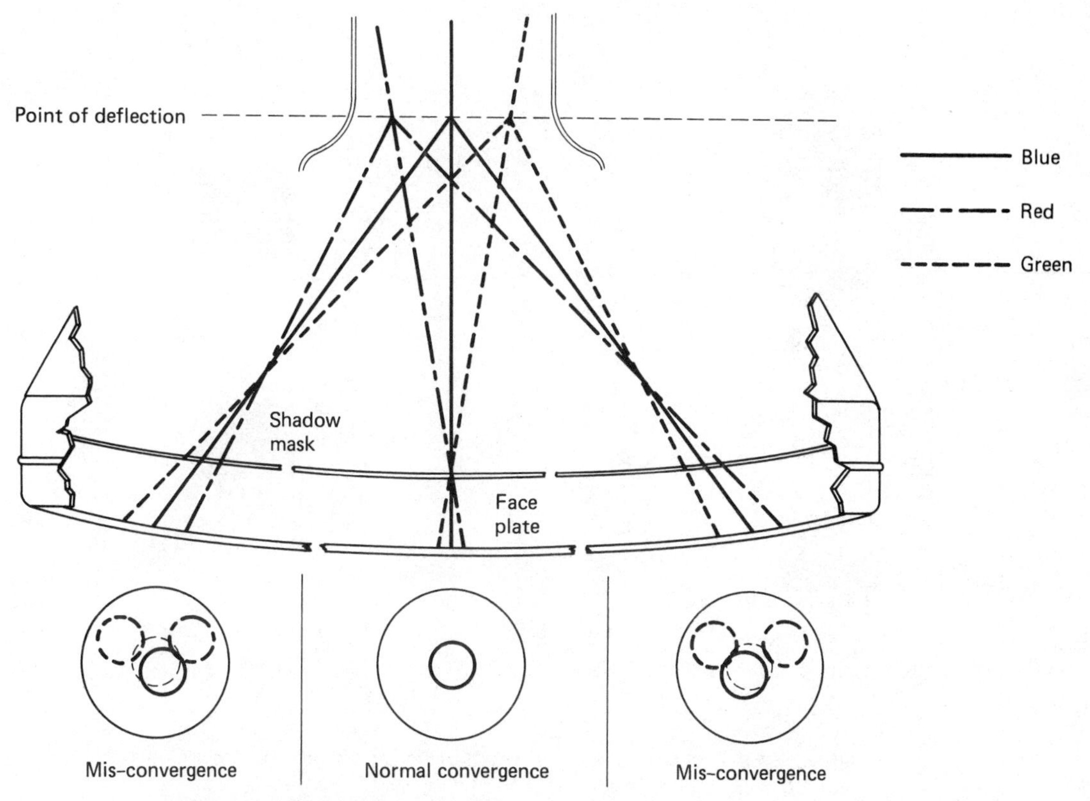

Figure 18-6 Electron beams from each CRT gun must meet at the same place on the face of the tube.

12 used for these adjustments controls the interrelation of the three electron beams. Adjustments are made for top, bottom, and both sides of the CRT. One set of controls is used for each color, thus the 12 adjustments.

ALTERNATE OUTPUT STAGES

Another circuit in use for the vertical output stage is shown in Figure 18-7. This system uses the sawtooth wave created in the vertical oscillator. The vertical signal is coupled through capacitor C_{505} to the base of Q_{503}. It is amplified by this transistor. The next stage to receive the signal is Q_{504}. This is an emitter-follower circuit. Its output is connected to the base of transistor Q_{503}. During the first half of the scanning this transistor is used to amplify the signal. Its output is connected to the deflection yoke and to the base of transistor Q_{506} through diode CR_{510}. During the second half of the vertical scan this transistor acts as an output transistor in order to make current flow through the yoke circuit.

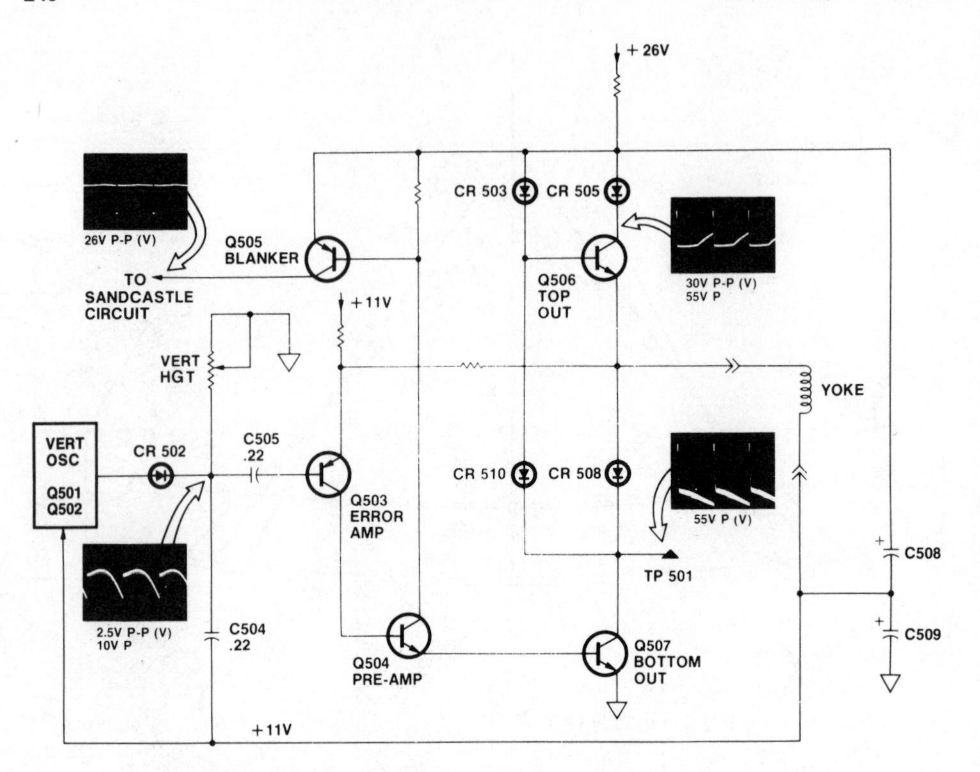

Figure 18-7 Another version of the vertical output system. (Courtesy of RCA Consumer Electronics.)

The circuits that are used for vertical sweep output are but two of several used for this purpose. The specific circuit in use is a factor in the design of the receiver. The technician has to understand that these circuits are used to develop an electric current and its resulting magnetic field in the deflection yoke. This varying magnetic field develops the vertical position of the electron beam as it scans horizontally.

TROUBLESHOOTING VERTICAL SWEEP

The initial step required in any troubleshooting procedure is to isolate the trouble area. Typical problems associated with vertical sweep include:

1. Loss of all sweep
2. Partial loss of sweep
3. Nonlinearity of picture
4. Vertical rolling

A total loss of vertical sweep results in one horizontal line appearing on the center of the CRT. This is the result of a lack of current in the vertical windings of the yoke. It may be caused by a failure anywhere in the vertical system. Waveform observation with an oscilloscope will aid in the location of the failure area. Be sure to check for proper operating voltages when localizing the problem area.

A partial loss of sweep may appear as two types of visual images. One of these is seen as a picture that is reduced in height at both the top and the bottom of the screen. This is caused by a lack of full amplification in the vertical system. Any stage may cause this problem. A check for proper amplitude of the waveforms will determine which stage is not performing properly.

The second type of partial loss will appear as a loss of only one part of the raster. Either the top or the bottom of the picture will be reduced in height. This problem is often associated with a distorted image. It often is described as a nonlinear picture. This is due to an improperly shaped wave in the vertical sweep circuit. The vertical linearity control circuit is a feedback type of circuit. A failure of one of the components in the feedback path will produce this type of problem. In any case, check control operations before making other tests. Often, an adjustment of a control will restore the receiver to working order.

The final problem encountered in vertical sweep systems is a vertical roll. This is due to a sync signal not reaching the oscillator. Check for the presence of a sync pulse first. Adjusting the vertical hold control should make the picture "lock" when the sync pulse and oscillator are on frequency.

When troubleshooting a horizontal/vertical countdown circuit a similar approach is used. In many receivers a failure in the horizontal circuits will shut down all the low-voltage power sources. This may be checked with a voltmeter. An oscilloscope will show if the oscillator is developing the proper waveforms. In checking this, a frequency counter will quickly indicate if the oscillator is operating on the proper frequency.

One symptom observed when the countdown circuit fails is a lack of any sync. The operation of the master oscillator should be checked when this occurs. Waveforms and the presence of proper operating frequency both must be checked. The use of service training literature will aid in the proper testing and repair of these systems.

QUESTIONS

18-1. How many lines of vertical raster are used in the receiver?

18-2. What is the basic shape of the vertical signal?

18-3. Why is this shape used for scanning?

18-4. Exactly what does the vertical signal's output do in the receiver?
18-5. What is "blanking" and what does it do?
18-6. Describe "pincushion" and its correction.
18-7. Describe convergence and tell why it is necessary.
18-8. Static convergence corrections are made at what part of the CRT screen?
18-9. Describe vertical system problems.
18-10. What type of signal flow system is used in the vertical section?

Chapter 19

Horizontal Sweep

The second part of the television sweep system is used to create the horizontal portion of the scan. This system is also used to develop the high voltage required for CRT operation. A block diagram of the various sections of the horizontal scanning system is shown in Figure 19-1. The system starts with an oscillator. This oscillator, like its vertical counterpart, is free running. Its frequency is 15,734 Hz. An AFC block is used to synchronize the oscillator frequency with that of the transmitted sync signal.

The output of the horizontal oscillator is a constant frequency signal. It is amplified and continues on its path until it reaches the horizontal output stage. The horizontal output amplifier is used to develop a high-voltage pulse. This pulse is used to drive the horizontal output transformer. This transformer has several outputs. These include high voltage, yoke windings, low voltage, and convergence circuits. In addition, there is a damper section. Each of these blocks is essential to the operation of the receiver. They have equal importance to the correct operation of the receiver. Each block is discussed in the next section.

HORIZONTAL AFC

The horizontal AFC circuit is described in detail in Chapter 16. Its purpose is to compare two signals in the receiver. These are the differentiated horizontal sync pulse and the feedback pulse from the horizontal output stage. A phase comparator circuit is used to develop a dc correction voltage at the output of

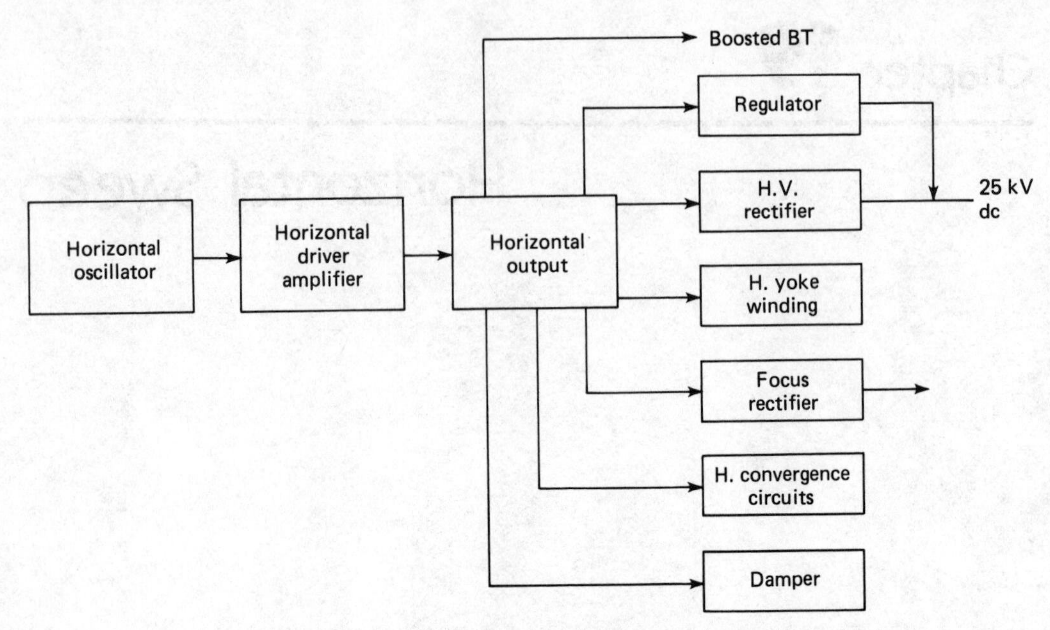

Figure 19-1 Block diagram of a horizontal scanning system.

this block. This dc correction voltage is fed to a voltage-controlled oscillator. Its purpose is to keep the oscillator on the correct frequency.

HORIZONTAL OSCILLATOR

The horizontal oscillator develops the required waveform for circuit operation. This block may use a voltage-controlled oscillator in order to maintain the proper frequency. In some receivers the horizontal sawtooth signal is developed in the horizontal/vertical countdown section. Regardless of where the signal is developed it is used to drive the horizontal output section of the receiver. The sawtooth-shaped waveform is amplified to a level that is required for proper horizontal output operation.

The output of the AFC circuit is often fed to a reactance circuit. This and the oscillator are shown in Figure 19-2. The dc correction voltage is developed at the junction of R_8 and R_9. The reactance control transistor, Q_2, is used to modify the total amount of capacitance in the oscillator tuned circuit. This tuned circuit consists of L_1, C_{13}, C_{12}, and C_{16}. In effect, the amount of conduction of the reactance transistor is controlled by the dc output voltage of the AFC circuit. The reactance circuit charges capacitor C_{10}. The amount of charge on this capacitor determines its capacitive value. The capacitor is connected as a part of the resonant "tank" circuit of the oscillator.

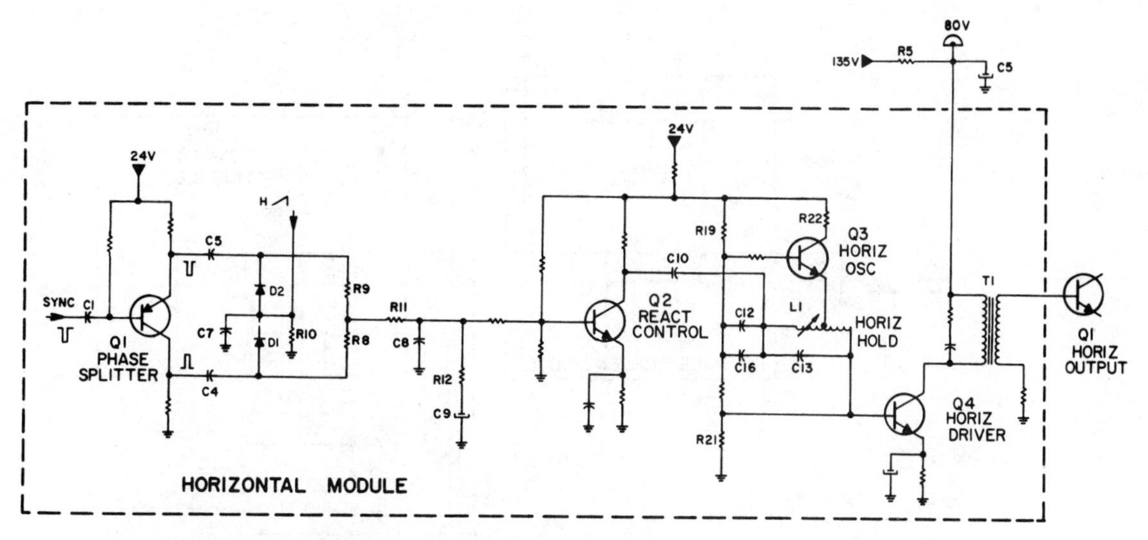

Figure 19-2 Schematic diagram of a horizontal oscillator circuit. (Courtesy of Magnavox Consumer Electronics.)

The output of the oscillator is connected to the base of the driver transistor Q_4. The value of the oscillator signal forces Q_4 into either cutoff or saturation. As a result of this the output of this transistor is a square wave. The output of the driver is coupled to the base of the horizontal output transistor through transformer T_1.

HORIZONTAL OUTPUT

The horizontal output stage is used to develop a strong signal. The output of this stage is a multipurpose transformer. One winding on this transformer is used to develop the high voltage required for CRT operation. Another winding is used to develop deflection yoke current and the resulting magnetic field. Other windings are used to develop low-voltage power sources. These often include filament voltage for the CRT. In addition, the transformer has a feedback winding. This winding is used for the keying pulses required for AFC, color, and AGC circuits. Other circuits associated with the horizontal output transformer (HOT) include high-voltage hold-down circuits, dampers, and focus voltage sources. The critical role this transformer has in the operation of the receiver cannot be overemphasized. This is a general overview of the horizontal scanning system. Details of the operation of each block follow.

The output stage is shown in Figure 19-3. The transistor is used to develop pulses of voltage at the horizontal rate. To do this, it has to act as a switch. When the transistor is turned on, the screen of the CRT is scanned from its center to the right-hand side. The output transistor is then turned

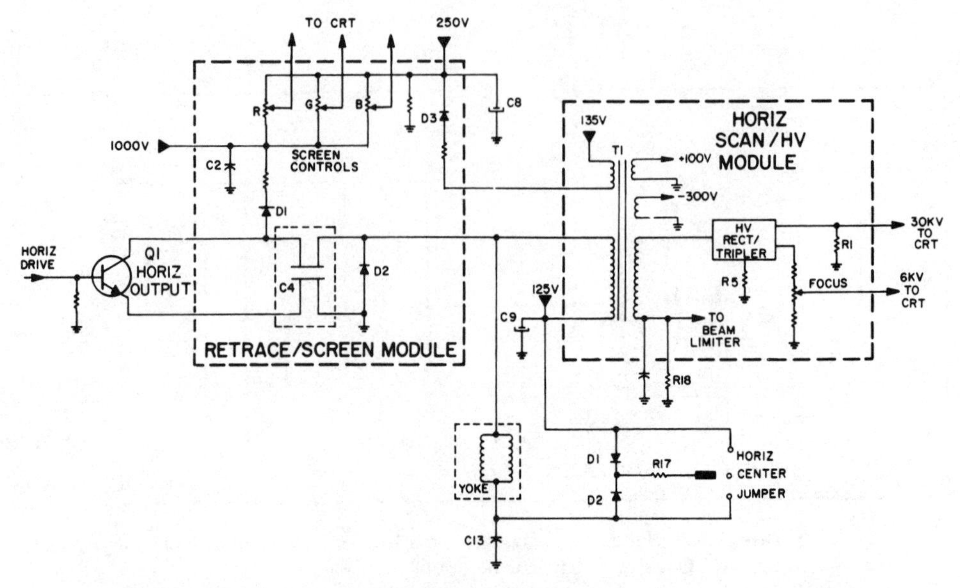

Figure 19-3 Schematic diagram of a horizontal output system. (Courtesy of Magnavox Consumer Electronics.)

off. During this period the beam retraces to the left of the screen. It also scans from the left side to the center. How this is developed is interesting.

The scanning system uses magnetic fields to position the electron beam in the CRT. The magnetic field is developed as a result of energizing a resonant circuit. The output transistor is required to develop a large current during a part of its duty cycle. To do this, the circuits used are often class C amplifiers. The class C amplifier is biased well beyond its cutoff point. A resonant tank circuit is used in the output circuit. When the transistor conducts, energy is developed in the tank circuit. When the transistor is not conducting, the energy in the tank circuit is used to complete the cycle.

The information shown in Figure 19-4 will help in understanding how this works. Another component is required in this circuit. This is the damper diode. The damper diode is used as a polarity sensitive switch in the circuit. It reduces undesired oscillation in the output circuit. During the diode conduction period the electron beam is moved from left to center on the screen.

A circuit that is equal to the output circuit is shown in this figure. The top of the illustration shows a parallel resonant circuit, its power source, and two switches. The parallel resonant circuit represents the deflection yoke and flyback components. Switch S_1 represents the horizontal output circuit. It is in series with the power source and the resonant circuit. Switch S_2 represents the damper diode. It is in parallel with the tank circuit $C_1 L_1$. This circuit is used to develop a sawtooth waveform for deflection purposes. The illustration begins with no deflection. This positions the beam at the center of the

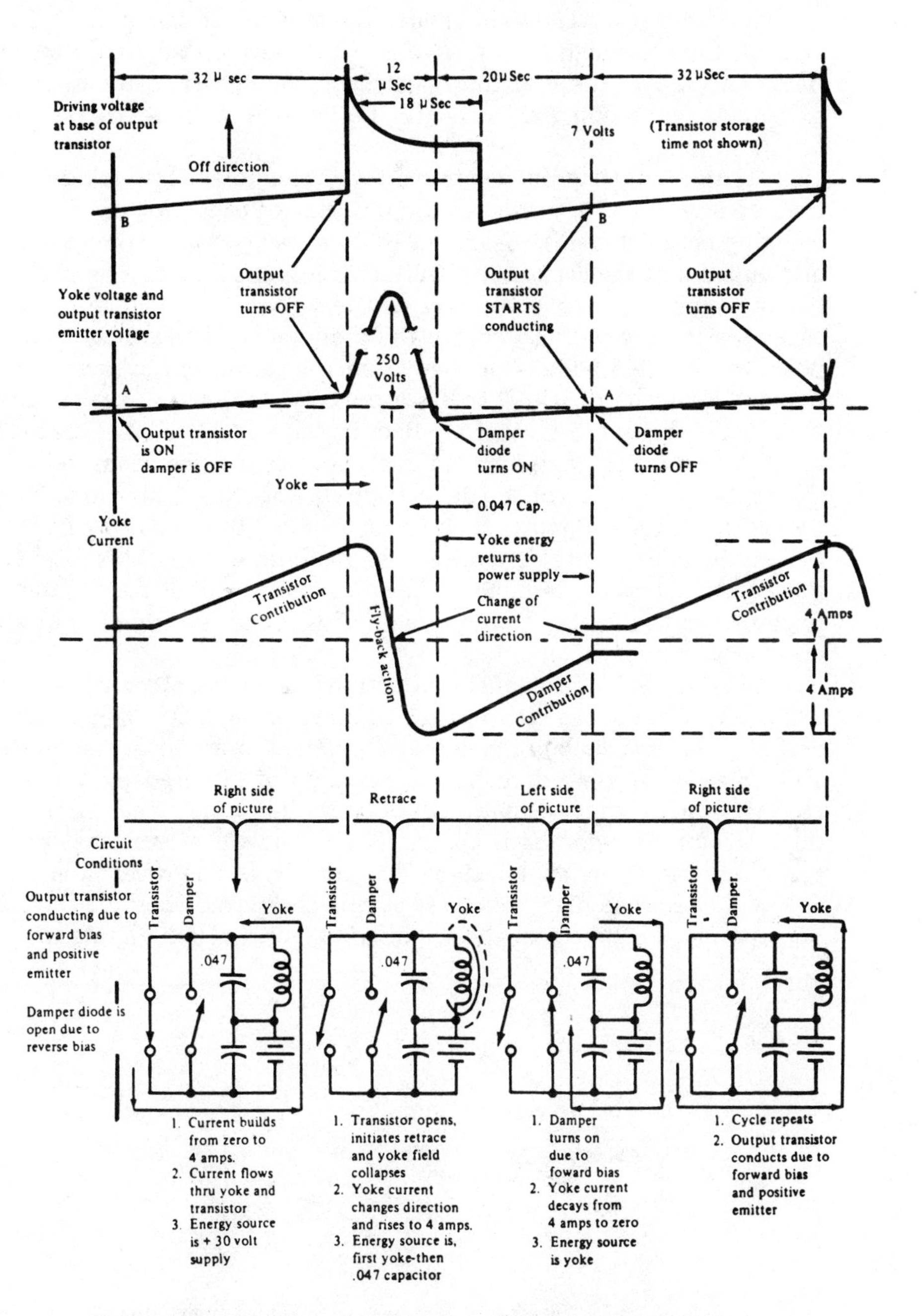

Figure 19-4 Horizontal output and damper action. The combination of these two circuits is necessary to develop full horizontal beam deflection. (Courtesy RCA Electronic Components Division.)

screen. The output transistor starts to conduct at this point. The electron current flow through the yoke and HOT develop an expanding magnetic field. This moves the electron beam from the center to the right side of the screen. At maximum field strength the beam is at the extreme right side of the CRT.

At this time the output transistor is shut off. This makes the magnetic field collapse. It also results in the charging of capacitor C_1. It is during this charging period that the beam starts its retrace period. When the inductor is discharged and the capacitor is fully charged, the lack of a magnetic field has returned the beam to the center of the screen. The voltage built up in the capacitor is now used to charge the inductance. The voltage is of opposite polarity to that originally applied from the output stage power supply. The result is a magnetic field that has a polarity opposite to that of the original field. This causes the beam to deflect to the left side of the screen. The only thing left to do to complete the cycle is to return the beam to the center of the screen. The action of the damper diode does this. This diode is connected across the parallel circuit. It is turned on by the polarity of the magnetic field. It acts as a short circuit across the tank circuit. This discharges both the capacitor and the inductor. The collapsing magnetic field forces the beam to return to the center of the screen. This is but one of the outputs of the horizontal output system.

In addition to the deflection action the square wave is used to develop a magnetic field in the high-voltage winding of the HOT. The principle of the development and collapse of a magnetic field in a coil is used in this circuit also. A secondary winding that is made up of hundreds of turns of wire is used to develop the required high voltage. In a color receiver the output of this transformer winding is about 10 kV. The waveforms that produce this action are shown in Figure 19-5. The portion of the horizontal output that is used to produce high voltage is during the retrace period. During this period the magnetic field in the winding collapses at a high rate of speed. The

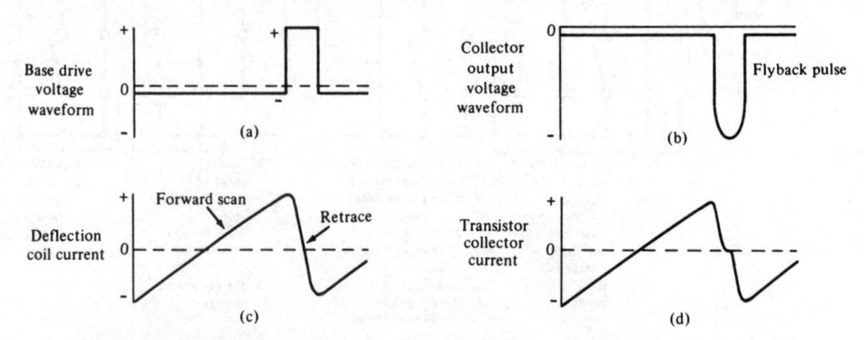

Figure 19-5 Waveforms required to produce high voltage in the receiver. (Clyde N. Herrick, *Television Theory and Servicing: Black/White and Color, 2nd ed.*, 1976. Reprinted with permission of Reston Publishing Co., a Prentice-Hall Co., 11480 Sunset Hills Road, Reston, VA 22090.)

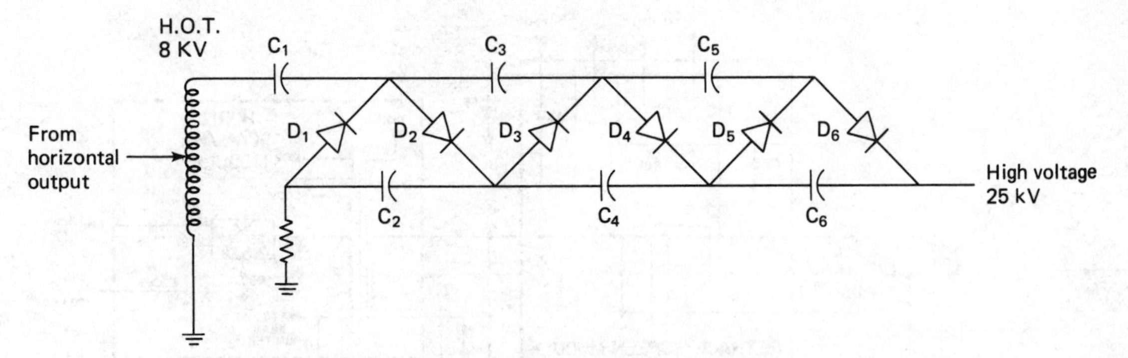

Figure 19-6 Schematic diagram for a high-voltage tripler circuit. The charging of the capacitor helps develop the necessary voltage values.

fast-moving lines of force that cut across the wires in this coil develop a high voltage. The voltage is often processed further by a high-voltage multiplier module. Its output is triple the input voltage. Tripler action is based upon the charging ability of a capacitor. The schematic diagram of a high-voltage tripler is shown in Figure 19-6. It works by alternately charging and discharging the capacitors in the circuit. Action of a voltage-doubling circuit is quite similar. During the first half of the duty cycle, diode D_1 conducts. This charges capacitor C_1 to the voltage value applied to it. During the second half of the duty cycle the voltage developed across C_1 is added to the output of the transformer. This makes diode D_2 conduct. It now charges capacitor C_2. As the pulses from the HOT continue, the process also continues. Capacitors C_3 and C_5 are charged during one half of the cycle. Capacitors C_4 and C_6 charge during the other half of the cycle. The voltage charge in the capacitors is about 8 to 10 kV. The output is taken from the three series-connected capacitors C_2, C_4, and C_6. The development of this circuit allows output voltages that are three, four, and five times the value of the input voltage. All that is required is to add additional diode and capacitor pairs.

The circuit also shows a tap-off point at the high end of the first diode set. This voltage is used to develop the required focus voltage for the CRT. This is on the order of 5 kV. A voltage divider network is used to drop the output of the tripler/rectifier to the required operating voltage. A variable resistance is used to adjust the focus voltage level for the clearest picture.

A circuit that uses these features is shown in Figure 19-7. An additional output is shown at the collector of the output transistor. This circuit is used to develop the required 1 kV for other circuits in the receiver. Diode D_1 rectifies the output of the transistor. The voltage is filtered by capacitor C_2. Capacitor C_4 is a special unit. It has four leads. If it should fail, it shuts down the high-voltage circuits in the receiver. This is used as a fail-safe device in order to protect the consumer. This is a critical value component and should be replaced with an exact value replacement. Diode D_2 is the damper diode. Its action in the chain of events that produce scanning has already been explained.

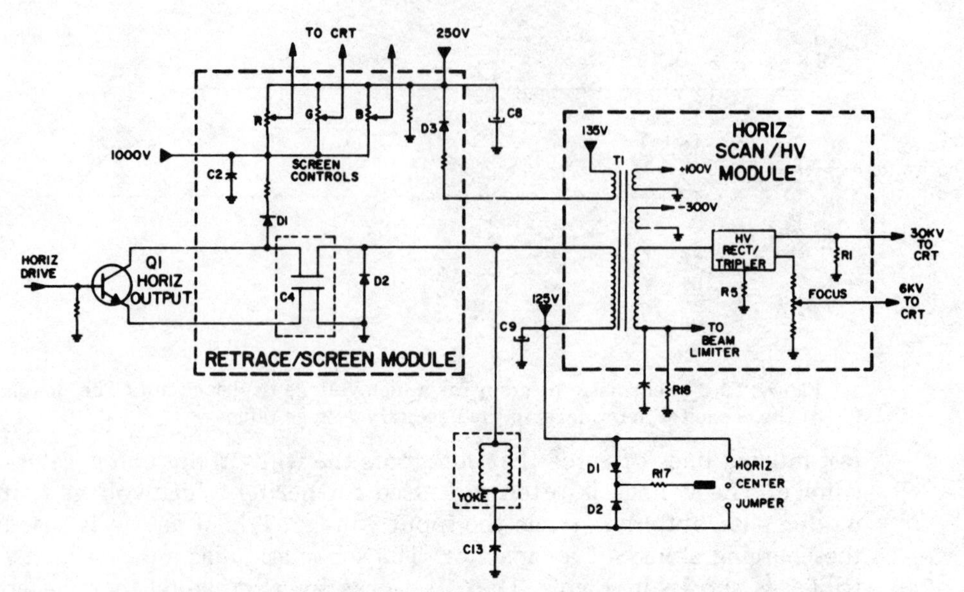

Figure 19-7 Schematic diagram of the high voltage/horizontal output section of a color receiver. (Courtesy of Magnavox Consumer Electronics.)

The secondary of transformer T_1 is used to develop several operating voltage sources. In this receiver these include -100 V, -300 V, as well as the 6 kV and 30 kV required for CRT operation. These secondary voltages will vary depending upon the needs and design of a specific receiver.

SCR SWEEP

Some receivers have used silicon-controlled rectifiers (SCRs) in place of the usual oscillator and amplifier circuits. A simplified schematic of one such circuit is shown in Figure 19-8. In this circuit yoke current during the trace time is developed by the switching action of diode D_1 and SCR_1. The other diode and SCR control beam deflection during the retrace period.

This system has operating features that are similar to those described previously. The major difference is that the SCRs form a pair of switches. These switches are used in order to direct current flow through the deflection circuits. Each SCR acts as a switch to develop the proper polarity magnetic field in the yoke. During the period of time when the beam moves from the center to the right side of the screen, a current flows through the yoke. This current is developed by the voltage charge on the yoke capacitor C_y.

After the beam is at its full right-hand position, the magnetic field collapses. This is during the retrace time period. The result is that the beam is returned to the center of the screen. While this is occurring, a current that

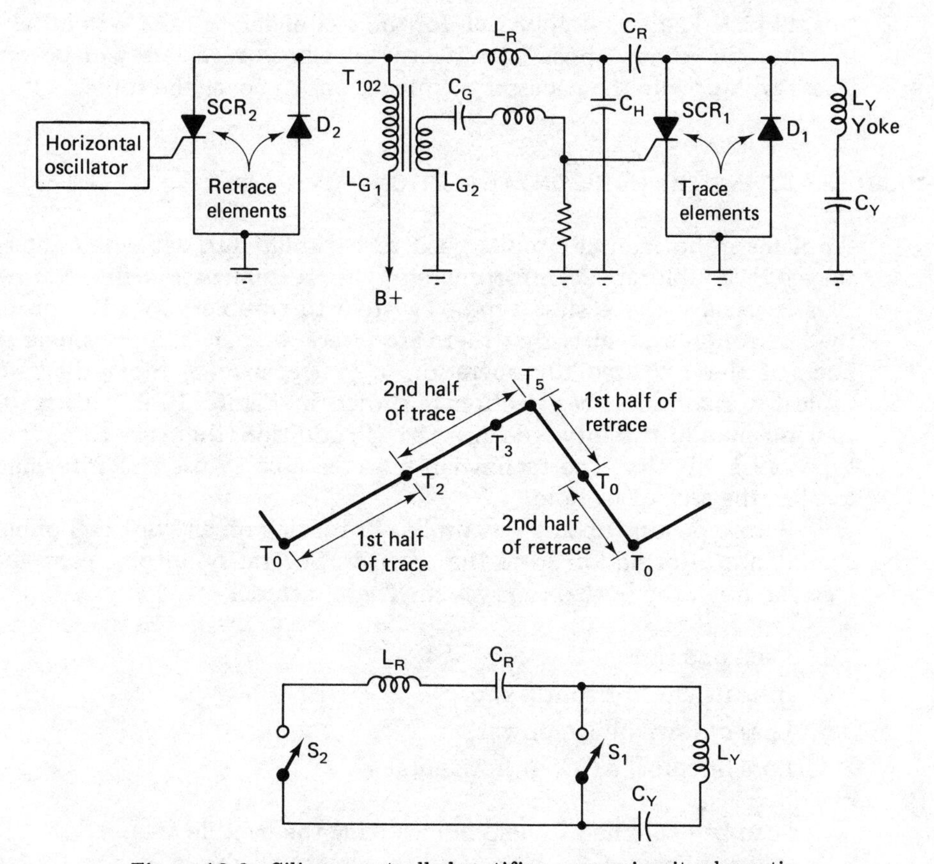

Figure 19-8 Silicon-controlled rectifier sweep circuit schematic.

has a polarity opposite to the original is developing in the yoke. This is because the second SCR is now conducting and the first SCR is shut off. This drives the electron beam to the left-hand side of the screen. This field collapses when the first SCR turns on and the second SCR shuts down. The result is the collapse of the magnetic field. The electron beam returns to its center screen position. The process repeats itself at the horizontal rate.

SHUTDOWN CIRCUITS

A problem that presents itself when the television system high voltage becomes too high is x-ray radiation. Concern about this hazard on the part of the U.S. government has forced receiver manufacturers to incorporate shutdown protection circuits in receivers. There are several different circuits used for this purpose. Each manufacturer is permitted to develop circuits to work in that manufacturer's receivers. As a result, a book could be written about this subject. The wisest approach for the technician to take is to obtain training literature from those manufacturers whose receivers will be repaired. These will provide the necessary explanations to cover the topic.

TROUBLESHOOTING THE HORIZONTAL SECTION

Problems in horizontal circuits need to be isolated to determine the specific stage, or trouble area. Unfortunately, this system uses a linear flow path. This is usually the easiest type of system to troubleshoot. The problem in the horizontal system is that there are several branches to this linear system. Each of these is used for operation of the receiver. A block diagram for a typical horizontal sweep section is shown in Figure 19-9. This system has four outputs in the forward flow path. In addition, there are three feedback-type pulse circuits. The technician must be able to use this information to localize the area of trouble.

Before picking up any test probes or turning on any test equipment, one should use information from the receiver. Several symptoms can be caused by a failure in the horizontal system. These include:

1. Lack of a raster
2. Insufficient horizontal sweep
3. Loss of low-voltage power
4. Loss of color, AGC, or horizontal sync

These can be examined to help in localizing the trouble area.

A loss of raster can be caused by almost all the blocks in this section. The raster development depends upon high voltage. This, in turn, is dependent upon a horizontal signal. Any stage from the oscillator to the high-voltage rectifier can fail and cause a loss of this voltage. In addition to this, in some receivers the video output stage is directly coupled to the CRT. A failure in this stage raises the dc voltage at the cathode of the CRT to the blanking level. A failure of this type can be identified by rotating the channel selector. Flashes of raster as the receiver is tuned between channels is an indication of this problem.

The identification of a specific block in the receiver horizontal section can be aided when one considers the following statements. Presence of a raster indicates that all horizontal sections are working properly. One should review the schematic diagram when attempting to localize the problem. If the sections that obtain power from these sources are working, the section of the horizontal up to the power source is also working.

After reviewing this information and analyzing the symptoms in the receiver, the next best thing to do is to pick up the proper piece of test equipment. This should be the oscilloscope. Use the basic rule for a linear flow

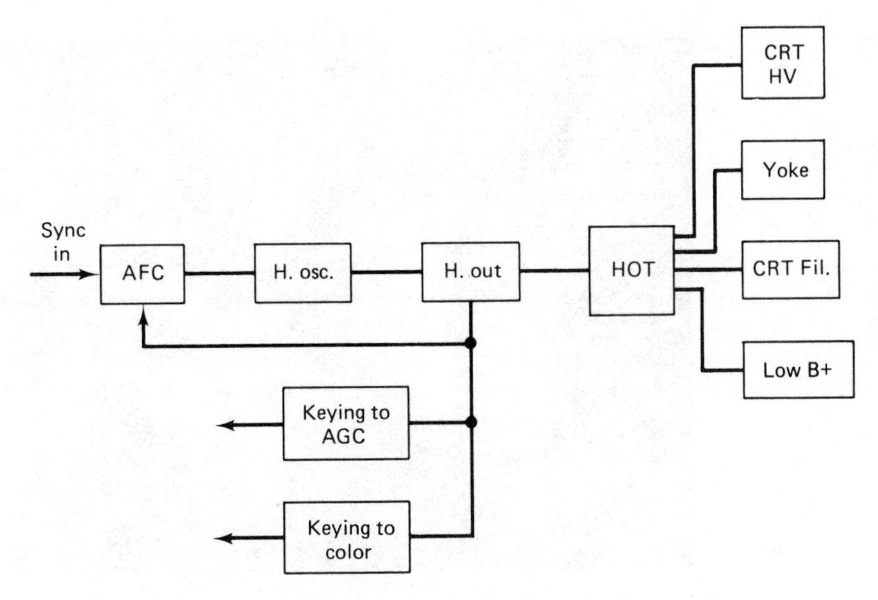

Figure 19-9 Block diagram of a horizontal sweep section.

path system. Make the first test at a point near the middle of the system. A good place to start is the input to the horizontal output stage. This test will tell if the signal is being created and if it is the proper amplitude to drive the output stage.

Further tests are made using a lot of caution. The level of voltages present in the HOT may be very high. A level of 30 kV is not uncommon in vacuum-tube receivers. Review the service literature. Many manufacturers tell you not to measure the high voltages at the output of the horizontal output stage or at the high-voltage winding of the transformer. Other manufacturers provide waveform and amplitude information to make valid measurements when the information is given.

Measuring the output voltages in the scan-derived power supply will tell if the output stage is working properly. A high-voltage probe is used to measure the CRT anode voltage. The information obtained from these measurements will aid in locating the problem area. Do not underestimate the importance of the feedback pulses in the receiver. These pulses are used with the presence of the proper signal in order to turn on, or key, a stage. Without the presence of these keying pulses, the stage will not operate. The keying pulse and the incoming signal are used in the converging, or meeting, signal flow path system.

Another area of trouble that has been common is the failure of the voltage divider resistors in the focus bleeder circuit. These resistors often break down, or fail, when in operation. Often, their resistance is so high that it cannot be correctly measured with the ohmmeter. When the high voltage is

Figure 19-10 Shorted windings on the yoke will produce this uneven horizontal scan, called a "keystone" effect.

correct but the focus voltage is off its correct value, one must suspect and replace the resistive network in the focus bleeder.

Another problem in the horizontal section of the receiver is associated with the deflection yoke. This problem is shown in Figure 19-10. It is called the keystone effect. It is caused when some of the turns in the deflection yoke winding short together. When this occurs, the full-strength magnetic field cannot be developed in the yoke. The raster shown in the illustration is a classic example of this problem.

QUESTIONS

19-1. What is the basic shape of the horizontal signal?
19-2. What outputs are found in the horizontal system?
19-3. Describe how high voltage is produced.
19-4. Why is horizontal AFC used in the receiver?
19-5. What is the frequency of the horizontal signal?
19-6. Describe damper diode action.
19-7. What is a shutdown circuit and why is it used?
19-8. What is the basic signal flow path in the horizontal system?
19-9. What are typical horizontal system symptoms?
19-10. How does one troubleshoot this system?

Chapter 20

Miscellaneous Circuits

Receiver manufacturers are attempting to make the television receiver almost foolproof. This has resulted in several automatic circuits. These are basically comparator-type circuits. They compare the signal in the receiver to a preset reference level. If the correct output signal from the comparator circuit is not present, a correction signal or voltage is developed. This changes the incorrect input signal to maintain the correct output. There are several of these automatic signals. These include color control, tint control, fine tuning, and degaussing. In addition, the receiver has a mandated safety shutdown circuit. Some receivers have a "preset" switch. This, when engaged, will establish predetermined levels of color, tint, brightness, and luminance. Each of these circuits is discussed in this chapter.

PRESET CONTROLS

The purpose of a preset control system is to permit the viewer to have all of the common controls set to a specific value, or level, when one switch is engaged. A multisection switching arrangement is required for this purpose. In effect, there are two sets of controls used in this system. One set is used by the viewer. The second set is adjusted by the service/installation technician when the set is delivered. Schematic diagrams for a select group of preset controls are shown in Figure 20-1. A group of two pole switch contact assemblies is used. The schematics show the controls switched to the manual position. The heavy bar between terminals 1 and 2 represents the switch

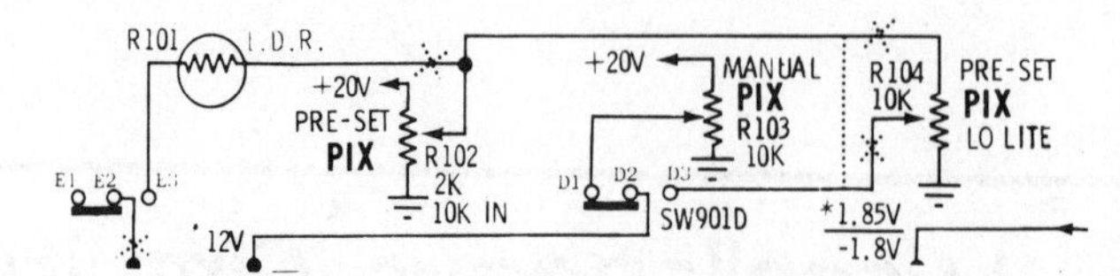

Figure 20-1 Preset controls shown on the schematic diagram. These controls are activated by a switch.

arm assembly. It is shown connecting these two terminals. When the switch is in the preset position, terminals 2 and 3 are in use. Samples of hue, brightness, luminance, and color intensity-level circuits are shown in this illustration. These types of control circuits are normally found in color receivers.

AUTOMATIC DEGAUSSING

The term "gauss" refers to the strength of a magnetic field. The color CRT is sensitive to the influence of external magnetic forces. A magnetic influence on the CRT will result in a picture that has poor convergence. In some instances the CRT cannot be converged at all. A device called a degaussing coil is used to neutralize any magnetic fields that develop around the face of the color CRT. The degaussing circuit consists of a large coil of wire and a thermistor circuit. One of these circuits is shown in Figure 20-2. The degaussing

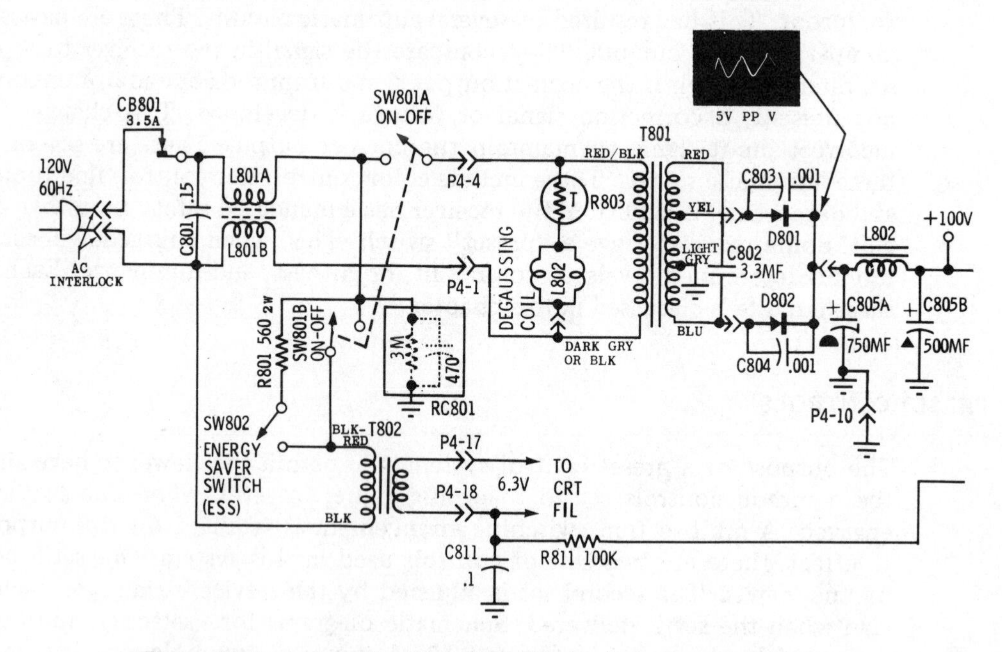

Figure 20-2 Automatic degaussing circuit used to neutralize magnetic fields in the color CRT. (Courtesy of Quasar Electronics Corp.)

coils are wired in parallel to each other. They are identified as L_{802} in this schematic. A thermistor is connected in series with the coils. The thermistor is a heat-sensitive resistor. Its ohmic value will increase as it heats. The flow of electrons through the thermistor produces heat. When the receiver is first turned on, the thermistor is cold. It has a very low resistance value. An ac current flows through the degaussing circuit. The varying magnetic field established by the ac voltage will neutralize any magnetic fields on the face of the CRT. As the current flows the thermistor heats. This gradually increases its resistance. The resistance increases until there is very little current flow in the circuit.

A second automatic degaussing circuit is shown in Figure 20-3. This circuit uses two special resistors. One of these is a thermistor. This style of thermistor has a negative temperature coefficient. Its resistance decreases as it heats. In this circuit the thermistor is connected in parallel with the degaussing coils. The other special resistor is called a *varistor*. Its resistance will decrease as the voltage across its terminals increases. The varistor is connected in series with the degaussing coils. The circuit works in the following manner. When power is applied to the cold receiver, the thermistor is cold and has a high resistance. Current, taking the path of least resistance, will flow through the coils and the varistor. This establishes an alternating magnetic field for a few seconds. The result is a neutral magnetic field around the CRT. As the current flows, it will heat the thermistor. This action causes its resistance to drop to a very low value. At the same time the resistance of the varistor increases because of the drop in voltage across its terminals. The result is that almost all the current flows around the degaussing coils. This circuit works only when the temperature-sensitive components are at normal, or room, temperature.

Troubleshooting of automatic degaussing circuits is relatively simple. The symptoms of a failure appear as blotches of color on the CRT. Often, the varistor and thermistor failure is evident because the components get very hot. The insulation around the varistor will even melt! Use of an ohmmeter to measure the resistance of these components when they are *cold* will show if they are good.

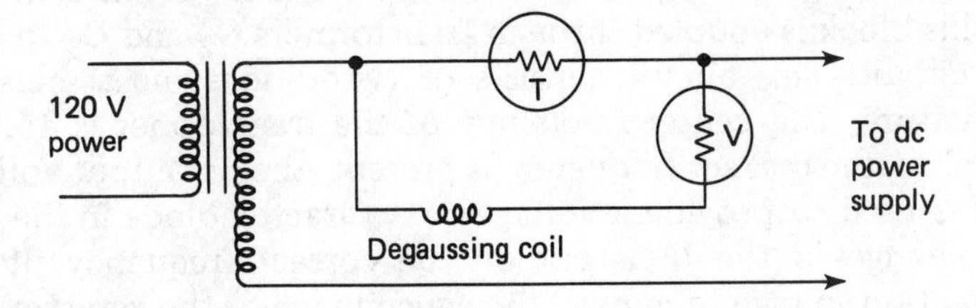

Figure 20-3 A schematic diagram of a degaussing circuit that uses a thermistor and a varistor to shift current flow.

AUTOMATIC CONTROL CIRCUITS

There may be several automatic control circuits used in television receivers. These have one basic purpose. It is to maintain the proper signal levels. The requirements of the color receiver almost dictate the need for each of these. Typical circuits are described in this section of the chapter.

Automatic color control (ACC). This circuit acts in a manner similar to AGC. The difference is that ACC is used only in the color section of the receiver. Its basic function is to maintain a constant level of chrominance signal under a variety of input signal levels. A block diagram of the ACC section of a receiver is shown in Figure 20-4(a). The ACC circuit is used to develop a dc control voltage. The control voltage is used to adjust the bias, and thus the gain, in the chroma bandpass amplifiers. The ACC circuit receives its signal from the level of the color burst signal.

A schematic diagram of this circuit is shown in Figure 20-4(b). This circuit uses the output of the ACC amplifier transistor to control the bias on the bandpass amplifier. The level of the collector voltage of the ACC amplifier is dependent upon the amount of conduction of this transistor. This voltage is totally dependent upon the level of the ACC signal received from the input to its base.

Troubleshooting this circuit is best done by dividing it into problem areas. Use a dc power supply to clamp the bias voltage at the base of the bandpass amplifier. If this restores color levels, the problem is in the ACC amplifier transistor. Further troubleshooting follows traditional patterns in order to locate a defective component.

Automatic fine tuning (AFT). The purpose of this section is to automatically establish the proper local oscillator frequency in the tuner. This is necessary in order to receive a proper level of color information. Without this circuit each channel has to be manually fine tuned. Incorrect tuning will produce either a no-color picture or sound bars in the picture. A block diagram for an IC AFT circuit is shown in Figure 20-5. An IF signal from the third IF amplifier stage is coupled into the IF amplifier block in the IC. The output of this block is coupled through transformers C_2 and C_3 to a balanced detector circuit. This circuit consists of two diodes and associated resistors and capacitors. The center frequency of the transformer is 45.75 MHz. When a signal of the correct frequency is present, the dc output voltage from the circuit is used to provide a voltage to a varactor diode in the tuners. This voltage changes if the IF is not on the correct frequency. It uses the varying correction voltage to control the capacitance of the varactor diodes. These, in turn, are used to tune the tuner oscillator. Correction voltages are generated that change the oscillator frequencies and keep it at the proper frequency.

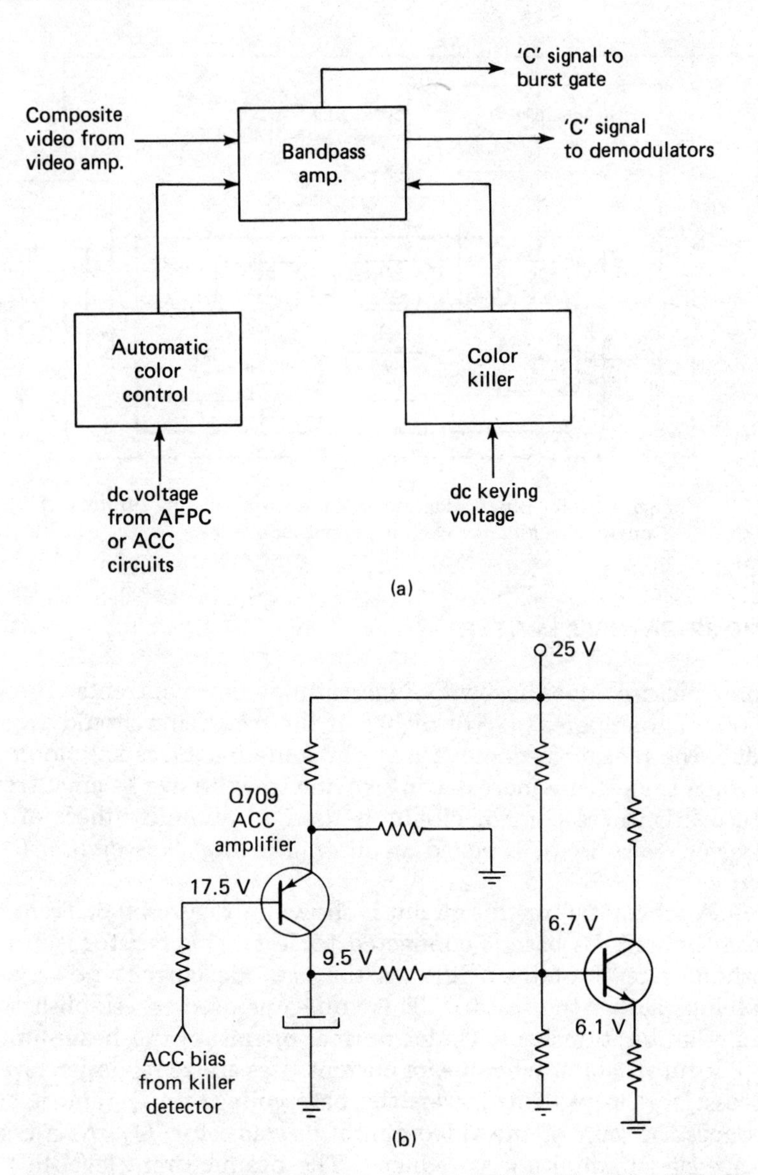

Figure 20-4 Automatic color control (ACC) circuits: (a) block diagram, and (b) schematic diagram.

Troubleshooting this circuit may require alignment of the detector transformer. A place to start is to measure the control voltage at the varactor diode. If this varies as the fine tuning control is rotated, the problem is located in the varactor diode circuit. If no change in voltage occurs, the problem is developed in the ACC tuning circuit.

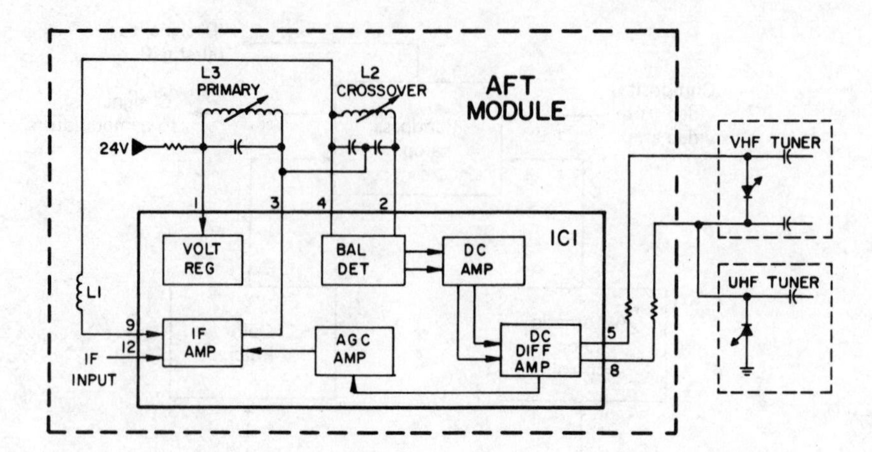

Figure 20-5 Block diagram of an automatic fine tuning (AFT) circuit. (Courtesy of Magnavox Consumer Electronics.)

AUTOMATIC BRIGHTNESS LIMITER

Color picture tubes have very high cathode beam currents. These often reach 1000 microamperes (μA) or higher. If these currents should exceed the design value, the picture becomes fuzzy. In some instances it "blooms." Blooming is when the picture increases in size due to excessive beam current. The automatic brightness control circuit is used to minimize these effects. In some receivers this circuit is called an *automatic brightness limiter* (ABL) or *beam limiter.*

A schematic for this circuit is shown in Figure 20-6. Transistor Q_6 is the beam limiter. Its base is connected to R_{18}. This resistor is in series with the cathode circuit of the CRT. As the cathode current flows, a voltage drop develops across the resistor. This voltage is used to establish the bias in the beam-limiter transistor. Under normal operation the beam-limiter transistor is in a cutoff state. When beam current rises above its design level, the voltage across R_{18} drops. This lowers the base voltage on Q_6 and it conducts. This reduces the gain of the video amplifier transistor, Q_4. As a result, the luminance signal amplitude is reduced. The beam-current level in the CRT is dependent upon the amplitude of the luminance signal. The reduction of this signal also reduces beam current to a safe level.

Troubleshooting of this circuit is relatively simple. Varying the brightness control in the receiver should produce a change in the voltage developed across R_{18}. This should produce a change in the voltage at the collector of the beam-limiter transistor. Measuring the collector voltage as the receiver brightness control is rotated will quickly determine if this circuit is working properly.

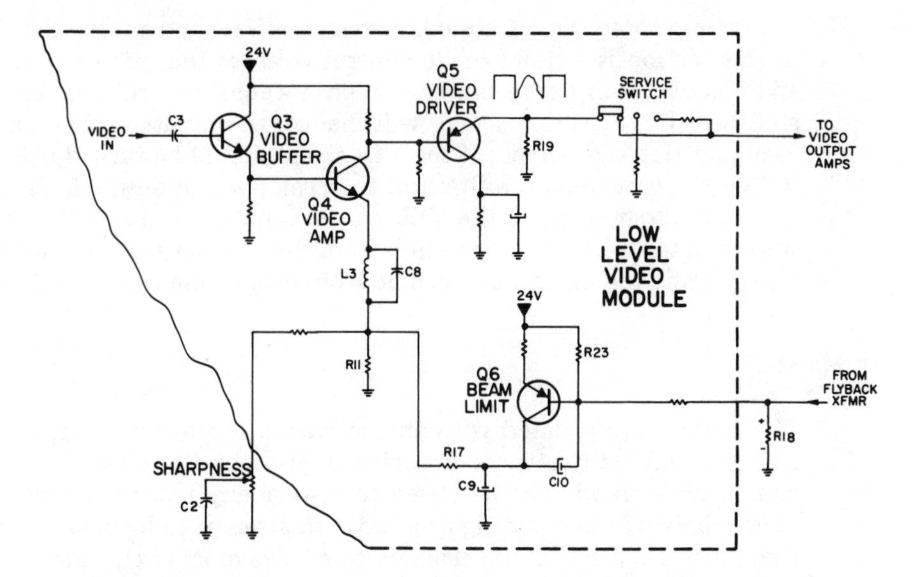

Figure 20-6 Schematic diagram of an automatic brightness limiter (ABC) circuit. (Courtesy of Magnavox Consumer Electronics.)

Vertical interval reference (VIR). The purpose of this circuit is to automatically adjust chroma level and chroma tint. This signal originates at the broadcast station. Its transmission is optional. It is transmitted on line 19 during the vertical blanking interval. It may be observed on the receiver by adjusting the vertical hold control so that the blanking bar is viewed. On a color receiver the VIR appears as a line of various colors. The signal is shown in Figure 20-7. It contains three reference signals. These are a chrominance reference of 3.579545 MHz. This signal is in phase with the color burst signal. A second signal is a luminance or white-level signal. The last reference is the black level.

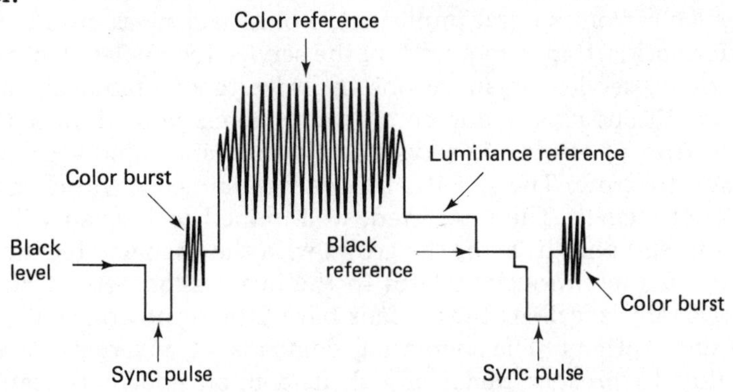

Figure 20-7 Line information for a VIR signal. This signal is used for automatic adjustment of color circuits in the receiver.

These three signals are compared in a VIR control section. The output of this section is a series of dc control voltages that are used to correct tint and color level in the receiver. A receiver's color controls can be misadjusted and the VIR correction signals will change the circuits so that a normal color level and tint is observed. If the VIR signal should be turned off at the transmitter, the two circuits will return to develop the undesirable display.

Troubleshooting of the VIR system is rather complex. The technician is wise to refer to specific literature from the receiver manufacturer. Little service information on this is available from other sources.

SUMMARY

The problems associated with any television receiver may appear to be difficult to resolve. Circuits, at first glance, also appear to be complex and difficult to understand. The best way to resolve any receiver problems is to use your senses of smell and sight in order to attempt to localize a problem area. One must use a schematic diagram to resolve most circuit problems. The use of proper test equipment is essential to be successful in the repair of the receiver. The technician has to learn to use the oscilloscope to measure signal levels and form. The proper piece of test equipment used at the proper time will aid in the correct diagnosis of the problem.

One must also be willing to apply the basic rules for electrical circuits to locate defective components. The rules developed by Ohm and Kirchhoff play an important role in defect diagnosis and repair. Analysis of circuits based on the rules identified by these men will make the diagnosis of the problem easier.

Do not be afraid of becoming a professional. The professional technician attends school periodically. This school may be a one-day workshop put on by the receiver manufacturer. It may also be a regular educational course at a local community college or technical school. Professional literature is available from several publishers. Books and magazines contain a wealth of information that is required by the service technician. Get involved with professional service organizations in order to communicate with others in the field. Discussions about common problems will often aid in solving them. The field of electronic service is dynamic and rapidly changing. It will continue to grow. The specific service requirements will change. Basic theories do not change. The interested, well-trained technician will survive and prosper if, and only if, he or she grows with the changing technology.

One last thought relates to the business aspects of electronic servicing. Too many excellent technicians have little or no concept of how to organize business affairs. The increasing demands of government reporting and the ability to prosper financially depend upon good business management. Include courses and readings on business-related subjects in your plans for success both today and tomorrow.

QUESTIONS

20-1. How do "preset" controls function?
20-2. Describe automatic degaussing.
20-3. What is a thermistor? a varistor?
20-4. What blocks are controlled by ACC?
20-5. Describe ACC action.
20-6. What blocks are controlled by AFT?
20-7. Describe AFT action.
20-8. What blocks are controlled by ABL?
20-9. Describe ABL action.
20-10. Describe VIR action.

Appendix

TV Channel Frequency Allocations

Channel Number	Frequency Band, MHz	Picture Carrier Frequency, MHz	Sound Carrier Frequency, MHz	Channel Number	Frequency Band, MHz	Picture Carrier Frequency, MHz	Sound Carrier Frequency, MHz
2	54–60	55.25	59.75	43	644–650	645.25	649.75
3	60–66	61.25	65.75	44	650–656	651.25	655.75
4	66–72	67.25	71.75	45	656–662	657.25	661.75
5	76–82	77.25	81.75	46	662–668	663.25	667.75
6	82–88	83.25	87.75	47	668–674	669.25	673.75
7	174–180	175.25	179.75	48	674–680	675.25	679.75
8	180–186	181.25	185.75	49	680–686	681.25	685.75
9	186–192	187.25	191.75	50	686–692	687.25	691.75
10	192–198	193.25	197.75	51	692–698	693.25	697.75
11	198–204	199.25	203.75	52	698–704	699.25	703.75
12	204–210	205.25	209.75	53	704–710	705.25	709.75
13	210–216	211.25	215.75	54	710–716	711.25	721.75
14	470–476	471.25	475.75	55	716–722	717.25	721.75
15	476–482	477.25	481.75	56	722–728	723.25	727.75
16	482–488	483.25	487.75	57	728–734	729.25	733.75
17	488–494	489.25	493.75	58	734–740	735.25	739.75
18	494–500	495.25	499.75	59	740–746	741.25	745.75
19	500–506	501.25	505.75	60	746–752	747.25	751.75
20	506–512	507.25	511.75	61	752–758	753.25	757.75
21	512–518	513.25	517.75	62	758–764	759.25	763.75

Channel Number	Frequency Band, MHz	Picture Carrier Frequency, MHz	Sound Carrier Frequency, MHz	Channel Number	Frequency Band, MHz	Picture Carrier Frequency, MHz	Sound Carrier Frequency, MHz
22	518–524	519.25	523.75	63	764–770	765.25	769.75
23	524–530	525.25	529.75	64	770–776	771.25	775.75
24	530–536	531.25	535.75	65	776–782	777.25	781.75
25	536–542	537.25	541.75	66	782–788	783.25	787.75
26	542–548	543.25	547.75	67	788–794	789.25	793.75
27	548–554	549.25	553.75	68	794–800	795.25	799.75
28	554–560	555.25	559.75	69	800–806	801.25	805.75
29	560–566	561.25	565.75	70	806–812	807.25	811.75
30	566–572	567.25	571.75	71	812–818	813.25	817.75
31	572–578	573.25	577.75	72	818–824	819.25	823.75
32	578–584	579.25	583.75	73	824–830	825.25	829.75
33	584–590	585.25	589.75	74	830–836	831.25	835.75
34	590–596	591.25	595.75	75	836–842	837.25	841.75
35	596–602	597.25	601.75	76	842–848	843.25	847.75
36	602–608	603.25	607.75	77	848–854	849.25	853.75
37	608–614	609.25	613.75	78	854–860	855.25	859.75
38	614–620	615.25	619.75	79	860–866	861.25	865.75
39	620–626	621.25	625.75	80	866–872	867.25	871.75
40	626–632	627.25	631.75	81	872–878	873.25	877.75
41	632–638	633.25	637.75	82	878–884	879.25	883.75
42	638–644	639.25	643.75	83	884–890	885.25	889.75

Index